STUDIES IN NUMERICAL ANALYSIS

Studies in Mathematics

Paul Concus
University of California

Iain Duff
AERE Harwell

Walter Gautschi
Purdue University

Gene H. Golub
Stanford University

Peter Henrici
Zurich, Switzerland

Dennis C. Jespersen
Oregon State University

Jorge Moré
Argonne National Laboratory

Dianne P. O'Leary
University of Maryland

Victor Pereyra
Universidad Central de Venezuela

J. Barkley Rosser
University of Wisconsin — Madison

Danny Sorensen
Argonne National Laboratory

Paul Swarztrauber
National Center for Atmospheric Research

James H. Wilkinson
Stanford University

Studies in Mathematics

Volume 24

STUDIES IN NUMERICAL ANALYSIS

Gene H. Golub, editor
Stanford University

Published and distributed by
The Mathematical Association of America

PREFACE

In this volume, we have collected a number of papers which describe some of the current research in numerical analysis. The papers illustrate the wide diversity of topics that are covered in numerical analysis; even so, we have failed to adequately cover some important areas such as approximation theory and initial value problems.

The last several years has seen the range of computational facilities expand enormously. In particular, the hand held calculator is a tool of every modern engineer. At the same time, more powerful computers are available than was imagined only a decade ago. Different numerical techniques are required in each situation and some of these papers illustrate the different possibilities.

These papers can be read independently of one another. It is our hope that advanced undergraduate and new graduate students will feel the excitement of this dynamic and relevant area of mathematics.

I wish to thank the following persons for having read the preliminary copies of the papers and having made helpful comments: R. E. Bank, M. Berger, J. Bunch, P. Concus, A. Feldstein, P. Gill, M. Heath, R. LeVeque, G. Meurant, E. Ng, M. Overton, R. D. Russell, P. Saylor, R. Sweet, R. Tewarson, J. Varah.

GENE H. GOLUB

vii

CONTENTS

THE PERFIDIOUS POLYNOMIAL

James H. Wilkinson

1. INTRODUCTION

The problem of finding the roots of polynomial equations has played a key role in the history of mathematics and indeed our very concept of numbers has been steadily broadened by consideration of it. The part it has played might be summarized as follows.

Starting with the positive integers which man, with uncharacteristic generosity, ascribes to the Almighty, negative integers were introduced so that the equation $x + p = 0$ has a solution. The requirement that $px - q = 0$ should have a solution then led to the introduction of the field of rationals q/p. This provided a very rich number system but, although it has the impressive property that there are an infinite number of rationals between any two of them, consideration of the equation $x^2 = 2$ showed it to be inadequate. There are no integers p and q such that $p^2/q^2 = 2$. This led to the concept of irrational numbers, though it must be admitted that there was still a long way to go before the concept of the real continuum was put on a firm basis. Finally the equation $x^2 + 1 = 0$ led to the introduction of the 'number' i such that $i^2 = -1$ and, thence, to the concept of complex numbers $a + ib$.

1

There then came what is perhaps the most momentous 'discovery' in the history of mathematics. With the field of complex numbers we are at the end of this particular line of development. There is no need to extend our number system further in order that polynomial equations of higher degree should have solutions. The Fundamental Theorem of Algebra asserts that every polynomial equation with complex coefficients has a complex root. All early proofs of this theorem made some appeal to intuition though a proof by Gauss is particularly convincing. It depends on the assumption that if a closed curve encircling the origin one or more times is shrunk continuously to a point it must at some stage pass through the origin. Attempts to put the intuitive element in these proofs on a firm basis provided a valuable stimulus to mathematicians.

In parallel with these developments there were attempts to find explicit expressions for the roots of polynomial equations. For linear and quadratic equations this was trivial and the problem was solved for the cubic and quartic by the sixteenth century. Related results for the quintic proved to be elusive and, finally, it was shown by Abel and Galois that it is impossible, in general, to express the solution of equations of the fifth degree and higher in terms of radicals. Again work in this area yielded particularly rich dividends for mathematics in general.

In classical analysis too, polynomial functions have for long occupied a key position. In many respects they may be regarded as the simplest functions having any real character; they can be differentiated and integrated an arbitrary number of times. In spite of their simplicity they have the remarkable property expressed in the approximation theorem of Weierstrass—"Any continuous function may be approximated to any desired accuracy by a polynomial of sufficiently high degree."

The history of mathematics has, in a sense, contrived to make mathematicians feel 'at home' with polynomials. This is well illustrated by Weierstrass' theorem. Why should one be interested in the fact that continuous functions can be satisfactorily approximated by polynomials? The assumption is implicit in the theorem that this is a desirable thing to do since polynomials are such agreeable functions.

2. NUMERICAL EVALUATION OF POLYNOMIALS

The cosy relationship that mathematicians enjoyed with polynomials suffered a severe setback in the early fifties when electronic computers came into general use. Speaking for myself I regard it as the most traumatic experience in my career as a numerical analyst.

The early electronic computers had no hardware facilities for computation in floating-point arithmetic and operation in this mode was effected by means of subroutines. Having constructed such a set of subroutines for our first electronic computer, PILOT ACE, I looked for a simple problem on which to give them a field trial. Computing the largest real root of a real polynomial equation seemed tailor-made for the exercise. Moreover since polynomials of even quite a modest degree can assume values covering a very wide range over the interval including their zeros, the use of floating-point arithmetic is particularly appropriate. I was well aware that there were practical difficulties associated with the problem but imagined that they were mainly confined to equations with multiple roots or pathologically close roots. I felt that it was not the time to concern myself with these difficulties and I decided, therefore, to make two programs with quite limited objectives.

The first took n prescribed real numbers $z_1 > z_2 > \cdots > z_n$ and constructed the *explicit* polynomial $\prod(x - z_i)$ by multiplying the factors together. This in itself provided a simple test of the basic subroutines.

The second took the resulting polynomial $f(x)$ and, starting with the initial approximation $z_1 + 1$, used the Newton Raphson iterative method

$$x_{r+1} = x_r - f(x_r)/f'(x_r) \qquad (2.1)$$

to locate the largest root.

It was my intention to avoid difficulties by choosing well separated z_i. An attractive feature of this problem is that the background theory is trivial. With exact computation the approximations tend *monotonically* to the root z_1 and convergence is ultimately quadratic.

After using these programs with complete success on examples of order three, I decided to give them their first genuine test. I took

$n = 20$ and the polynomial with roots $20, 19, \ldots, 2, 1$ and submitted it to the Newton Raphson program expecting immediate success. To my disappointment the iteration continued indefinitely and no root emerged. Since the polynomial was clearly devoid of all guile, I assumed that there was an error either in the floating-point subroutines or the two polynomial programs.

Failing to find an error in the subroutines and verifying that the computed polynomial was correct to working accuracy, I returned to the zero finder and examined the successive iterates x_r and the computed values $f(x_r)$. The x_r appeared to be virtually random numbers in the neighbourhood of 20 while the computed values were of arbitrary sign and were in any case much larger than the true values. (Notice that the true values of $f(x_r)$ could be computed directly via $\Pi(x_r - z_i)$ since we were in the privileged position of knowing the roots *a priori*.)

Now the interesting thing about this behaviour is that it would not have been difficult to predict, but it was so unexpected that it was some time before I could bring myself to attempt the analysis; I was convinced that either one of the programs or the computer was at fault.

3. ELEMENTARY ERROR ANALYSIS

We wish to avoid any detailed rounding error analysis in this article. Fortunately a very simplified analysis is quite adequate to demonstrate that the observed behaviour is to be expected. Let us concentrate on the evaluation of the polynomial

$$f(z) = z^n + a_{n-1}z^{n-1} + \cdots + a_1 z + a_0 \qquad (3.1)$$

at $z = x$. This was achieved via the algorithm

$$s_n = 1$$

$$s_p = xs_{p+1} + a_p \ (p = n-1, \ldots, 1, 0), \qquad (3.2)$$

$$f(x) = s_0,$$

usually known as *nested multiplication*. We note in passing that $z^{n-1} + s_{n-1}z^{n-2} + \cdots + s_2 z + s_1$ is the quotient polynomial and s_0 is the remainder when $f(z)$ is divided by $(z - x)$.

Consider now a hypothetical computation in which every operation is performed *exactly* except the addition involved in the computation of s_k. We assume that an error ε is made at this stage. This error could be the result of a 'blunder' or a mere rounding error. The computed $\bar{s}_i$ satisfy *exactly* the relations

$$\bar{s}_n = 1,$$

$$\bar{s}_p = x\bar{s}_{p+1} + a_p \qquad (p \neq k), \tag{3.3}$$

$$\bar{s}_k = x\bar{s}_{k+1} + a_k + \varepsilon.$$

(Of course the $\bar{s}_i$ will be the same as the s_i until we reach $\bar{s}_k$; after that they will differ.) Clearly

$$\bar{f}(x) = f(x) + \varepsilon x^k. \tag{3.4}$$

This is an exact relation and is in no sense dependent on ε being small. The performance is clearly dependent on the relative sizes of $|f(x)|$ and $|\varepsilon x^k|$. It will appear later that when x is in the neighbourhood of 20, the most dangerous term, in the case when ε arises from a rounding error, is s_{15}, so we concentrate on the case $k = 15$.

Suppose $x = 20 + y$ where $|y| \ll 1$, then

$$f(x) = y(1+y)(2+y) \cdots (19+y) \doteq 19!y \doteq 1.22 \times 10^{17}y \tag{3.5}$$

while

$$\varepsilon x^{15} \doteq \varepsilon(20)^{15} \doteq 3 \times 10^{19}\varepsilon. \tag{3.6}$$

We now have to decide the order of magnitude of ε. This error arises when we attempt to compute

$$x\bar{s}_{16} + a_{15} \tag{3.7}$$

and the coefficient a_{15} in our polynomial lies between 10^9 and 10^{10}. In floating-point arithmetic on the PILOT ACE the mantissa had 30 binary digits, roughly the equivalent of 9 decimal digits. Hence, since a_{15} requires ten digits for its representation, the rounding error involved in its mere representation may be as large as 5 units! If $|x\bar{s}_{16}| < |a_{15}|$ (as, in fact, it is), the rounding error made in this addition is also of that order. Taking ε to be 5 then, we have

$$\varepsilon x^{15} \doteqdot 1.5 \times 10^{20}. \qquad (3.8)$$

If y really *is* appreciably less than unity, then from (3.5) and (3.8) $|f(x)|$ is far smaller than $|\varepsilon x^{15}|$. Even when $y \doteqdot 0.1$ the error is some 10^4 times as large as the true value. The computed value of $f(x)$ is completely dominated by this single rounding error. Since it could be positive or negative we would expect the computed values to be of random sign and far larger in value than the true values. In general there will be rounding errors in the computation of each s_p and on a nine-digit decimal computer no accuracy can be achieved in a computed value for any x in the range of, say, 12 to 22. It may readily be verified that when $y = 10^{-4}$, seventeen decimals are required in the computation in order to guarantee that even the sign of $\bar{f}(x)$ is correct. This is a very disappointing result and is all the more surprising since twenty is quite a modest degree; for a polynomial equation of, say, degree 100 with a similar distribution of roots the computational requirements are prohibitive. However the implications of the trivial error analysis we have just given are even more severe, as we shall show in later sections.

4. BACKWARD ERROR ANALYSIS

Let us return now to the equations defining the s_p and consider the true computing process with rounding errors made in every arithmetic operation. The computed values $\bar{s}_p$ satisfy the relations

$$\bar{s}_n = 1,$$
$$\bar{s}_p = x\bar{s}_{p+1} + (a_p + \varepsilon_{1p} + \varepsilon_{2p}), \qquad (4.1)$$
$$\bar{s}_0 = \bar{f}(x),$$

where ε_{1_p} is the rounding error made in the multiplication and ε_{2_p} is the error made in the addition. Equations (4.1) are *exact* relations; in practice we shall not know ε_{1_p} and ε_{2_p} but we shall have upper bounds for them in terms of the computed precision, $|x\bar{s}_{p+1}|$ and $|a_p|$. We have bracketed a_p with ε_{1_p} and ε_{2_p} in order to highlight a simple property of the $\bar{s}_p$. Equations (4.1) state that the computed sequence is an *exact* sequence corresponding to a polynomial with coefficients $a_p + \varepsilon_{1_p} + \varepsilon_{2_p}$ for precisely the relevant argument. Although the observation is almost trivial, it is of profound practical importance. It shows that all the errors made in the computation are precisely equivalent in their effect to perturbing the original coefficients by $\varepsilon_{1_p} + \varepsilon_{2_p}$ and then performing the computation exactly. This is one of the simplest examples of a *backward error analysis*; it is so named since it reflects back all the rounding errors made in the course of the computation as equivalent errors in the data. Although basically a simple concept, many are uneasy when first introduced to it. There is a tendency to worry about the effect of errors in one $\bar{s}_p$ on the computation of subsequent values. From the point of view of a backward rounding error analysis this is irrelevant; such an analysis makes no comment on the errors in the $\bar{s}_p$ though it may be used subsequently to obtain bounds for them. Once we have admitted the perturbations in the data then the computed values are *exact*. There are several advantages in this form of analysis.

(i) It puts the rounding errors made during the computation on the same footing as the original errors in the data. In practice the data will seldom be exact so that one must of necessity consider the effect of errors in them.

(ii) It reduces the problem of determining the errors in the solution to that of the effect of perturbations in the data. Perturbation theory has been extensively studied in classical mathematics and hence there is a rich body of experience on which to draw.

(iii) The third advantage may be described as follows. Suppose we have a problem with data elements c_i and we can show that the computed solution is exact for data $c_i(1 + \varepsilon_i)$ and give bounds for $|\varepsilon_i|$. The size of these bounds for the $|\varepsilon_i|$

provides a basis for assessing the performance of the algorithm in the presence of rounding errors.

In connexion with (iii), the following considerations are of particular relevance. In any algorithm all of the data elements c_i must be used at least once, i.e., they must take part in some arithmetic operation. In general that operation will involve a rounding error and the effect of that error will be that we have effectively used $c_i(1 + \eta_i)$ rather than c_i where $|\eta_i|$ is bounded by the relative precision ε of the basic computer operations. Hence it is virtually impossible for any practical algorithm to do better *in general* than to give the exact results for data $c_i(1 + \varepsilon_i)$ where the $|\varepsilon_i|$ can take values up to ε. In so far as each c_i and its successors take part in a large number of operations, even a very stable algorithm is likely to give much larger values of ε_i than ε.

Now the remarkable thing about nested multiplication is that in spite of the fact that it can provide such poor computed values of $f(x)$ it is extremely stable in the sense we have described and is particularly so on precisely the example we discussed in section 2. In fact we shall show that each $\bar{f}(x)$ is exact for a polynomial with coefficients $a_i(1 + \varepsilon_i)$ with bounds on $|\varepsilon_i|$ only marginally greater than the computer precision. Let us assume for the moment that $|a_p| > |x\bar{s}_{p+1}|$ really is true. It is not our intention in this paper to indulge in any detailed rounding error analysis since this tends to be indigestible. Instead we shall make our point by illustrating what happens on a six-decimal digit computer. Let us assume that the exponent of a_p is k and the exponent of $x\bar{s}_{p+1}$ is, say, $k - 3$. Then we may write

$$a_p = 10^k(.u_1 u_2 u_3 u_4 u_5 u_6) \tag{4.2}$$

$$x\bar{s}_{p+1} = 10^k(.000\, v_1 v_2 v_3 v_4 v_5 v_6 | v_7 v_8 v_9 v_{10} v_{11} v_{12}) \tag{4.3}$$

where the u_i denote the six decimal digits in a_p. The exact value of $x\bar{s}_{p+1}$ (note that in a backward error analysis we are not concerned with the fact that $\bar{s}_{p+1}$ is in error) is a twelve-digit number with an exponent three less (by our assumption); hence when aligned with

a_p it starts with three zeros. The v_i represent the twelve digits. In practice the product $x\bar{s}_{p+1}$ will be rounded, giving an error bounded by $\frac{1}{2}$ in the digital position of v_6. The addition then takes place and the sum is rounded. The rounding at this stage is bounded by $\frac{1}{2}$ in the digital position of u_6, or just possibly one digit lower if cancellation takes place (N.B., one digital position at most can be lost by cancellation). Hence the bound for $|\varepsilon_{p1}| + |\varepsilon_{p2}|$ is only marginally greater than $\frac{1}{2}$ in the last digit of the computer representation of a_p. This must be regarded as a best possible result.

It is not difficult to justify the assumption that $|x\bar{s}_{p+1}| < |a_p|$ for an x in the region of interest. This is because (with exact computation) the s_p are the coefficients of the quotient polynomial when $f(z)$ is divided by $z - x$.

Our conclusion is that there is no alternative algorithm which is likely to give better results than nested multiplication since a relative perturbation of up to $\frac{1}{2}$ in the last digit of each a_i is virtually inevitable.

In Table 1 we give the computed values of $f(z) = (z-1)(z-2) \cdots (z-12)$ for a range of arguments in the neighbourhood of 10. These were obtained using floating-point computation with a mantissa of 46 binary digits. The polynomial of order 12 rather than 20 was used because some of the coefficients of the latter have too many digits to be represented exactly on a 46 binary digit computer.

Notice that the computed values of $f(z)$ for values of z from $10 + 2^{-42}$ to $10 + 2^{-23}$ are all roughly the same size and of unpre-

TABLE 1

z	Computed $f(z)$		True $f(z)$	
$10 + 1 \times 2^{-42}$	$+0.63811$		0.16501×10^{-6}	
$10 + 2 \times 2^{-42}$	$+0.57126$	Dominated	0.33003×10^{-6}	
$10 + 3 \times 2^{-42}$	-0.31649	by	0.49505×10^{-6}	Almost linear
$10 + 4 \times 2^{-42}$	-0.45823	rounding	0.66007×10^{-6}	over whole of
$10 + 2^{-28} + 7 \times 2^{-42}$	$+0.29389$	errors	0.27048×10^{-2}	this interval
$10 + 2^{-23} + 7 \times 2^{-42}$	$+0.70396$		0.86518×10^{-1}	
$10 + 2^{-18} + 7 \times 2^{-42}$	$+0.33456 \times 10^{1}$		0.27685×10^{1}	
$10 + 2^{-13} + 7 \times 2^{-42}$	$+0.89316 \times 10^{2}$		0.88608×10^{2}	

dictable sign. The contribution made by the rounding errors swamps
the true value. It is not until we reach $z = 10 + 2^{-13}$ that we are
sufficiently far from the root at 10 for the value of $f(x)$ itself to
dominate the rounding errors. The size of the computed values is
perfectly predictable. The most dangerous term turns out to be
$a_8 z^8$. In fact, $a_8 = 749463$ and hence the contribution made by the
rounding error when we compute s_8 is bounded by

$$2^{-46} \times 749643 \times 10^8 \text{ (approximately)} \doteq \frac{7.50 \times 10^{13}}{2^{46}} \doteq \frac{7.50 \times 10^{13}}{7.04 \times 10^{13}}$$

$$\doteq 1.00. \qquad (4.4)$$

Accordingly, we can expect the rounding errors to contribute
something of the order of magnitude of unity to the computed
value of $\bar{f}(z)$ and it is therefore only when the true $f(z)$ is
substantially larger than this that we have any correct significant
figures. The table deserves careful study. Since 10 requires 4 binary
digits for its representation, 2^{-42} is the value of the least significant
digit on a 46-digit computer for any number in the neighbourhood
of 10. Evaluation at $x = 10$ gives no rounding errors so the *computed* value is equal to the true value, i.e., zero. Accordingly the
first four arguments were taken to be the four successive computer
numbers following 10 itself. For these $f(z)$ is of the order of 10^{-6}
and this is completely swamped. Next we chose arguments which
were essentially $10 + 2^{-28}$, $10 + 2^{-23}$, $10 + 2^{-18}$, $10 + 2^{-13}$ but in
order to avoid the danger of atypical roundings each of these was
augmented by 7×2^{-42}. With the third of these arguments $f(z)$ is
just large enough to rise above the level of the rounding errors and
$\bar{f}(z)$ has one correct decimal digit; with the fourth argument $f(z)$
has reached 10^2 and the computed $\bar{f}(z)$ now has rather better than
two decimals of accuracy. It may be verified that for all eight values
of the argument the error is of order unity with the sort of variation
one might expect from the random fluctuations. Had we attempted
the exercise with the polynomial $(x - 1)(x - 2) \cdots (x - 20)$ on the
same computer, the rounding errors would have dominated the
computed value over most of the range $z + 12$ to $z = 22$.

5. SENSITIVITY OF THE ROOTS

So far we have concentrated on what might be regarded as the rather unexciting problem of evaluating $f(z)$, though we are primarily concerned with finding the roots of $f(z) = 0$. However, if changes in the least digit in the coefficients a_i completely change the exact value of the polynomial for a given argument, it is obvious that the implications for the roots must be very serious. At this point we turn to perturbation theory and ask ourselves "What is the effect on the roots of a perturbation of a_i by εa_i?"

Let us denote the roots of the original polynomial by x_i. If ε is sufficiently small the root x_j becomes $x_j + \delta x_j$ where

$$|\delta x_j| \doteqdot |\varepsilon a_i x_j^i / f'(x_j)|$$

$$\doteqdot \left| \varepsilon a_i x_j^i / \prod_{i \neq j} (x_i - x_j) \right|. \tag{5.1}$$

For the polynomial we have discussed it can be verified that the greatest sensitivity is that of the root 15 (which is x_6) with respect to a perturbation in a_{15}. For this we have

$$|\delta x_6| = |\varepsilon a_{15} \cdot 15^{15} / 5! 14!|, \tag{5.2}$$

and since $a_{15} \doteqdot 1.67 \times 10^9$, this gives

$$|\delta x_6| \doteqdot 0.7\varepsilon \times 10^{14}. \tag{5.3}$$

Even with ε as small as 10^{-14} the perturbation is 0.7 and this is so large that the first order theory is obviously inadequate.

The implications of this are quite calamitous. All coefficients of our polynomial from a_{15} to a_0 require more than 9 decimal digits for their representation. Hence on a nine-digit computer even the best possible representations of each of these coefficients will involve a rounding error and this will correspond to a value of ε between $\pm \frac{1}{2} 10^{-9}$. In fact the PILOT ACE program for computing the polynomial performed extremely well; whenever all the given values x_i were of the same sign the computed coefficients were, in general, correct to working accuracy. Unfortunately, the exact roots of the computed polynomial differed wildly from the given x_i. We illustrate this in Table 2 by giving the roots of the polynomial equation $(z-1)(z-2)\cdots(z-20)-2^{-23}z^{19}=0$, which differs from the original polynomial only in the coefficient of z^{19}. The true a_{19} is 210, which requires 8 binary digits for its representation; on a 30-binary digit computer the perturbation 2^{-23} therefore represents a unit in the first position beyond the end of the computer representation of a_{19}. It will be seen from the table that the smaller roots are scarcely affected but by the time we reach the root $x=7$ the perturbation has become substantial, while the ten roots from 10 to 19 have become five complex conjugate pairs; the roots 18 and 19, for example, have become $19.5\cdots\pm(1.9\cdots)i$. Perturbations in a_{15} and a_{16} are even more devastating in their effect.

6. IMPLICATIONS OF THE ERROR ANALYSIS

We now turn to the implications of the results of sections 2 to 5 for practical numerical analysts. Polynomial equations arise in very many branches of 'applied mathematics', using that term in its widest sense. However, these equations do not present themselves directly in explicit polynomial form. Perhaps the most important illustration of this is provided by the algebraic eigenvalue problem which arises in connexion with vibration problems in physics, chemistry and engineering. Here the polynomial equation is in the determinantal form

$$\det(\lambda I - A) = 0, \tag{6.1}$$

TABLE 2
Roots of $(z-1)(z-2)\cdots(z-20)-2^{-23}2^{19}=0$ correct to 9 decimal places

1.00000 0000	6.00000 6944	$10.09526\ 6145 \pm 0.64350\ 0904i$
2.00000 0000	6.99969 7234	$11.79363\ 3881 \pm 1.65232\ 9728i$
3.00000 0000	8.00726 7603	$13.99235\ 8137 \pm 2.51883\ 0070i$
4.00000 0000	8.91725 0249	$16.73073\ 7466 \pm 2.81262\ 4894i$
4.99999 9928	20.84690 8101	$19.50243\ 9400 \pm 1.94033\ 0347i$

which for an $n \times n$ matrix gives a monic polynomial equation of degree n; however, the primary data are the elements of A. Now it was perfectly natural for mathematicians to approach this problem by devising algorithms which would give coefficients of the *explicit* polynomial corresponding to $\det(\lambda I - A)$, i.e., the characteristic polynomial of A. This had the advantage of reducing the volume of data from the n^2 elements of A to the n elements of the characteristic polynomial. However, the real incentive to the development of such algorithms was that it reduced the problem to the solution of an explicit polynomial equation, and this was felt to be a highly desirable transformation. Although attempts were made to analyse the effect of the rounding errors made in the algorithm on the accuracy of the computed coefficients of the explicit polynomial form, the desirability of this form does not seem to have been questioned. Almost all of the algorithms developed before the 1950s for dealing with the unsymmetric eigenvalue problem (i.e., problems in which A is unsymmetric) were based on some device for computing the explicit polynomial equations.

So far we have based our exposition of the numerical problems associated with computing the roots of polynomial equations on a specific polynomial of degree twenty. This polynomial now enjoys a certain notoriety; it is sometimes referred to as 'Wilkinson's remarkable polynomial' and the opinion is quite widely held that there is something unusual about it. This is a particularly disappointing development. The polynomial was first used by me quite by chance; it was in fact the first polynomial of order 20 that came to mind; so, far from trying to invent a particularly intractable polynomial, I used it precisely because I imagined (quite wrongly) that it would be free from any numerical difficulties. It has the advantage that the orders of magnitude of $f(x)$ and $f'(x)$ may easily be determined in terms of factorials.

The really disturbing fact is that it is quite typical of polynomials with real coefficients and real roots and, indeed, many polynomials which arise in practice behave much worse than this. As an experiment I once took some 100 eigenvalue problems of order 25 for which I had found the eigenvalues by an algorithm that is known to be very stable with respect to eigenvalues. Knowing the eigenvalues, I then computed the minimum accuracy to which the explicit polynomial would need to be computed in order to ensure that it would give the eigenvalues correct to 5 decimals. (This would be a reasonable requirement in practice.) In every single case it would have been necessary to obtain the coefficients correct to more than 20 decimals and for many of the examples a much higher accuracy would have been necessary.

Again it should be emphasized that the polynomial $(z - 1)(z - 2)$ $\cdots (z - 20)$ is not a 'difficult' polynomial *per se*. I have sometimes been asked what precision of computation was necessary in order to determine the roots in Table 2 to 9 decimal places. This question is asked in the mistaken belief that this must be a difficult numerical problem. The 'difficulty' with the polynomial $\Pi(z - i)$ is that of evaluating the *explicit* polynomial accurately. If one already knows the roots, then the polynomial can be evaluated without any loss of accuracy. The roots in Table 2 were computed using a zero finder and evaluating the function directly by means of the expression $\Pi(z - i) - 2^{-23}z^{19}$. Eleven decimal computation was perfectly adequate to determine the roots quoted.

Practical computer users sometimes find it difficult to believe that there can be any justification for determining the coefficients of an explicit polynomial to high accuracy when the primary data are known to much lower accuracy. The fallacy in this argument may be exposed by quite a simple example. Consider the determinantal equation

$$\det \begin{bmatrix} 1 + \varepsilon_1 - \lambda & \varepsilon_3 \\ \varepsilon_4 & 1 + \varepsilon_2 - \lambda \end{bmatrix} = 0. \qquad (6.2)$$

The explicit polynomial form is

$$\lambda^2 - (2 + \varepsilon_1 + \varepsilon_2)\lambda + (1 + \varepsilon_1)(1 + \varepsilon_2) - \varepsilon_3\varepsilon_4 = 0 \qquad (6.3)$$

and the roots are

$$\frac{(2 + \varepsilon_1 + \varepsilon_2) \pm \left((\varepsilon_1 - \varepsilon_2)^2 + 4\varepsilon_3\varepsilon_4\right)^{1/2}}{2}. \qquad (6.4)$$

Observe that under the square root we have only second-order terms in the ε_i, the potential first-order terms cancel out. If the ε_i are of the order of 10^{-6}, the two roots differ from unity by quantities of order 10^{-6}. Random perturbations of order 10^{-6} in the ε_i make changes of order 10^{-6} in these roots. However, random independent perturbations of order 10^{-6} in the coefficients of the explicit polynomial make perturbations of order 10^{-3} in the roots. Perturbations in the ε_i in (6.2) produce perturbations in the explicit polynomial (6.3) which are so correlated as to make their effect on the roots far less important. Unfortunately, rounding errors made in the coefficients are inevitably uncorrelated in general.

Determinantal equations of the type $\det(A - \lambda I) = 0$ are commonly such that the roots are relatively insensitive to independent perturbations in the a_{ij}. This may be true even when the roots include close clusters. In fact when A is symmetric the roots are well-conditioned in this respect, however sensitive the roots of the corresponding explicit polynomial equation may be. The numerical instability introduced in transforming to the explicit polynomial equation is an induced instability.

7. AMELIORATION OF ILL-CONDITIONING

The sensitivity of the roots of explicit polynomial equations is something which is inherent in that mode of representation. All algorithms which involve rounding errors are, in general, doomed to give results of disappointing accuracy relative to the computer precision when used to deal with an ill-conditioned polynomial.

It is natural to ask whether the ill-condition can be removed by some simple transformation of the variables. The answer is that sometimes it can but the transformation itself must be performed to high accuracy because, as we have remarked, the transformed polynomial will be exactly related to a polynomial $\Sigma(a_i + \varepsilon_i)z^i$ with ε_i at best of the order of the precision used.

An interesting example in this respect is provided by the polynomial equation

$$z^3 - 3.0000002z^2 + 2.9999997z - 1.0000004 = 0. \qquad (7.1)$$

This polynomial is very close to $(z-1)^3$ and, hence, the roots can be expected to be very sensitive to random perturbations in the coefficients such as will occur in the evaluation of the polynomial. It is attractive to make the transformation $z-1 = w$ and thus to obtain a cubic equation in w which no longer has close roots. Now the standard algorithm for determining the coefficients of the transformed polynomial uses repeated division by $z-1$, the successive remainders providing the coefficients. As we have remarked, the process of dividing by $z - \alpha$ is precisely the same as evaluating the polynomial at $z = \alpha$. Now if the transformation is performed using standard 8-decimal digit, floating-point arithmetic the resulting polynomial equation is

$$w^3 - 2\times 10^{-7}w^2 - 7\times 10^{-7}w - 9\times 10^{-7} = 0, \qquad (7.2)$$

and this provides accurate roots. This would seem to contradict the claim that the transformation must be performed in higher precision but that is an illusion. The fact is that no rounding errors occur in this reduction, so we may regard it as having been done to arbitrarily high precision even though we appeared to be using only 8-decimal digit computation. Had we been foolish enough to make, say, the transformation $w = z - 1.0000123$, then rounding errors would have occurred and these would have been just as damaging as those arising in the use of the original polynomial.

Representation of (7.1) in the form

$$(z-1)^3 = 10^{-7}(2z^2 + 3z + 4) \qquad (7.3)$$

also enables one to compute the roots accurately via the iterative procedure

$$(z_{r+1} - 1)^3 = 10^{-7}(2z_r^2 + 3z_r + 4), \qquad (7.4)$$

provided care is taken to select the same branch throughout for iteration to a specific root. Again this is possible because the transformation from (7.1) to (7.3) has been performed *exactly*.

8. MULTIPLE ROOTS

The equation $(z-1)(z-2)\cdots(z-20)=0$ was adopted initially precisely because it has 'well separated' roots. The objective was to avoid the known difficulties associated with multiple roots or root clusters. Since the numerical problems proved to be unexpectedly severe, it is pertinent to ask whether the selected polynomial equation can be thought of as in any sense related to one with multiple roots.

In fact very small relative perturbations in the coefficients can give an equation with multiple roots. To see this consider, for example, the perturbed equation

$$(x-1)(x-2)\cdots(x-20)-\varepsilon x^{15}=0. \qquad (8.1)$$

The roots of this equation occur at the points of intersection of the curves $y=f(x)\equiv(x-1)(x-2)\cdots(x-20)=0$ and $y=g(x)\equiv +\varepsilon x^{15}$. The first curve is symmetric about the point $x=10.5$, crosses the axis at $x=i$ $(i=1,\dots,20)$ and has peaks in between these values, alternatively positive and negative, at the zeros of the derivative. Consider now values of ε of order 10^{-5}. For values of x from, say, 0 to 6, $g(x)$ is *very* close to the x-axis compared with the peaks of $f(x)$, and the points of intersection of $f(x)$ with $g(x)$ are virtually the same as the points of intersection of $f(x)$ with the x-axis. Hence (as we have already seen) the early roots are scarcely affected. However, as x increases, $g(x)$ increases quite rapidly while $f(x)$ continues to dissipate its energies by oscillating in sign. When we reach values of x, say, from 14 to 17, $g(x)$ is fully comparable with $|f(x)|$ even at the peaks of the latter. This is illustrated in Fig. 1 for $\varepsilon=\pm(0.5)10^{-5}$ and $\varepsilon=\pm(1.0)10^{-5}$. For $\varepsilon=+(0.5)10^{-5}$ the roots 14 and 15 have moved towards each other to a substantial extent as have also the roots 16 and 17. For $\varepsilon=(1.0)10^{-5}$ the roots originally at 14 and 15 have become complex conjugate pairs and the roots originally at 16 and 17 have almost coalesced. For an ε between these two values, $y=g(x)$ will just touch $y=f(x)$, giving a double root between 14 and 15. A slightly larger value of ε will produce a double root between 16 and 17. Similarly, for negative ε, a value between $-(0.5)10^{-5}$ and

$-(1.0)10^{-5}$ will give a double root between 15 and 16. Since $a_{15} \doteq (1.67)10^9$ these ε represent relative perturbations of the order of 10^{-14}. It is an instructive exercise to determine perturbations which give two pairs of double roots and other combinations. A more difficult question is the following.

Suppose we have a computer with an inadequate word length for the exact representation of the polynomial $(x-1)(x-2)\cdots$ $(x-20)$, for example, a ten-digit decimal floating-point computer. The correctly rounded version of the polynomial equation has roots which are very different from the true values. Determine the ten-digit decimal polynomial such that its roots z_i give the mini-

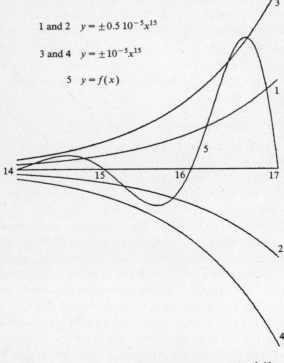

1 and 2 $y = \pm 0.5\,10^{-5}x^{15}$

3 and 4 $y = \pm 10^{-5}x^{15}$

5    $y = f(x)$

$$y = (x-1)(x-2)\cdots(x-20) \quad \text{and} \quad y = k(10^{-5}x^{15})$$
$$\text{for} \quad x = 14 \text{ to } 17$$

Fig. 1.

mum value of $\sum |z_i - i|$ (or any other appropriate measure of the difference between the two sets of roots).

9. WELL-CONDITIONED POLYNOMIALS

Although it is important to realize that explicit polynomial equations with ill-conditioned roots are remarkably common, not all polynomials of high degree are like that. The polynomial equations

$$x^n \pm 1 = 0 \qquad (9.1)$$

are extremely well-conditioned for all values of n. A perturbation ε in any coefficient gives a perturbation of order ε/n in all the roots.

Of particular interest for the following section are polynomial equations with roots $\alpha, \alpha^2, \ldots, \alpha^n$. If we take $\alpha = 1/2$, $n = 20$, we find that the coefficients vary enormously in order of magnitude. In fact, in terms of orders of magnitude, the polynomial is

$$x^{20} + x^{19} + 2^{-1}x^{18} + 2^{-4}x^{17} + 2^{-8}x^{16} + 2^{-13}x^{15} + 2^{-19}x^{14}$$

$$+ 2^{-26}x^{13} + 2^{-34}x^{12} + 2^{-43}x^{11} + 2^{-53}x^{10} + 2^{-64}x^9$$

$$+ 2^{-76}x^8 + 2^{-89}x^7 + 2^{-103}x^6 + 2^{-118}x^5 + 2^{-134}x^4$$

$$+ 2^{-151}x^3 + 2^{-169}x^2 + 2^{-189}x + 2^{-210} \qquad (9.2)$$

At first sight this appears to be an unsatisfactory state of affairs; moreover there is a natural tendency to regard the roots as dangerously close with a cluster near zero. Such fears prove to be completely unfounded. Small relative perturbations in all coefficients make correspondingly small *relative* perturbations in each of the roots. Even the smallest root behaves well. Since rounding errors made in evaluating a polynomial are always equivalent in their effect to small relative perturbations in the coefficients, one can predict that this polynomial equation will behave well. In fact even the rather unsophisticated program used on the PILOT ACE was capable of determining any one of the roots almost to full working accuracy.

To establish this result we observe that for a perturbation εa_k in a_k, the relative change in root x_i is

$$|\delta x_i / x_i| \sim \varepsilon |a_k| 2^{-ki} / (2^{-i} |PQ|), \qquad (9.3)$$

where

$$P = (2^{-i} - 2^{-1})(2^{-i} - 2^{-2}) \cdots (2^{-i} - 2^{-i+1})$$

$$Q = (2^{-i} - 2^{-i-1})(2^{-i} - 2^{-i-2}) \cdots (2^{-i} - 2^{-20}). \qquad (9.4)$$

We may write

$$|P| = 2^{-\frac{1}{2}i(i-1)} \left[(1 - 2^{-1})(1 - 2^{-2}) \cdots (1 - 2^{-i+1}) \right]$$

$$|Q| = 2^{-i(20-i)} \left[(1 - 2^{-1})(1 - 2^{-2}) \cdots (1 - 2^{-20+i}) \right] \qquad (9.5)$$

and the bracketed expressions are both convergents to the infinite product

$$(1 - 2^{-1})(1 - 2^{-2})(1 - 2^{-3}) \cdots. \qquad (9.6)$$

All convergents certainly lie between $1/2$ and $1/4$ and hence

$$|\delta x_i / x_i| < \varepsilon |a_k| 2^{s+4} \quad \text{where } s = \tfrac{1}{2}i[41 - 2k - i]. \qquad (9.7)$$

For a fixed value of k this takes its maximum where $i = 20 - k$, so that

$$|\delta x_i / x_i| < \varepsilon |a_k| 2^{4 + \frac{1}{2}(20-k)(21-k)}. \qquad (9.8)$$

However from (9.2) it may be verified that all a_k satisfy

$$|a_k| \leqslant 4.2^{-\frac{1}{2}(20-k)(21-k)} \qquad (9.9)$$

and hence

$$|\delta x_i / x_i| < 2^6 \cdot \varepsilon. \qquad (9.10)$$

A relative perturbation of ε in any coefficient can be amplified at worst by the factor 2^6 though for most i and k this is an overestimate since the above inequalities are quite crude. This polynomial then is well-behaved and for smaller values of α its behaviour is even better. However, such polynomials proved to have a sting in their tails.

10. POLYNOMIAL DEFLATION

The Fundamental Theorem of Algebra asserts that every polynomial equation over the complex field has a root. It is almost beneath the dignity of such a majestic theorem to mention that in fact it has precisely n roots. The latter result is regarded as trivial and is usually mentioned only as an aside. When a root x_1 has been found, the polynomial $f(z)$ may be divided by $z - x_1$ to give a quotient polynomial of degree $n - 1$

$$f(z) = (z - x_1)q(z). \tag{10.1}$$

By the Fundamental Theorem $q(z)$ has a root and the process can be continued to give the n roots.

I must have been subconsciously influenced by this attitude because when I turned to polynomial deflation (as the above process is called), I was not expecting substantial difficulties. Since I knew by this time that the polynomial with roots 2^{-i} was well-conditioned, I used this polynomial as my first test. The largest root 2^{-1} was found first and this had an error of only 1 in its least significant digit. The approximation to the root 2^{-2} derived from the deflated polynomial was of disappointing accuracy and later roots proved to have virtually no relation to the correct roots even as regards their orders of magnitude.

Again backward error analysis is an invaluable tool for demonstrating the cause of this difficulty. Suppose we have accepted a value $x_1 + \varepsilon$ as an approximation to x_1. Let us consider the result of performing the deflation process exactly using this approximation. We have

$$f(z) = (z - (x_1 + \varepsilon))q(z) + r \tag{10.2}$$

where $q(z)$ is the deflated polynomial and r is the remainder. In practice, of course, r is ignored and work is continued with $q(z)$. Now, by our assumption, equation (10.2) is exact and hence

$$(z-(x_1+\varepsilon))q(z) = f(z)-r \text{ (exactly).} \qquad (10.3)$$

This equation implies that $x_1 + \varepsilon$ and the exact roots of $q(z)$ are those of $f(z)-r$, not of $f(z)$ itself. However, from the Remainder Theorem, $r = f(x_1 + \varepsilon)$ and since

$$f(z) = (z - x_1)(z - x_2)\cdots(z - x_n), \qquad (10.4)$$

we have

$$r = f(x_1 + \varepsilon) = \varepsilon(x_1 - x_2 + \varepsilon)(x_1 - x_3 + \varepsilon)\cdots(x_1 - x_n + \varepsilon).$$

$$(10.5)$$

Consider now the implications of this for a polynomial equation with roots $x_i = 2^{-i}$ when the accepted approximation to x_1 is $2^{-1} + 10^{-10}$ and, therefore, has an error only in the tenth decimal. From (10.5)

$$r = \varepsilon(2^{-1} - 2^{-2} + \varepsilon)(2^{-1} - 2^{-3} + \varepsilon)\cdots(2^{-1} - 2^{-20} + \varepsilon)$$

$$(\varepsilon = 10^{-10}). \quad (10.6)$$

Each of the terms in parenthesis is certainly greater than 2^{-2} and hence

$$r > \varepsilon 2^{-38} = 10^{-10} 2^{-38} > 2^{-72}. \qquad (10.7)$$

With exact deflation the roots of $q(z) = 0$ will be roots of $f(z) - r$, which has the constant term $2^{-210} - r$ instead of 2^{-210}. However, r is larger than 2^{-210} by a factor greater than $2^{138} \doteq 10^{41}$. Since the product of the roots of a monic polynomial equation is equal to the constant term and this has been changed by an enormous factor, the roots of $q(z)$ cannot possibly give good approximations to the remaining roots of $f(z)$. It is hardly likely that when rounding errors are made in the deflation process itself the results will be superior to those resulting from exact deflation with the approximate root.

On the other hand, suppose an approximation $2^{-20} + \varepsilon$ to 2^{-20} with a low relative error has been accepted. We have now

$$r = f(2^{-20} + \varepsilon) = (2^{-20} + \varepsilon - 2^{-1})(2^{-20} + \varepsilon - 2^{-2})$$

$$\cdots (2^{-20} + \varepsilon - 2^{-19})\varepsilon \qquad (10.8)$$

and we assume that $\varepsilon = 10^{-16}$, which again corresponds to a relative error of approximately one part in 10^{10}. We require an approximation to $|r|$ and, fortunately, quite a crude bound will suffice. We have certainly

$$|r| < 2^{-1} \cdot 2^{-2} \cdots 2^{-19} \cdot 10^{-16} = 2^{-190} \times 10^{-16} \doteq 2^{-210} \times 10^{-10}.$$

$$(10.9)$$

Hence the constant term 2^{-210} is changed by less than one part in 10^{10}, and we know that this has little effect on any of the roots.

Although this is a very well-conditioned polynomial, we see that division by an approximate factor corresponding to the largest root is disastrous even if performed exactly; division by an approximate factor corresponding to the smallest root is completely harmless. This analysis is entirely born out in practice. A similar result is true in general for polynomials having roots of widely varying magnitudes.

The above argument suggests that roots should be found in terms of increasing absolute magnitude. Unfortunately, with many of the more efficient algorithms it is not easy to ensure that roots *are* found in this order. It is natural to ask whether there is some stratagem which overcomes this problem. In fact there is a very simple solution.

It is a well-established tradition when dividing a polynomial of degree n by an (exact) linear factor to obtain the coefficient of z^{n-1} first and then those of $z^{n-2}, z^{n-3}, \ldots$ in succession. With an exact factor and exact division there is, of course, no remainder. However, we could just as well start at the other end and find the constant term first, followed by those of $z, z^2, \ldots$ in succession;

again there will be no remainder. The quotient polynomials will, of course, be exactly equal.

If α is merely an approximate root, these quotient polynomials will not be exactly equal; the remainder will be a constant with the conventional mode of division and a multiple of z^n when the division is performed in the reverse direction. Now we may think of division in the reverse direction in the following terms. Suppose we write $z = w^{-1}$, then

$$f(z) = f(1/w) = \left(a_0 w^n + a_1 w^{n-1} + \cdots + a_n\right)/w^n = g(w)/w^n$$

$$(10.10)$$

$$(z - \alpha) = (1 - \alpha w)/w = (\alpha^{-1} - w)\alpha/w \qquad (10.11)$$

and if we write $\beta = \alpha^{-1}$, we can think of dividing $(w - \beta)$ into

$$\left(a_0 w^n + a_1 w^{n-1} + \cdots + a_n\right).$$

If α is the largest root of $f(z)$, then β is the smallest root of $g(w)$. If a large root of $f(z)$ has been found, then clearly it should be divided out in the reverse direction.

However, this still leaves us with the necessity of finding either the largest root or the smallest at each stage. If we find one of the inner ones, such as 2^{-10} in our example, then neither quotient is satisfactory. Suppose now we denote the two quotients by

$$b_{n-1}z^{n-1} + b_{n-2}z^{n-2} + \cdots + b_1 z + b_0, \qquad (10.12)$$

and

$$c_{n-1}z^{n-1} + c_{n-2}z^{n-2} + \cdots + c_1 z + c_0, \qquad (10.13)$$

respectively. We could use a composite quotient using coefficients $b_{n-1}, b_{n-2}, \ldots, b_r$ and $c_{r-1}, c_{r-2}, \ldots, c_0$. Note that in dividing forward, b_{n-1} is found first and in dividing backwards, c_0 is found first; hence it seems appropriate to use those terms which are computed earlier in the two processes. Will there always be a good choice of r, presumably different according to which root one happens to have found?

The answer to this question is a resounding 'yes'. However, we must be clear what we mean by this. There is a limitation on the accuracy with which any root can be found even when iterating in the original polynomial. This is a fundamental property of the polynomial with respect to that root. The irreducible error in a root is that corresponding to relative perturbations in the original coefficients which are a modest multiple of the computer precision. It would be unreasonable to expect that the deflation process has positive virtues that enable us to find ill-conditioned roots more accurately after several deflations than by iteration in the original polynomial, assuming a fixed precision of computation throughout. In fact, by using the appropriate value of r when each approximate root has been accepted, the accuracy obtained in each root is almost equal to that attainable by iteration for that root in the original polynomial.

It should be appreciated that if the original polynomial equation is very ill-conditioned, the successive deflated polynomials will usually become steadily better conditioned. When, for example, the roots $1, 2, \ldots, 15$ of our notorious polynomial equation have been found, the remaining quintic with roots $16, \ldots, 20$ is very much better conditioned even though this includes several of the worst conditioned roots of the original. This improvement in the condition avails us little. When the first approximate root has been accepted and the appropriate decomposition (with rounding errors) has been performed, the deflated polynomial will be related to a slightly perturbed version of the original. These perturbations, small as they are, will already have severely damaged ill-conditioned roots. Even if the remaining $n - 1$ roots and deflations are performed exactly from that point onwards we can never recover from the loss of accuracy already inflicted.

However, there are circumstances in which this loss of accuracy may not be serious. This is when our primary objective is to produce a set of approximate roots $\bar{x}_i$ such that $\prod(z - \bar{x}_i)$ is 'close' to the given polynomial. A good iterative method for finding a single root plus the *composite* deflation we have described achieves just this. Though the errors in the ill-conditioned roots will be severe (except when the rounding errors happen, fortuitously, to be benign) the errors in the complete set will be correlated in such a way as to reproduce the original polynomial remarkably accurately. For a detailed explanation of this and the details of composite deflation the reader is referred to Peters and Wilkinson [7]. The error analysis involved is not particularly complex but its presentation here would violate the policy adopted in this article. However, a trivial example illustrates how this comes about.

Consider the polynomial equation

$$z^2 - 2z + 1 = 0, \tag{10.14}$$

which has $z = 1$ as a double root. This root is accordingly fairly ill-conditioned. If iteration is performed using six-decimal floating-point computation then, say, $x = 1.00024$ will be accepted as a root since the computed value of $f(x)$ will be exactly zero. Notice that the smallest attainable computed value (other than zero) is 10^{-6}, since the last step is to add xs_1 to a_0, i.e., to 1 to give s_0. The smallest value is attained when computed $xs_1 = -0.999999$. However, when deflation is performed (composite deflation is irrelevant here) the computed quotient is

$$z - 0.999760, \tag{10.15}$$

and the second computed root is 0.999760. Both roots have substantial errors but

$$(z - 1.00024)(z - 0.999760) = z^2 - 2z + 0.9999999424 \tag{10.16}$$

and the original polynomial is given to much greater accuracy than are the individual factors. Such a phenomenon is referred to as 'backwards stability' and is of great importance in numerical analysis.

11. CONCLUSIONS

The title of this article reflects my feelings in early encounters with solving polynomial equations but perhaps it is a misnomer. Polynomial equations may be said to have played an 'instructive' role in the history of mathematics and could have continued to play this role for numerical analysts in the period immediately following the advent of electronic computers.

There *was* an intense interest in the effect of rounding errors at that time but, unfortunately, it was focused primarily on matrix problems, particularly the solution of systems of linear equations. These problems had a superficial formal complexity which had the effect of making their analysis seem rather difficult. This formal complexity is completely absent from the problem of evaluating a polynomial or even of locating a simple real root of a real polynomial equation. Viewed in retrospect it is interesting that I used backward error analysis both in polynomial evaluation and in polynomial deflation almost, as it were, by accident. (The term backward error analysis had not been coined at that time and it was certainly not in use as a recognized tool.) Although I was astonished, indeed affronted, by my experiences with simple polynomials, I was rather pleased with the effectiveness of the analysis. Nevertheless it did not impress me sufficiently to make me adopt similar policies in the analysis of matrix problems. My conversion to backward error analysis in a matrix context did not occur until several years later. I had observed that for stable methods of solving the eigenvalue problem the residual vector r defined by

$$r = Ax - \lambda x \qquad (11.1)$$

was of the order of magnitude of the computer precision times $\|A\|$. Quite suddenly it occurred to me that this implied that

$$(A - rx^T)x = \lambda x \qquad (11.2)$$

(assuming x is normalized so that $x^Tx = 1$) and that λ and x were therefore exact for $A - rx^T$, which is a 'very close neighbour' of A. This *did* indeed ultimately persuade me consciously to adopt backward error analysis as a working tool and it yielded rich dividends almost immediately. By this time Givens [3] had already used it quite specifically in his analysis of the reduction of a real symmetric matrix to tri-diagonal form by orthogonal similarities. It is interesting that this did not lead to its explicit adoption by others or indeed by Givens himself in the analysis of other algorithms.

For accidental historical reasons therefore backward error analysis is always introduced in connexion with matrix problems. In my opinion the ideas involved are much more readily absorbed if they are presented in connexion with polynomial equations. Perhaps the fairest comment would be that polynomial equations narrowly missed serving once again in their historical didactic role and rounding error analysis would have developed in a more satisfactory way if they had not.

REFERENCES

1. G. Birkhoff and S. Mac Lane, *A Survey of Modern Algebra*, 4th Edition, Macmillan, New York, 1977.
2. B. Dejon and P. Henrici (Editors), *Constructive Aspects of the Fundamental Theorem of Algebra*, Proceedings of a symposium at the IBM Research Laboratory, Zurich-Ruschlikon, June 1967, Interscience, London, New York, 1969.
3. J. W. Givens, "Numerical computation of the characteristic values of a real symmetric matrix," Oak Ridge National Laboratory, ORNL-1574, 1954.
4. A. S. Householder and F. L. Bauer, "On certain methods for expanding the characteristic polynomial," *Numer. Math.*, 1 (1959), pp. 29–39.
5. M. A. Jenkins and J. F. Traub, "A three-stage variable shift iteration for polynomial zeros and its relation to generalized Rayleigh iteration," *Numer. Math.*, 14 (1970), pp. 252–263.
6. F. W. J. Olver, "Evaluation of zeros of high degree polynomials," *Phil. Trans. Roy. Soc.*, 244 (1952), pp. 385–415.
7. G. Peters and J. H. Wilkinson, "Practical problems arising in the solution of polynomial equations," *J. Inst. Math. Appl.*, 8 (1971), pp. 16–35.
8. G. W. Stewart, "Some iterations for factoring a polynomial," *Numer. Math.*, 13 (1969), pp. 458–471.
9. J. H. Wilkinson, "The evaluation of zeros of ill-conditioned polynomials," Part I, *Numer. Math.*, 1 (1959), pp. 150–166, Part II, *Numer. Math.*, 1 (1959), pp. 167–180.
10. _____, "Error analysis of floating point computation," *Numer. Math.*, 2 (1960), pp. 319–340.
11. _____, *Rounding Errors in Algebraic Processes*, Notes on Applied Sciences No. 32, Her Majesty's Stationery Office, London, 1963.
12. _____, *The Algebraic Eigenvalue Problem*, Oxford University Press, London, 1965.

NEWTON'S METHOD

Jorge J. Moré and D. C. Sorensen

1. INTRODUCTION

Many fundamental problems in science, engineering, and economics can be phrased in terms of minimizing a scalar valued function of several variables. Problems that arise in these practical settings usually have constraints placed upon the variables. Special techniques are required to handle these constraints but eventually the numerical techniques used must rely upon the efficient solution of unconstrained minimization problems.

Newton's method plays a central role in the development of numerical techniques for optimization. One of the reasons for its importance is that it arises very naturally from considering a Taylor approximation to the function. Because of its simplicity and wide applicability, Newton's method remains an important tool for solving many optimization problems. In fact, most of the current practical methods for optimization (e.g., quasi-Newton methods) can be viewed as variations on Newton's method. It is therefore important to understand Newton's method as an algorithm in its own right and as a key introduction to the most recent ideas in this area.

One of the aims of this paper is to present and analyze two main approaches to Newton's method for unconstrained minimization: the line search approach and the trust region approach. The other aim is to present some of the recent developments in the optimization field which are related to Newton's method. In particular, we explore several variations on Newton's method which are appropriate for large-scale problems, and we also show how quasi-Newton methods can be derived quite naturally from Newton's method.

We assume familiarity with some of the basic notions from computational linear algebra (see, for example, Stewart [51]), and the calculus of functions of several variables (see, for example, Chapter 7 of Bartle [1] or Chapter 3 of Ortega and Rheinboldt [39]), but otherwise the background necessary for this paper is minimal. We begin our development by reviewing some standard definitions and results.

Given a function $f: R^n \to R$ defined in an open set D, the unconstrained minimization problem is to find $x^* \in D$ such that

$$f(x^*) \leqslant f(x), \qquad x \in N(x^*), \tag{1.1}$$

for some open neighborhood $N(x^*)$ of the *local minimizer* x^*. If x^* is the only minimizer of f in $N(x^*)$ then x^* is an *isolated* minimizer of f. If $N(x^*)$ is all of D then x^* is a *global* minimizer of f in D.

The properties of local minimizers are better understood if we focus our attention on a reasonable class of functions. For our purposes, it is reasonable to assume that f is twice continuously differentiable. Under this assumption, the properties of local minimizers can be expressed in terms of the quadratic function

$$\psi(w) = \nabla f(x)^T w + \tfrac{1}{2} w^T \nabla^2 f(x) w$$

where $\nabla f(x)$ is the *gradient* of f at x and $\nabla^2 f(x)$ is the *Hessian matrix* of f at x. Recall that the ith component of the gradient is $\partial_i f(x)$ and that the (i, j) element of the Hessian matrix is $\partial_{i,j} f(x)$. Since

$$f(x + w) = f(x) + \psi(w) + o(\|w\|^2), \tag{1.2}$$

the quadratic ψ is the *local quadratic model* at x of the possible reduction in f. Unless otherwise stated, in this paper $\|\cdot\|$ is the Euclidean norm on R^n, or the induced operator norm.

THEOREM (1.3). *Let $f: R^n \to R$ be twice continuously differentiable in an open set D. If $x^* \in D$ is a local minimizer of f then $\nabla f(x^*) = 0$ and $\nabla^2 f(x^*)$ is positive semidefinite. If $\nabla f(x^*) = 0$ and $\nabla^2 f(x^*)$ is positive definite for some $x^* \in D$, then x^* is an isolated local minimizer of f.*

Proof. Let ψ be the local quadratic model at x^* of the possible reduction in f. If x^* is a local minimizer for f then (1.2) shows that

$$0 \leqslant \psi(\alpha p) + o(\alpha^2) = \alpha \nabla f(x^*)^T p + \alpha^2 p^T \nabla^2 f(x^*) p + o(\alpha^2,)$$

for each $p \in R^n$ and all α sufficiently small. This implies that $\nabla f(x^*)^T p = 0$ and that $p^T \nabla^2 f(x^*) p \geqslant 0$. Since p is arbitrary, we must conclude that $\nabla f(x^*) = 0$ and that $\nabla^2 f(x^*)$ is positive semidefinite. On the other hand, if $\nabla f(x^*) = 0$ and $\nabla^2 f(x^*)$ is positive definite then

$$\psi(w) = \tfrac{1}{2} w^T \nabla^2 f(x^*) w \geqslant \tfrac{1}{2} \lambda \|w\|^2,$$

where $\lambda > 0$ is the smallest eigenvalue of $\nabla^2 f(x^*)$. Now it follows from (1.2) that x^* must be an isolated local minimizer for f. ∎

A point $x^* \in R^n$ such that $\nabla f(x^*) = 0$ is a *critical point* of f. Critical points can be divided into local minimizers, local maximizers, and *saddle points*. Theorem (1.3) shows, in particular, that if x^* is a critical point of f and $\nabla^2 f(x^*)$ is indefinite then x^* is a saddle point of f. If, however, $\nabla^2 f(x^*)$ is semidefinite and singular then Theorem (1.3) does not provide any information on the nature of the critical point. This gap between the necessary and sufficient conditions of Theorem (1.3) is illustrated by the 2-dimensional function

$$f(\xi_1, \xi_2) = \xi_1^3 + \xi_2^2.$$

Note that $(0,0)$ is a critical point of f and that the Hessian matrix at $(0,0)$ is positive semidefinite. However, $(0,0)$ is a saddle point of f and not a local minimizer.

Algorithms for the unconstrained minimization of a function f: $R^n \to R$ are usually descent methods. Given an initial starting point x_0, a *descent method* generates a sequence of approximations $\{x_k\}$ to a local minimizer with the property that

$$f(x_{k+1}) < f(x_k), \qquad k \geqslant 0. \tag{1.4}$$

This descent condition alone is not sufficient to guarantee that the iterates $\{x_k\}$ approach a local minimizer. Stronger conditions are required to actually force the sequence into a neighborhood of a local minimizer. Once the iterates are in such a neighborhood, descent methods usually allow a rapidly convergent local method to determine the iterates. In this paper, the local method is Newton's iteration

$$x_{k+1} = x_k - \nabla^2 f(x_k)^{-1} \nabla f(x_k), \qquad k \geqslant 0,$$

and our concern here is with modifications to this local method that will provide a general purpose algorithm.

An algorithm that is designed for general use should be analyzed as thoroughly as possible. The purpose of a convergence analysis is to predict the behavior of the sequences produced by the algorithm. This involves establishing properties of limit points and rates of convergence. These features, together with requirements of storage and computational effort, aid in the selection of an algorithm for a specific application. At the very least, we expect an unconstrained minimization algorithm to produce sequences which satisfy

$$\lim_{k \to +\infty} \nabla f(x_k) = 0. \tag{1.5}$$

This condition guarantees that any limit point x^* of $\{x_k\}$ is a critical point of f. For algorithms which only use gradient information this is all that can be expected. If an algorithm requires Hessian information, then it is reasonable to expect that the second-order necessary conditions of Theorem (1.3) will be satisfied.

This can be done by ensuring that

$$\lim_{k \to +\infty} \inf \lambda_1 \left(\nabla^2 f(x_k) \right) \geqslant 0, \tag{1.6}$$

where $\lambda_1(A)$ is the smallest eigenvalue of a symmetric matrix A. If (1.5) and (1.6) hold then any limit point x^* of $\{x_k\}$ satisfies the necessary conditions of Theorem (1.3).

In the remainder of this paper we shall derive Newton's method in its basic form and then introduce various modifications which have been devised to ensure that (1.4), (1.5), and (1.6) are satisfied by the sequences $\{x_k\}$ produced by the method. Techniques for forcing convergence from poor starting points are the subject of two sections. We discuss line search methods and trust region methods in Sections 4 and 5, respectively. Both approaches are important and can be applied to other optimization problems. Variations on Newton's method are discussed in Sections 6 and 7. Since the techniques for forcing convergence are all designed to bring the sequence into a neighborhood of a local minimizer and then switch automatically to Newton's method, it is most appropriate to begin, in Section 2, with a discussion of the unmodified local algorithm.

It will be worthwhile to have a specific problem in mind in order to appreciate some of the concerns we express with respect to implementation of the methods. The problem we consider is the simplest problem in the calculus of variations. An excellent introduction to this problem may be found in Fleming and Rishel [19]. The problem is to minimize the functional

$$J(u) \equiv \int_0^1 L(\tau, u, \dot{u}) \, d\tau \tag{1.7}$$

over the set W of piecewise continuously differentiable functions u on the interval $[0,1]$ with specified endpoints $u(0)$ and $u(1)$. We assume that L is twice continuously differentiable. Two classical problems of this form in which L is independent of τ are the brachistochrone for which $L(\tau, u, v) = (u - \alpha)^{-1/2}(1 + v)^{1/2}$ for some constant α, and the minimal surface of revolution for which $L(\tau, u, v) = u(1 + v)^{1/2}$. An accessible introduction to the many applications related to the minimization of J may be found in Smith [47].

The solution techniques available for minimizing J directly are very limited. In most practical settings one would almost surely need to resort to numerical techniques. One such technique is to discretize the continuous problem and then construct an approximate solution by solving the discrete problem. To see how this might be accomplished, consider a family of n-dimensional subspaces $W_n \subset W$. If $\{\varphi_k\}$ is a basis for W_n, then we can determine the minimum of J on W_n by setting

$$f(x) = f(\xi_1, \xi_2, \ldots, \xi_n) \equiv J\left(\sum_{k=1}^{n} \xi_k \varphi_k\right),$$

and minimizing f. If Newton's method is used to minimize f, then we must be able to compute ∇f and $\nabla^2 f$. These derivatives can be computed by noting that the ith component of ∇f is

$$\partial_i f(x) = \int_0^1 \left((\partial_2 L)\varphi_i + (\partial_3 L)\dot{\varphi}_i\right) d\tau,$$

and that the (i, j) element of the Hessian $\nabla^2 f(x)$ is

$$\partial_{i,j} f(x) = \int_0^1 \left((\partial_{2,2} L)\varphi_i \varphi_j + (\partial_{2,3} L)(\varphi_i \dot{\varphi}_j + \dot{\varphi}_i \varphi_j) + (\partial_{3,3} L)\dot{\varphi}_i \dot{\varphi}_j\right) d\tau,$$

where the partial derivatives of L are all evaluated at $(\tau, u(\tau), \dot{u}(\tau))$ and

$$u(\tau) \equiv \sum_{k=1}^{n} \xi_k \varphi_k(\tau). \tag{1.8}$$

Once a solution x^* is found, the components of x^* can be used in (1.8) to construct an approximate minimizer u_n^* of J. Some analysis must be carried out to ensure that u_n^* is near a minimizer of J. An introduction to the type of analysis that is necessary may be found

in Daniel [10]. We shall only be concerned with the finite dimensional minimization process that occurs once n and a particular
basis is selected.

In principle this is all that is required to apply Newton's method
to f. However, some important practical considerations remain.
First of all, a reasonable starting point x_0 may be difficult to
provide so it becomes important to have an algorithm which
converges from arbitrary starting points. Also, since the dimension
of the approximating subspace W_n must increase in order to
guarantee that u_n^* is close to a minimizer of J, it may be necessary
to solve large dimensional problems. This means that the storage of
the Hessian and that solutions of linear systems involving the
Hessian matrix may become prohibitively costly. There is also the
matter of evaluating the integrals to sufficient accuracy as well as
the concern over how inaccuracies might affect the performance of
the optimization algorithm. We shall take up some of these problems in this paper.

2. THE LOCAL ALGORITHM

Newton's method can be studied from a local point of view in
which we assume that the starting point x_0 is close to a local
minimizer. This point of view is helpful because it provides information about the ultimate behavior of Newton's method. In Sections 4 and 5 we shall study Newton's method from a global
viewpoint.

Let $f: R^n \to R$ be a twice continuously differentiable function.
Newton's method for the unconstrained minimization problem can
be derived by assuming that we have an approximation x_k to a
local minimizer of f and that in a neighborhood of x_k the approximation

$$f(x_k + w) \approx f(x_k) + \psi_k(w)$$

is appropriate, where

$$\psi_k(w) = \nabla f(x_k)^T w + \tfrac{1}{2} w^T \nabla^2 f(x_k) w$$

is the local quadratic model at x_k of the possible reduction in f. If this approximation is appropriate, then a presumably better approximation $x_{k+1} = x_k + s_k$ can be found by requiring that the step s_k be a minimizer of ψ_k. Theorem (1.3) shows that s_k must then satisfy

$$\nabla \psi_k(s_k) = \nabla f(x_k) + \nabla^2 f(x_k) s_k = 0.$$

Thus Newton's method takes an approximation x_0 and attempts to improve it through the iteration

$$x_{k+1} = x_k - \nabla^2 f(x_k)^{-1} \nabla f(x_k), \qquad k \geqslant 0. \qquad (2.1)$$

Note that in this derivation the only restriction on the step s_k is that it satisfy the system of linear equations $\nabla \psi_k(w) = 0$. In other words, we only require that s_k be a critical point of ψ_k. As a consequence, the Newton iteration (2.1) has the same behavior in the neighborhood of any critical point of f regardless of its type. This seems undesirable since we would like our algorithms to have a predilection towards local minimizers.

As it turns out, however, this behavior is just a consequence of the fact that iteration (2.1) is Newton's method for the solution of the system of nonlinear equations $\nabla f(x) = 0$. Since the local properties of iteration (2.1) only depend on the mapping $F(x) = \nabla f(x)$, let us consider Newton's method in this more general setting.

Let $F: R^n \to R^n$ be a mapping with range and domain in R^n and consider the problem of finding the solution to the system of n equations in n unknowns $F(x) = 0$, or equivalently,

$$f_i(\xi_1, \ldots, \xi_n) = 0, \qquad 1 \leqslant i \leqslant n.$$

where f_i is the ith component function of F. Newton's method for this problem can be derived by assuming that we have an approximation x_k to the solution of the system of nonlinear equations $F(x) = 0$, and that in a neighborhood of x_k the approximation

$$F(x_k + w) \approx L_k(w) \equiv F(x_k) + F'(x_k) w$$

is appropriate where $F'(x)$ is the *Jacobian matrix* of the mapping

F at x. The next approximation $x_{k+1} = x_k + s_k$ can then be obtained by requiring that the step s_k satisfies the system of linear equations $L_k(w) = 0$. Thus Newton's method attempts to improve x_0 by the iteration

$$x_{k+1} = x_k - F'(x_k)^{-1}F(x_k), \qquad k \geq 0. \qquad (2.2)$$

In comparing iterations (2.1) and (2.2), note that (2.1) is a special case of the Newton iteration (2.2) applied to the mapping $F(x) = \nabla f(x)$. Because of this relationship, it suffices to study the local behavior of iteration (2.2). The most important aspects of this local behavior are summarized in the following two theorems.

THEOREM (2.3).   *Let $F: R^n \to R^n$ be a continuously differentiable mapping defined in an open set D, and assume that $F(x^*) = 0$ for some x^* in D and that $F'(x^*)$ is nonsingular. Then there is an open set S such that for any x_0 in S the Newton iterates (2.2) are well defined, remain in S, and converge to x^*.*

Proof. Let α be a fixed constant in $(0, 1)$. Since F' is continuous at x^* and $F'(x^*)$ is nonsingular, there is an open ball S and a positive constant μ such that

$$\|F'(x)^{-1}\| \leq \mu, \qquad \|F'(y) - F'(x)\| \leq \frac{\alpha}{\mu},$$

for every x and y in S. Suppose that $x_k \in S$. Since x_{k+1} satisfies (2.2) and $F(x^*) = 0$ we have that

$$x_{k+1} - x^* = -F'(x_k)^{-1}\big(F(x_k) - F(x^*) - F'(x_k)(x_k - x^*)\big),$$

and hence

$$\|x_{k+1} - x^*\| \leq \mu\|F(x_k) - F(x^*) - F'(x_k)(x_k - x^*)\|.$$

Now, the fundamental theorem of integral calculus implies that

$$F(x_k) - F(x^*) - F'(x_k)(x_k - x^*)$$
$$= \int_0^1 \big[F'(x^* + \xi(x_k - x^*)) - F'(x_k)\big](x_k - x^*)\,d\xi,$$

and hence,

$$\|x_{k+1} - x^*\|$$

$$\leqslant \mu \left\{ \max_{0 \leqslant \xi \leqslant 1} \|F'(x^* + \xi(x_k - x^*)) - F'(x_k)\| \right\} \|x_k - x^*\|.$$

(2.4)

Thus,

$$\|x_{k+1} - x^*\| \leqslant \alpha \|x_k - x^*\|$$

as long as $x_k \in S$. Since $\alpha < 1$, this last inequality implies that if $x_0 \in S$ then $x_k \in S$ for $k = 1, 2, \ldots$, and that $\{x_k\}$ converges to x^*. ∎

Theorem (2.3) states that Newton's method is *locally convergent* in the sense that if the starting point x_0 is sufficiently close to a solution x^* then Newton's method converges to x^*. Unfortunately, for many important problems the *domain of attraction S* guaranteed by Theorem (2.3) is quite small, and much research has gone into developing techniques to overcome this weakness of Newton's method. For systems of nonlinear equations this is a particularly active field, and much interest has been generated by the recent global Newton methods. For an account of some of this work see, for example, Keller [34]. For the unconstrained minimization problem, the situation is in much better shape, and we examine two main approaches to globalizing Newton's method in Sections 4 and 5.

Although Theorem (2.3) is undeniably important, it does not tell the full story. It is not enough to know that a sequence converges if the rate is so slow that we could not afford to see it converge. Generally, when analyzing an iterative method we are also interested in saying as much as possible about the expected rate of convergence of a sequence produced by the method. A reasonable optimization algorithm should be able to generate *linearly convergent* sequences $\{x_k\}$ in the sense that

$$\|x_{k+1} - x^*\| \leqslant \alpha \|x_k - x^*\|, \qquad k \geqslant 0, \qquad (2.5)$$

for some constant α in $(0,1)$. If α is small then (2.5) is adequate, but if α is close to unity, say $\alpha \geqslant 0.9$, then (2.5) is not reassuring.

For many optimization algorithms which use second-order information, it is possible to establish a stronger result than (2.5). A sequence $\{x_k\}$ converges *quadratically* to x^* if

$$\|x_{k+1} - x^*\| \leqslant \beta \|x_k - x^*\|^2, \qquad k \geqslant 0, \qquad (2.6)$$

for some constant $\beta > 0$. Since

$$\frac{\|x_{k+1} - x^*\|}{\|x^*\|} \leqslant (\beta\|x^*\|) \left(\frac{\|x_k - x^*\|}{\|x^*\|} \right)^2,$$

quadratic convergence implies that the number of significant digits of x_k as an approximation to x^* doubles at each iteration. Typically, as soon as two significant digits are obtained, the next three iterations will produce roughly sixteen significant digits.

There is a middle ground between (2.5) and (2.6). A sequence $\{x_k\}$ converges *superlinearly* to x^* if

$$\|x_{k+1} - x^*\| \leqslant \beta_k \|x_k - x^*\|, \qquad k \geqslant 0, \qquad (2.7)$$

for some sequence $\{\beta_k\}$ which converges to zero. It should be clear that a superlinearly convergent sequence is linearly convergent, and that a quadratically convergent sequence is superlinearly convergent. Also note that since

$$\left| \|x_{k+1} - x_k\| - \|x_k - x^*\| \right| \leqq \|x_{k+1} - x^*\|,$$

it follows that

$$\lim_{k \to +\infty} \frac{\|x_{k+1} - x_k\|}{\|x_k - x^*\|} = 1$$

when $\{x_k\}$ converges superlinearly to x^*. This is an important property because it implies that $\|x_{k+1} - x_k\|$ can be used to estimate the distance of x_k from x^*.

An iterative method is assigned a rate of convergence if it is possible to show that every convergent sequence produced will have at least this rate. Usually reasonable restrictions are imposed upon the domain of application for a method in order to obtain a useful assessment of this rate. Newton's method is usually quadratically convergent, as we now demonstrate.

THEOREM (2.8).   *Let F: $R^n \to R^n$ satisfy the assumptions of Theorem (2.3). Then the sequence $\{x_k\}$ produced by iteration (2.2) converges superlinearly to x^*. Moreover, if*

$$\|F'(x) - F'(x^*)\| \leqslant \kappa \|x - x^*\|, \qquad x \in D, \qquad (2.9)$$

for some constant $\kappa > 0$, then the sequence converges quadratically to x^*.

Proof.   Convergence of the sequence $\{x_k\}$ was established in Theorem (2.3), so it only remains to establish the rate of convergence. To this end define

$$\beta_k \equiv \mu \left\{ \max_{0 \leqslant \xi \leqslant 1} \|F'(x^* + \xi(x_k - x^*)) - F'(x_k)\| \right\},$$

and assume that $x_0 \in S$ with μ and S defined in the proof of Theorem (2.3). The hypothesis on F' at x^* and the convergence of the sequence to x^* implies that $\{\beta_k\}$ converges to zero. Since inequality (2.4) shows that

$$\|x_{k+1} - x^*\| \leqslant \beta_k \|x_k - x^*\|,$$

this proves that $\{x_k\}$ converges superlinearly to x^*. Moreover, if (2.9) holds then

$$\beta_k \leqslant 2\mu\kappa \|x_k - x^*\|,$$

and hence $\{x_k\}$ converges quadratically to x^*. ∎

Note that the Lipschitz condition (2.9) is necessary in order to guarantee that Newton's method is quadratically convergent. For example, Newton's method applied to the 1-dimensional problem defined by

$$f(\xi) = \xi\left[1 + \left(\frac{1}{\log(|\xi|)}\right)\right], \qquad \xi \neq 0, \qquad f(0) = 0,$$

is precisely superlinearly convergent to $\xi^* = 0$; that is, if $\{\xi_k\}$ is a sequence generated by Newton's method, then the ratio

$$\frac{|\xi_{k+1}|}{|\xi_k|^{1+\rho}}$$

is unbounded for any $\rho > 0$.

Rate of convergence results are sometimes used to compare algorithms by claiming that the superior algorithm is the one with the highest rate of convergence. Claims of this type should be made with care because these results are asymptotic. It is usually not possible to establish the magnitude of the constants that appear in expressions like (2.5), (2.6) and (2.7). Moreover, rate of convergence results do not measure the work necessary to compute x_{k+1} from x_k, and in many cases this information is decisive in the choice of algorithm. For example, consider the class of quasi-Newton methods as described in Section 7. Sequences generated by these methods are known to be superlinearly convergent, and usually not quadratically convergent. However, since they do not require the computation of the Hessian matrix, quasi-Newton methods are often regarded as being superior to the quadratically convergent Newton methods.

3. PROPERTIES OF QUADRATIC FUNCTIONS

Quadratic functions play an important role in the development of algorithms for optimization problems. For example, we have seen in Section 2 that in a neighborhood of a local minimizer of a function $f: R^n \to R$, Newton's method can be derived by requiring that the step be the minimizer of the local quadratic model

$$\psi_k(w) = \nabla f(x_k)^T w + \tfrac{1}{2} w^T \nabla^2 f(x_k) w \tag{3.1}$$

of the expected reduction in f. It is therefore important to understand the properties of quadratic functions and to provide numerically stable algorithms for minimizing them.

Our first result completely describes the unconstrained minimization of quadratic functions.

LEMMA (3.2).   *Let $\psi: R^n \to R$ be the quadratic function*

$$\psi(w) = g^T w + \tfrac{1}{2} w^T B w \tag{3.3}$$

where $g \in R^n$ and $B \in R^{n \times n}$ is a symmetric matrix.

a) *The quadratic ψ has a minimizer if and only if B is positive semidefinite and g is in the range of B.*
b) *The quadratic ψ has a unique minimizer if and only if B is positive definite.*
c) *If B is positive semidefinite then every solution to the equation $Bp = -g$ is a global minimizer.*

Proof.   Suppose B is positive semidefinite with g in the range of B. Then $Bp = -g$ has a solution p, and thus

$$\psi(p + w) = \psi(p) + (Bp + g)^T w + \tfrac{1}{2} w^T B w$$

$$= \psi(p) + \tfrac{1}{2} w^T B w \geqslant \psi(p) \tag{3.4}$$

for every $w \in R^n$. On the other hand, if p is a minimizer of ψ then Theorem (1.3) implies that $Bp + g = \nabla \psi(p) = 0$, and that $B = \nabla^2 \psi(p)$ is positive semidefinite. To establish b) and c) note that (3.4) holds whenever $Bp = -g$ and B is positive semidefinite, and that strict inequality holds for $w \neq 0$ if and only if B is positive definite. ∎

Given the quadratic ψ, there is an excellent numerical procedure for finding its minimizer. First an attempt is made at computing the Cholesky factorization of B. This factorization exists if and only if B is positive semidefinite, and in this case it leads to an upper triangular matrix R such that

$$B = R^T R.$$

If a negative diagonal is encountered during the factorization process, then B is not positive semidefinite and hence Lemma (3.2) shows that the quadratic ψ has no minimizer. If the factorization is successful and R is nonsingular, then the minimizer is computed by solving the system $Bp = -g$, or equivalently,

$$R^T v = -g, \qquad Rp = v.$$

If the factorization is successful but R is singular then B is positive semidefinite and singular. It still may be possible to compute a solution p, but from a numerical point of view, this computation is unstable because arbitrarily small perturbations can transform B into a positive definite matrix or an indefinite matrix.

Theorem (1.3) shows that in a neighborhood of a local minimizer of f, we can expect the Hessian matrix to be positive definite and then Lemma (3.2) shows that the local quadratic model (3.1) has a unique minimizer. Thus, in this case, the minimizer of the local quadratic model is a reasonable step for a minimization algorithm. However, away from a local minimizer the Hessian matrix $\nabla^2 f(x)$ may have a negative eigenvalue and then Lemma (3.2) tells us that this local quadratic model does not have a minimizer. In fact, the model is not even bounded below. There are several remedies to this difficulty. One possibility is to modify the quadratic model by adding a positive semidefinite matrix $E(x)$ so that

$$\nabla^2 f(x) + E(x)$$

is positive definite. When B is replaced by this matrix the step calculation can proceed as described above. An algorithm based upon this step calculation is described in Section 4.

Another possible remedy is to restrict the region in which we assume that the local quadratic model is appropriate. Locally the model still provides an excellent approximation to the expected reduction in f, so it is reasonable to restrict ψ to a ball $\{ w : \|w\| \leqslant \Delta \}$ for some $\Delta > 0$, and to compute a step as the minimizer of ψ on this ball. An algorithm based upon this step calculation is described in Section 5. The following result characterizes the global solutions to the problem of minimizing a quadratic function on this restricted region.

LEMMA (3.5). *Let $\psi: R^n \to R$ be the quadratic function (3.3) and let $\Delta > 0$ be given. A point $p \in R^n$ solves the problem*

$$\min\{\psi(w): \|w\| \leqslant \Delta\} \tag{3.6}$$

if and only if $\|p\| \leqslant \Delta$ and there is $\lambda \geqslant 0$ such that

$$(B + \lambda I)p = -g, \qquad \lambda(\Delta - \|p\|) = 0, \tag{3.7}$$

with $B + \lambda I$ positive semidefinite.

Proof. Suppose that λ and p satisfy (3.7) with $\|p\| \leqslant \Delta$ and $B + \lambda I$ positive semidefinite. Then Lemma (3.2) implies that p minimizes the quadratic function

$$\hat{\psi}(w) = g^T w + \tfrac{1}{2} w^T (B + \lambda I)w.$$

Thus $\hat{\psi}(w) \geqslant \hat{\psi}(p)$, which implies that

$$g^T w + \frac{1}{2} w^T B w \geqslant g^T p + \frac{1}{2} p^T B p + \frac{\lambda}{2} \left(p^T p - w^T w \right) \tag{3.8}$$

for all $w \in R^n$. Since $\lambda p^T p = \lambda \Delta^2$ and $\lambda \geqslant 0$, it follows from (3.8) that $\psi(w) \geqslant \psi(p)$ whenever $\|w\| \leqslant \Delta$, so p must solve (3.6).

Now suppose that p solves (3.6). If $\|p\| < \Delta$ then p is an unconstrained minimizer of ψ, so Lemma (3.2) implies that (3.7) holds with $\lambda = 0$, and that B is positive semidefinite. If $\|p\| = \Delta$ then p must also solve the equality constrained problem $\min\{\psi(w): \|w\| = \Delta\}$. Therefore, the method of Lagrange ensures the existence of λ such that

$$\nabla L(p) = 0, \quad \text{where } L(w) = \psi(w) + \frac{\lambda}{2} (w^T w - \Delta^2).$$

This implies that (3.7) holds for this λ and p. Moreover, since p solves (3.6) we have that (3.8) is valid for this λ and p whenever $\|w\| = \|p\|$. Using (3.7) to replace g and then rearranging terms in (3.8) shows that

$$\tfrac{1}{2}(w - p)^T (B + \lambda I)(w - p) \geqslant 0$$

for every w with norm $\|p\|$. It follows readily from this inequality that $B + \lambda I$ is positive semidefinite. To show that $\lambda \geqslant 0$, note that Lemma (3.2) implies that (3.8) is valid for every $w \in R^n$. Now, if λ is not positive then (3.8) implies that $\psi(w) \geqslant \psi(p)$ whenever $\|w\| \geqslant \|p\|$. Since p solves (3.6) we must have that p is an unconstrained minimizer of ψ and then Lemma (3.2) implies that $\lambda = 0$. Hence, $\lambda \geqslant 0$ as claimed. ■

Those familiar with various multiplier rules associated with mathematical programming will of course recognize that most of Lemma (3.5) could be obtained through direct application of these rules. At the very least one could invoke the Karush-Kuhn-Tucker conditions to avoid having to argue that the λ obtained from the Lagrange theory for equality constraints must be non-negative. Unfortunately, the multiplier theory for inequality constraints is not treated in most advanced calculus texts. McShane [36] has an elementary treatment of the standard results, and Pourciau [41] surveys the most recent results. It is interesting, however, that Lemma (3.5) cannot be obtained through direct application of the standard second-order multiplier rules. The gap between necessary and sufficient second-order conditions precludes this possibility since there is no such gap in this result.

Computing a numerical approximation to a solution of (3.6) requires some care. One immediate complication is that, due to the nonlinear constraint, there cannot be any general direct method for solving (3.6). In fact, if $g = 0$ and B has a negative eigenvalue then a solution p to (3.6) must be an eigenvector of norm Δ corresponding to the smallest eigenvalue of B. Therefore, a general method for solving (3.6) must solve a symmetric eigenvalue problem in this special case. Since we are concerned with global solutions to (3.6) it may seem that an algorithm which solves (3.6) is bound to be quite costly. However, we now show that there is an algorithm which produces a nearly optimal solution to (3.6) in all cases and with only a few iterations.

The solution of (3.6) is straightforward if there are no solutions on the boundary of $\{w : \|w\| \leqslant \Delta\}$. In fact, it is not difficult to prove that (3.6) has no solution with $\|p\| = \Delta$ if and only if B is positive definite and $\|B^{-1}g\| < \Delta$.

If (3.6) has a solution on the boundary of $\{w : \|w\| \leqslant \Delta\}$, then Lemma (3.5) shows that it is reasonable to expect that the nonlinear equation

$$\|p_\alpha\| = \Delta, \tag{3.9}$$

where

$$p_\alpha = -(B + \alpha I)^{-1}g$$

has a solution $\alpha = \lambda \geqslant 0$ in $(-\lambda_1, \infty)$, where $\lambda_1 \leqslant 0$ is the smallest eigenvalue of B. Note that (3.9) is a 1-dimensional zero finding problem in α that can be solved, for example, by Newton's method. However, since each evaluation of p_α requires the solution of a system of linear equations, it is important to solve (3.9) with very few evaluations of p_α.

To solve (3.9), Reinsch [44], [45] and Hebden [33] observed independently that great advantage could be taken of the fact that the function $\|p_\alpha\|^2$ is a rational function in α with second order poles on a subset of the negatives of the eigenvalues of the symmetric matrix B. To see this consider the decomposition

$$B = Q\Lambda Q^T \qquad \text{with } \Lambda = \text{diag}(\lambda_1, \lambda_2, \dots, \lambda_n) \text{ and } Q^TQ = I,$$

and observe that

$$\|p_\alpha\|^2 = \|Q(\Lambda + \alpha I)^{-1}Q^Tg\|^2 = \sum_{j=1}^{n} \frac{\gamma_j^2}{(\lambda_j + \alpha)^2} \tag{3.10}$$

where γ_i is the ith component of Q^Tg. Knowledge of the functional form (3.10) shows that Newton's method may not be very efficient if it is applied to the function

$$\varphi_1(\alpha) = \|p_\alpha\| - \Delta.$$

A reason for this is that φ_1 has a pole at $-\lambda_1$, and thus Newton's method tends to perform poorly when the solution of (3.9) is near $-\lambda_1$. This point is clear from Figure 3.1 which shows a typical sketch of $\|p_\alpha\|$ with $\lambda_1 = -7.5$. Reinsch and Hebden suggested that

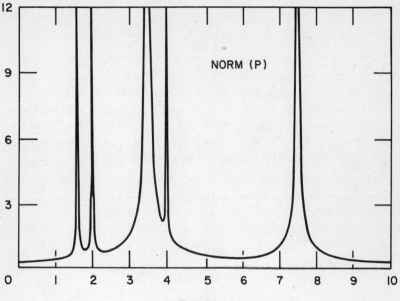

FIG. 3.1.

it is more efficient to apply Newton's method to the function

$$\varphi_2(\alpha) = \frac{1}{\Delta} - \frac{1}{\|p_\alpha\|}.$$

This function has no poles, and is almost linear near a solution of (3.9). This is illustrated quite well in Figure 3.2. This graph shows $(\|p_\alpha\|)^{-1}$ where $\|p_\alpha\|$ appears in Figure 3.1. Note that $(\|p_\alpha\|)^{-1}$ is almost linear for $\lambda > -\lambda_1$. It is clear from Figure 3.2 that Newton's method is bound to perform well on φ_2. The Newton iteration applied to finding a zero of φ_2 takes the following form.

Algorithm (3.11)

1) Let λ_0 and $\Delta > 0$ be given.
2) For $k = 0, 1, \ldots$ until "convergence"
   a) Factor $B + \lambda_k I = R_k^T R_k$.
   b) Solve $R_k^T R_k p_k = -g$.

c) Solve $R_k^T q_k = p_k$.

d) $\lambda_{k+1} = \lambda_k + \left(\dfrac{\|p_k\|}{\|q_k\|} \right)^2 \left(\dfrac{\|p_k\| - \Delta}{\Delta} \right)$.

If certain precautions are taken, then this basic iteration can be used to solve (3.6) in most cases. However, when B is indefinite there are cases in which the equation (3.9) has no solutions in $(-\lambda_1, \infty)$, and then Algorithm (3.11) fails. This happens, for example, when $g = 0$ and B is indefinite. It may also happen when $g \neq 0$, as illustrated by the following simple example. If

$$B = \begin{pmatrix} -1 & 0 \\ 0 & 1 \end{pmatrix}, \qquad g = \begin{pmatrix} 0 \\ 1 \end{pmatrix},$$

then $\lambda_1 = -1$, and if $\alpha > 1$ then $\|p_\alpha\|^2 < \frac{1}{2}$. In our examples g is orthogonal to the eigenspace of B corresponding to the smallest

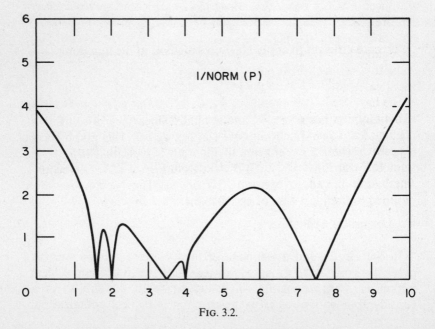

Fig. 3.2.

eigenvalue. This is typical; g must be orthogonal to the eigenspace

$$S_1 \equiv \{ z : Bz = \lambda_1 z, \ z \neq 0 \}$$

corresponding to the smallest eigenvalue of B whenever (3.9) has no solutions in $(-\lambda_1, \infty)$. To see this it suffices to note that if g is not orthogonal to S_1, then $\gamma_1 \neq 0$ in (3.10), and hence

$$\lim_{\alpha \to -\lambda_1} \| p_\alpha \| = \infty, \qquad \lim_{\alpha \to +\infty} \| p_\alpha \| = 0.$$

Since small perturbations of g lead to a nonzero γ_1, it is tempting to ignore the case when g is orthogonal to S_1. However, in many cases g is almost orthogonal to S_1, and in these cases an algorithm based completely on Newton's method would require a large number of iterations. This is not acceptable since a matrix factorization is required for each of these iterations.

Several algorithms have been proposed for the numerical solution of (3.6), but Gay [23] was the first to show that his algorithm produced a nearly optimal solution. Gay's algorithm, however, may require a large number of iterations when g is orthogonal to S_1, and fails when $g = 0$ and B is indefinite. Moré and Sorensen [38] have improved on Gay's algorithm, and their numerical results show that it is possible to produce a nearly optimal solution to (3.6) in all cases and with only a few iterations.

We have dealt with problem (3.6) at length because it arises in a variety of applications. For example, the solution of ill-posed problems in linear algebra usually requires the solution of (3.6) for a positive definite B. The literature on just this problem is extensive; for more information consult Eldén [18], Gander [22], and Varah [54].

4. LINE SEARCH METHODS

In Section 2 we mentioned the difficulty of providing a starting point x_0 for Newton's method which is sufficiently close to a local minimizer. Overcoming this difficulty has been the subject of a considerable amount of recent research in numerical optimization, and in this section we discuss the line search approach to this problem. In discussing this approach it is best to consider general line search methods first, and then specialize to Newton's method.

Given an iterate x_k, the basic idea of a line search method is to compute a direction p_k and a parameter $\alpha_k > 0$ such that the next iterate $x_{k+1} = x_k + \alpha_k p_k$ has a lower function value. Convergence of the iterates to a minimizer depends on the choice of p_k and α_k.

A direction $p \in R^n$ is a *descent direction* for a function $f: R^n \to R$ at a point $x \in R^n$ if there is a constant $\bar{\alpha} > 0$ such that

$$f(x + \alpha p) < f(x), \qquad \alpha \in (0, \bar{\alpha}]. \tag{4.1}$$

For differentiable functions, the easiest way to guarantee that (4.1) holds is to require that

$$\nabla f(x)^T p < 0. \tag{4.2}$$

In particular, the *steepest descent* choice $p = -\nabla f(x)$ satisfies (4.2). Condition (4.2) requires that the angle between $-\nabla f(x)$ and p be acute, and is equivalent to requiring that there is a positive definite matrix B such that

$$p = -B^{-1}\nabla f(x). \tag{4.3}$$

This is not difficult to prove. If (4.3) holds then certainly (4.2) follows. Conversely, if (4.2) holds then

$$B = I - \frac{pp^T}{p^Tp} - \frac{gg^T}{g^Tp}, \qquad g = \nabla f(x),$$

is positive definite and satisfies (4.3). Thus descent directions differ only in the choice of the positive definite matrix B in (4.3). The steepest descent method chooses B as the identity matrix and Newton's method chooses B as the Hessian matrix; the choice of B made in a quasi-Newton method (described in Section 7) is a compromise between these two choices.

Line search methods for differentiable functions assume that (4.2) holds. Note that if (4.2) does not hold then (4.1) can fail and then it may not be possible to make further reductions in f. Later on in this section we shall see that for convergence purposes it is necessary to require that p is not even nearly orthogonal to $\nabla f(x)$. This can be achieved by imposing a bound on the condition number of B in (4.3).

A *line search* algorithm examines points along the ray $\{x + \alpha p: \alpha \geqslant 0\}$ in search of a *steplength* α such that $f(x + \alpha p) < f(x)$. If p is a descent direction then such a point exists. In fact, the smallest positive local minimizer α^* of the univariate function

$$\varphi(\alpha) \equiv f(x + \alpha p), \qquad \alpha \geqslant 0, \tag{4.4}$$

is such an α. However, it would not be practical to search for this point. Indeed, a line search algorithm is usually an iterative scheme for 1-dimensional minimization, but the search process is usually terminated long before an accurate minimizer is found. Finding an accurate minimizer along a given ray usually does not yield a significantly larger reduction in f than a crude search, and better progress can often be made by making a reasonable reduction in the function f and then exploring other directions. These considerations have led to the development of stopping rules which terminate the line search process as soon as some minimal requirement is satisfied.

Given parameters $\mu \in (0, \tfrac{1}{2})$ and $\eta \in (\mu, 1)$, and a descent direction $p \in R^n$ which satisfies (4.2), let $\mathrm{SR}(\mu, \eta)$ be the set of steplengths $\alpha > 0$ such that

$$f(x + \alpha p) \leqslant f(x) + \alpha \mu \nabla f(x)^T p,$$

and

$$|\nabla f(x + \alpha p)^T p| \leqslant \eta |\nabla f(x)^T p|.$$

In other words, the set $\mathrm{SR}(\mu, \eta)$ specifies the stopping rule. In terms of the function φ defined by (4.4), a steplength α belongs to $\mathrm{SR}(\mu, \eta)$ if and only if

$$\varphi(\alpha) \leqslant \varphi(0) + \alpha \mu \varphi'(0), \tag{4.5}$$

and

$$|\varphi'(\alpha)| \leqslant \eta|\varphi'(0)|. \tag{4.6}$$

For example, for the function φ shown in Figure 4.1, the set $SR(\mu, \eta)$ consists of the intervals I_1 and I_2. The intuitive nature of these rules should be clear. If α is not too small, then the first condition of the stopping rule $SR(\mu, \eta)$ forces a sufficient decrease in the function. However, (4.5) allows arbitrarily small choices of $\alpha > 0$, so this condition is not sufficient to guarantee convergence. The second condition rules out arbitrarily small choices of $\alpha > 0$ and usually implies that α is near a local minimizer of φ.

We assume that $\mu < \frac{1}{2}$ because if φ is a quadratic with $\varphi'(0) < 0$ and $\varphi''(0) > 0$ then the global minimizer α^* of φ satisfies

$$\varphi(\alpha^*) = \varphi(0) + \tfrac{1}{2}\alpha^*\varphi'(0),$$

and thus α^* satisfies (4.5) if and only if $\mu \leqslant \frac{1}{2}$. The restriction $\mu < \frac{1}{2}$

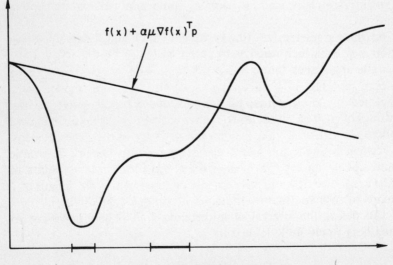

FIG. 4.1.

also allows $\alpha = 1$ to be ultimately acceptable to Newton and quasi-Newton methods; failure to take $\mu < \frac{1}{2}$ prevents these methods from converging superlinearly.

The restriction $\mu < \eta$ guarantees that under reasonable conditions $\mathrm{SR}(\mu, \eta)$ contains a non-trivial interval. For example, suppose that φ is continuously differentiable and bounded below. Then for sufficiently large $\beta > 0$

$$\varphi(\beta) \geqslant \varphi(0) + \mu\beta\varphi'(0). \tag{4.7}$$

Now let α^* be the smallest α in $(0, \beta]$ such that

$$\varphi(\alpha) = \varphi(0) + \mu\alpha\varphi'(0).$$

Then the mean value theorem shows that there is a τ such that

$$\varphi'(\tau) = \mu\varphi'(0) > \eta\varphi'(0), \qquad 0 < \tau < \alpha^*.$$

In particular, since $\varphi'(0) < 0$ we must have that $\varphi'(\tau) < 0$. Hence, τ satisfies (4.6). Moreover, $\tau < \alpha^*$ implies that

$$\varphi(\tau) < \varphi(0) + \mu\tau\varphi'(0),$$

and thus τ also satisfies (4.5). Continuity of φ' now shows that $\mathrm{SR}(\mu, \eta)$ contains a non-trivial interval.

The algorithms for selecting the steplength α are usually based upon minimizing a univariate quadratic or cubic model to φ defined by interpolation of function and first derivative at trial values of α. It is important to realize that it is possible to *safeguard* these algorithms so that they terminate in a finite number of steps.

Safeguarding a line search algorithm requires that we determine and update an *interval of uncertainty* I which contains points in $\mathrm{SR}(\mu, \eta)$. The updating process must guarantee that the length of I tends to zero and that eventually I is contained in $\mathrm{SR}(\mu, \eta)$.

To define the interval of uncertainty I, it is helpful to use an auxiliary function ψ defined by

$$\psi(\alpha) \equiv \varphi(\alpha) - \varphi(0) - \mu\alpha\varphi'(0),$$

and require that I be a closed interval with endpoints α_l and α_u such that

$$\psi'(\alpha_l)(\alpha_u - \alpha_l) < 0, \qquad \psi(\alpha_l) \leqslant \psi(\alpha_u), \qquad \psi(\alpha_l) \leqslant 0.$$

We now prove that $\psi'(\alpha^*) = 0$ and $\psi(\alpha^*) < 0$ for some α^* in I. As a consequence, if I is sufficiently small then

$$|\psi'(\alpha)| \leqslant (\eta - \mu)|\varphi'(0)|, \qquad \psi(\alpha) \leqslant 0,$$

for all $\alpha \in I$, and thus I is contained in SR(μ, η). If we let α^* be a global minimizer of ψ on I then α^* cannot be an endpoint of I because this contradicts the above requirements on α_l and α_u. Hence, α^* is interior to I and thus $\psi'(\alpha^*) = 0$. Moreover, since $\psi(\alpha_l) \leqslant 0$ we must also have that $\psi(\alpha^*) < 0$.

We now show how to update I. Given a trial value α_t in I, we can determine a new interval I_+ with endpoints α_l^+ and α_u^+ as follows:

If $\psi(\alpha_t) \geqslant \psi(\alpha_l)$ then $\alpha_l^+ = \alpha_l$ and $\alpha_u^+ = \alpha_t$.

If $\psi(\alpha_t) < \psi(\alpha_l)$ and $\psi'(\alpha_t)(\alpha_t - \alpha_l) < 0$ then $\alpha_l^+ = \alpha_t$ and $\alpha_u^+ = \alpha_u$.

If $\psi(\alpha_t) < \psi(\alpha_l)$ and $\psi'(\alpha_t)(\alpha_t - \alpha_l) > 0$ then $\alpha_l^+ = \alpha_t$ and $\alpha_u^+ = \alpha_l$.

It is straightforward to show that α_l^+ and α_u^+ still define an interval of uncertainty unless $\psi'(\alpha_t) = 0$ and $\psi(\alpha_t) < \psi(\alpha_l)$. Of course, in this case α_t belongs to SR(μ, η) and there is no need to update I. Also note that these updating rules can be used to determine an initial interval of uncertainty. If we set $\alpha_l = 0$ then $\alpha_t > 0$ defines an interval of uncertainty if $\psi(\alpha_t) \geqslant 0$ or if $\psi'(\alpha_t) < 0$. For α_t sufficiently large, we must have that $\psi(\alpha_t) \geqslant 0$ unless φ is not bounded below.

There are many ways to compute the trial value of α_t. For example, α_t could be the minimum of a cubic polynomial which interpolates φ and φ' at two previous trial values of α. The only requirement on α_t is that the length of I tends to zero. This can be done by monitoring the length of I, and if, say, the length of I is not reduced by a factor of 0.5 after two trials, then a bisection step can be used for the next trial α_t.

THEOREM (4.8). *Let* $f: R^n \to R$ *be continuously differentiable and bounded below on* R^n, *and assume that the starting point* x_0 *is such that* ∇f *is uniformly continuous on the level set*

$$\Omega \equiv \{ x \in R^n : f(x) \leqslant f(x_0) \}. \tag{4.9}$$

If the sequence $\{x_k\}$ *is defined by* $x_{k+1} = x_k + \alpha_k p_k$ *where* $\nabla f(x_k)^T p_k < 0$ *and* α_k *is any steplength in* $\mathrm{SR}(\mu, \eta)$ *then*

$$\lim_{k \to +\infty} \left(\frac{\nabla f(x_k)^T p_k}{\| p_k \|} \right) = 0. \tag{4.10}$$

Proof. Since $\nabla f(x_k)^T p_k < 0$ and since f is bounded below, the sequence $\{ x_k \}$ is well defined and lies in Ω. Moreover, $\{ f(x_k) \}$ is decreasing and hence converges.

The proof is by contradiction. If (4.10) does not hold then there is an $\varepsilon > 0$ and a subsequence with index set K such that

$$-\frac{\nabla f(x_k)^T p_k}{\| p_k \|} \geqslant \varepsilon, \qquad k \in K.$$

The first condition of the stopping rule $\mathrm{SR}(\mu, \eta)$ shows that

$$f(x_k) - f(x_{k+1}) \geqslant \mu \alpha_k \| p_k \| \left(-\frac{\nabla f(x_k)^T p_k}{\| p_k \|} \right) \geqslant \mu \alpha_k \| p_k \| \varepsilon,$$

$$k \in K,$$

and since $\{ f(x_k) \}$ is a convergent sequence, $\{ \alpha_k p_k : k \in K \}$ converges to zero. Now, the second condition of the stopping rule $\mathrm{SR}(\mu, \eta)$ yields the inequality

$$(1 - \eta)\left(-\nabla f(x_k)^T p_k \right) \leqslant \left(\nabla f(x_k + \alpha_k p_k) - \nabla f(x_k) \right)^T p_k,$$

$$k \geqslant 0,$$

and hence

$$\varepsilon \leqslant - \frac{\nabla f(x_k)^T p_k}{\|p_k\|} \leqslant \left(\frac{1}{1-\eta}\right) \|\nabla f(x_k + \alpha_k p_k) - \nabla f(x_k)\|,$$

$$k \in K.$$

However, since we have already shown that $\{\alpha_k p_k : k \in K\}$ converges to zero, this contradicts the uniform continuity of ∇f on Ω. ∎

Wolfe [55] proved Theorem (4.8) under various choices of steplength rules, while Gill and Murray [27] obtained a variation of (4.10) with the steplength chosen by a safeguarded algorithm designed for 1-dimensional minimization. Note, however, that 1-dimensional minimization algorithms must be modified in order to find points in $\text{SR}(\mu, \eta)$ because the first condition of this rule may exclude all 1-dimensional local minimizers. Also note that Theorem (4.8) holds under the hypothesis that $f: R^n \to R$ is continuously differentiable in an open set D and that the level set

$$\Omega = \{x \in D : f(x) \leqslant f(x_0)\} \tag{4.11}$$

is compact. The proof is almost identical to that of Theorem (4.8); the only difference occurs in proving that $\text{SR}(\mu, \eta)$ is not empty.

The specialization of Theorem (4.8) to algorithms of the Newton class is almost immediate. In this case,

$$p_k = - B_k^{-1} \nabla f(x_k), \tag{4.12}$$

where $\{B_k\}$ is a sequence of positive definite matrices with uniformly bounded condition numbers; that is, there is a constant $\kappa > 0$ such that

$$\|B_k\| \|B_k^{-1}\| \leqslant \kappa, \qquad k \geqslant 0. \tag{4.13}$$

Under this assumption we have that

$$-\frac{\nabla f(x_k)^T p_k}{\|p_k\|} \geqslant \left(\frac{1}{\kappa}\right)\|\nabla f(x_k)\|,$$

and thus (4.10) implies that $\{\nabla f(x_k)\}$ converges to zero. In particular, every limit point of $\{x_k\}$ is a critical point of f.

For a line search method with p_k chosen via (4.12), this is the strongest type of result possible. It is not possible to prove that the limit points of $\{x_k\}$ are local minimizers because, for example, if x_0 is any critical point of f then the line search method terminates at x_0. It should be pointed out, however, that it is possible to modify the search procedure so that it uses the curvature information contained in the Hessian matrix to avoid regions where the Hessian has a negative eigenvalue. In these modifications the search for a lower function value usually takes place in a 2-dimensional subspace. Various ideas along these lines may be found in Gill and Murray [28], McCormick [35], Moré and Sorensen [37], and Goldfarb [31].

The choice of B_k in (4.12) is guided by a desire to satisfy (4.13) and still guarantee a fast rate of convergence. In the steepest descent method B_k is the identity matrix. For this method (4.13) is satisfied but convergence can be quite slow. The convergence of Newton's method is quite rapid when it occurs, but since $B_k = \nabla^2 f(x_k)$ is not necessarily positive definite, there is no guarantee of convergence. Modifications to Newton's method have been designed to overcome this problem. They set

$$B_k = \nabla^2 f(x_k) + E_k, \tag{4.14}$$

where E_k is chosen so that B_k is positive definite and satisfies (4.13). There are many ways to do this, but one of the most effective methods is due to Gill and Murray [28].

Given a symmetric matrix A and parameters $\varepsilon \geqslant 0$ and $\beta > 0$, Gill and Murray's method produces an upper triangular matrix R and a diagonal matrix $E = \text{diag}(\varepsilon_i) \geqslant 0$ such that $A + E = R^T R$.

The ith step of the algorithm sets

$$\gamma_{ij} = a_{ij} - \sum_{k=1}^{i-1} r_{kj} r_{ki}, \qquad i \leqslant j \leqslant n,$$

$$\mu_i = \max\left\{ |\gamma_{ij}| : i < j \leqslant n \right\},$$

$$r_{ii} = \max\left\{ \varepsilon, |\gamma_{ii}|^{1/2}, \frac{\mu_i}{\beta} \right\},$$

$$r_{ij} = \frac{\gamma_{ij}}{r_{ii}}, \qquad i < j \leqslant n,$$

$$\varepsilon_i = r_{ii}^2 - \gamma_{ii}.$$

Note that if $\varepsilon = 0$ then it is possible that $r_{ii} = 0$, but in this case set $r_{ij} = 0$.

The idea behind the Gill and Murray algorithm is to increase the diagonal elements of A so that $A + E$ has a Cholesky decomposition. The increase in r_{ii} is designed to ensure that r_{ij} is bounded relative to $\|A\|$, and that if A is sufficiently positive definite then A is not modified. Note that the increase in r_{ii} due to the term μ_i/β forces $|r_{ij}| \leqslant \beta$ for $i < j$. Hence,

$$|\gamma_{ij}| \leqslant |\alpha_{ij}| + \beta^2 n, \qquad i \leqslant j.$$

This shows that μ_i/β is bounded in terms of β and $|\alpha_{ij}|$. It is sensible to choose β so that this bound on μ_i/β is as small as possible, and this leads to a choice of

$$\beta^2 = \frac{1}{n}\max\left\{ |\alpha_{ij}| : i \neq j \right\}.$$

This choice, however, may conflict with the desire to leave A unmodified whenever A is sufficiently positive definite. The definition of r_{ii} shows that in order to accomplish this β cannot be too small. It is sufficient to require that

$$\beta^2 \geqslant \max\left\{ |\alpha_{ii}| : 1 \leqslant i \leqslant n \right\}.$$

To establish this claim we first show that if A is positive definite then $\gamma_{jj} > 0$. The proof is easy. Given an index j, define $p \in R^n$ by letting $Rp = r_{jj}e_j$. Then $p_j = 1$ and

$$0 < p^T A p = r_{jj}^2 - p^T E p \leqslant r_{jj}^2 - \varepsilon_j = \gamma_{jj}.$$

Now, since $\gamma_{jj} > 0$, it follows that

$$\beta^2 \geqslant \alpha_{jj} > \sum_{k=1}^{j-1} r_{kj}^2 \geqslant r_{ij}^2, \qquad i < j,$$

and since $\gamma_{ij} = r_{ij}r_{ii}$, we have that $\mu_i^2 < \beta^2 r_{ii}^2$. Hence, if A is positive definite then $r_{ii} = \max\{\varepsilon, \gamma_{ii}^{1/2}\}$. This shows that if A is sufficiently positive definite then $r_{ii} = \gamma_{ii}^{1/2}$, and thus $E = 0$.

A reasonable way to guarantee that r_{ij} is bounded and that $E = 0$ whenever A is sufficiently positive definite, is to choose

$$\beta^2 = \max\left\{\frac{1}{n}\max\{|\alpha_{ij}| : i \neq j\}, \max\{|\alpha_{ii}| : 1 \leqslant i \leqslant n\}\right\}.$$

For this choice of β it is not difficult to prove that

$$\varepsilon \leqslant |r_{ii}| \leqslant \max\{\varepsilon, 2n\beta\}, \qquad |r_{ij}| \leqslant \beta, \qquad i < j.$$

For $\varepsilon > 0$, these inequalities show that if the Gill and Murray algorithm is applied to a bounded sequence $\{A_k\}$ of symmetric matrices, then (4.13) is satisfied for $B_k = A_k + E_k$.

We now have all the ingredients for a modified Newton's method with a line search. In this method, we compute p_k via (4.12) and (4.14), and determine E_k by applying Gill and Murray's modified Cholesky factorization to $\nabla^2 f(x_k)$ with some $\varepsilon > 0$.

THEOREM (4.15). *Let $f : R^n \to R$ be twice continuously differentiable on an open set D, and assume that the starting point x_0 is such that the level set (4.11) is compact. If the sequence $\{x_k\}$ is defined by $x_{k+1} = x_k + \alpha_k p_k$ where p_k is computed by the modified Newton's method and α_k is any steplength in $SR(\mu, \eta)$ then*

$$\lim_{k \to +\infty} \nabla f(x_k) = 0.$$

Proof. We have already noted that if $\varepsilon > 0$ and $\{A_k\}$ is bounded then (4.13) holds for $B_k = A_k + E_k$. In this case $A_k = \nabla^2 f(x_k)$, and since Ω is compact, $\{A_k\}$ is bounded. Thus our result is a consequence of Theorem (4.8). ∎

There is an interesting variation of Theorem (4.15) which shows that the iterates $\{x_k\}$ usually converge. In this variation the stopping rule $SR(\mu, \eta)$ is modified by the addition of an upper bound β on the steplength. We accept α as the steplength if $\alpha \leqslant \beta$ and α belongs to $SR(\mu, \eta)$, or if $\alpha = \beta$ and

$$f(x + \beta p) \leqslant f(x) + \beta \mu \nabla f(x)^T p.$$

Note that if this condition is not satisfied then (4.7) holds, and then there is an α in $(0, \beta)$ which satisfies $SR(\mu, \eta)$. It is not difficult to show that Theorem (4.15) holds with this variation, and since now we have

$$\|x_{k+1} - x_k\| \leqslant |\alpha_k| \left\| \left(\nabla^2 f(x_k) + E_k \right)^{-1} \right\| \|\nabla f(x_k)\| \leqslant \beta \gamma \|\nabla f(x_k)\|$$

for some constant γ dependent on ε, it follows that

$$\lim_{k \to +\infty} \|x_{k+1} - x_k\| = 0. \tag{4.16}$$

This shows that if $\{x_k\}$ has an isolated limit point x^* then $\{x_k\}$ converges to x^*. In particular, note that if $\nabla^2 f(x^*)$ is nonsingular at a limit point x^*, then x^* is an isolated solution of $\nabla f(x) = 0$ and hence x^* is also an isolated limit point of $\{x_k\}$. The structure of the set of limit points of $\{x_k\}$ is further restricted by a result of Ostrowski [40], page 203, which states that if $\{x_k\}$ is a bounded sequence and (4.16) holds, then the set of limit points of $\{x_k\}$ is connected.

To investigate the rate of convergence of the modified Newton's method, assume that the sequence $\{x_k\}$ converges to a point x^* at which $\nabla^2 f(x^*)$ is sufficiently positive definite in the sense that $E_k = 0$ for all k sufficiently large. Then

$$p_k = p_k^N \equiv -\nabla^2 f(x_k)^{-1} \nabla f(x_k), \tag{4.17}$$

and it can then be shown that there is a k_0 such that the steplength $\alpha_k = 1$ is in $SR(\mu, \eta)$ for $k \geqslant k_0$. With this choice of α_k, the rate of convergence is given by Theorem (2.8).

The above argument relies on the fact that $\alpha_k = 1$ is eventually in $SR(\mu, \eta)$. To establish this result it is only necessary to assume that $\{p_k\}$ tends to the Newton step in both length and direction, that is,

$$\lim_{k \to +\infty} \left(\frac{\|p_k - p_k^N\|}{\|p_k\|} \right) = 0,$$

where p_k^N is the Newton step (4.17). For a proof of this result, and a discussion of its relationship to quasi-Newton methods, see Dennis and Moré [16].

5. TRUST REGION METHODS

In Newton's method with a line search the Hessian is modified when it is not sufficiently positive definite. This modification to the quadratic model guarantees convergence but seems to ignore the role of the quadratic model as a local approximation to the objective function. We now consider an alternative approach in which the quadratic model is not modified but instead, the quadratic model is only considered in a restricted *trust* region. We mentioned this technique briefly in Section 3 as a motivation for Lemma (3.5); its use for globalizing Newton's method has resulted in reliable algorithms with strong convergence properties. In this section we introduce the main ideas of this approach and establish some of the basic convergence properties.

Let $f: R^n \to R$ be a twice continuously differentiable function. In Newton's method with a trust region strategy, each iterate x_k has a bound Δ_k such that

$$f(x_k + w) \approx f(x_k) + \psi_k(w), \qquad \|w\| \leqslant \Delta_k,$$

where

$$\psi_k(w) = \nabla f(x_k)^T w + \tfrac{1}{2} w^T \nabla^2 f(x_k) w$$

is the quadratic model of f within a neighborhood of the iterate x_k. This suggests that it may be desirable to compute a step s_k which approximately solves the problem

$$\min\{\psi_k(w) : \|w\| \leq \Delta_k\}. \tag{5.1}$$

If the step is satisfactory in the sense that $x_k + s_k$ produces a sufficient reduction in f, then Δ_k can be increased; if the step is unsatisfactory then Δ_k should be decreased. The following algorithm expresses these ideas in more detail.

Algorithm (5.2). Let $0 < \mu < \eta < 1$ and $0 < \gamma_1 < \gamma_2 < 1 < \gamma_3$ be specified constants.

1) Let $x_0 \in R^n$ and $\Delta_0 > 0$ be given.
2) For $k = 0, 1, \ldots$ until "convergence"
   a) Compute $\nabla f(x_k)$ and $\nabla^2 f(x_k)$.
   b) Determine an approximate solution s_k to problem (5.1).
   c) Compute $\rho_k = \dfrac{f(x_k + s_k) - f(x_k)}{\psi_k(s_k)}$.
   d) If $\rho_k \leq \mu$ then replace Δ_k by a member of the interval $[\gamma_1 \Delta_k, \gamma_2 \Delta_k]$ and go to b).
   e) Compute $x_{k+1} = x_k + s_k$.
   f) If $\rho_k \leq \eta$ then choose $\Delta_{k+1} \in [\gamma_2 \Delta_k, \Delta_k]$, else choose $\Delta_{k+1} \in [\Delta_k, \gamma_3 \Delta_k]$.

This is a basic form of a trust region Newton's method. An interesting variant of this algorithm includes a scaling matrix for the variables. In this variation sub-problem (5.1) is replaced by

$$\min\{\psi_k(w) : \|D_k w\| \leq \Delta_k\},$$

where D_k is a nonsingular matrix. We shall not discuss this generalization here; however, it is important to note that all of the results presented here hold for this variant if $\{D_k\}$ has uniformly bounded condition numbers. Such a modification can be very important in practice when the units of the variables are on widely different scales. Another variation is to use the hypercube $\{w: \|w\|_\infty \leqslant \Delta_k\}$ as the trust region in (5.1). In this variation subproblem (5.1) is replaced by the quadratic programming problem

$$\min\{\psi_k(w): |w^T e_i| \leqslant \Delta_k, \quad 1 \leqslant i \leqslant n\}. \tag{5.3}$$

A difficulty with the hypercube approach is that although a local solution to (5.3) can be obtained with a reasonable amount of computational effort, it may be quite expensive to obtain a global solution. This is unfortunate since providing a nearly optimal solution enhances the convergence properties of Algorithm (5.2). As noted in Section 3, it is indeed possible to provide nearly optimal solutions to (5.1) with a reasonable amount of computational effort, so this is the only method considered in the remainder of the section.

Just as in the case of a line search we are not interested in solving the model problem (5.1) with great accuracy. Instead, we are interested in providing relaxed conditions for accepting an approximate solution s_k to problem (5.1) which are sufficient to force convergence of the sequence $\{x_k\}$ generated by Algorithm (5.2). In fact, there are conditions which guarantee much more than convergence of the method. If ψ_k^* is the optimal value of (5.1), and if the approximate solution s_k to (5.1) satisfies

$$-\psi_k(s_k) \geqslant \beta_1|\psi_k^*|, \qquad \|s_k\| \leqslant \beta_2\Delta_k, \tag{5.4}$$

for specified constants $\beta_1 > 0$ and $\beta_2 > 0$, then it is possible to prove that under suitable conditions on f, the sequence $\{x_k\}$ is convergent to a point x^* with $\nabla f(x^*) = 0$ and $\nabla^2 f(x^*)$ positive semidefinite.

It is not difficult to obtain a vector s_k which satisfies (5.4), although as mentioned at the end of Section 3, this requires attention to a number of details. Given σ in $(0,1)$, the algorithm of Moré and Sorensen [38], for example, finds a vector s_k such that

$$\psi_k(s_k) - \psi_k^* \leqslant \sigma(2-\sigma)|\psi_k^*|, \qquad \|s_k\| \leqslant (1+\sigma)\Delta_k,$$

provided $\psi_k^* \neq 0$. Of course, if $\psi_k^* = 0$, then $\nabla f(x_k) = 0$ and $\nabla^2 f(x_k)$

is positive semidefinite, so Algorithm (5.2) terminates at x_k. It is also worthy of mention that if $\sigma = 0.1$ then the cost of this algorithm is quite reasonable. On the average the approximate solution of each model problem requires less than two factorizations of a symmetric positive definite matrix of order n.

Condition (5.4) can be expressed in an alternate form which is more convenient for proofs of convergence. If $p_k \in R^n$ is a solution to problem (5.1) then Lemma (3.5) implies that there is a parameter $\lambda_k \geqslant 0$ such that

$$\left(\nabla^2 f(x_k) + \lambda_k I \right) p_k = -\nabla f(x_k), \qquad \lambda_k(\Delta_k - \|p_k\|) = 0.$$

Now let $R_k^T R_k$ be the Cholesky factorization of $\nabla^2 f(x_k) + \lambda_k I$. Then

$$|\psi_k^*| = \tfrac{1}{2}\left(\|R_k p_k\|^2 + \lambda_k \Delta_k^2 \right). \tag{5.5}$$

This expression for ψ_k^* shows that if (5.4) holds then

$$-\psi_k(s_k) \geqslant \tfrac{1}{2}\beta_1\left(\|R_k p_k\|^2 + \lambda_k \Delta_k^2 \right), \tag{5.6}$$

and thus the iterates $\{x_k\}$ generated by Algorithm (5.2) satisfy

$$f(x_k) - f(x_{k+1}) \geqslant \tfrac{1}{2}\mu\beta_1\left(\|R_k p_k\|^2 + \lambda_k \Delta_k^2 \right). \tag{5.7}$$

These two inequalities are essential to the proof of our next result.

THEOREM (5.8). *Let* $f: R^n \to R$ *be twice continuously differentiable on an open set* D, *and assume that the starting point* x_0 *is such that the level set*

$$\Omega = \left\{ x \in D : f(x) \leqslant f(x_0) \right\}$$

is compact. If the sequence $\{x_k\}$ *is produced by Algorithm (5.2) where* s_k *satisfies (5.4), then either the algorithm terminates at* $x_l \in \Omega$ *because* $\nabla f(x_l) = 0$ *and* $\nabla^2 f(x_l)$ *is positive semidefinite, or* $\{x_k\}$ *has a limit point* x^* *in* Ω *with* $\nabla f(x^*) = 0$ *and* $\nabla^2 f(x^*)$ *positive semidefinite.*

Proof. If $\nabla f(x_l) = 0$ and $\nabla^2 f(x_l)$ is positive semidefinite for some iterate x_l in Ω then the algorithm terminates; otherwise (5.4) implies that $\psi_k(s_k) < 0$ for $k \geqslant 0$ and thus $\{x_k\}$ is well defined and lies in Ω.

Let us now prove the result under the assumption that $\{\lambda_k\}$ is not bounded away from zero. If some subsequence of $\{\lambda_k\}$ converges to zero then since Ω is compact we can assume, without loss of generality, that the same subsequence of $\{x_k\}$ converges to some x^* in the level set Ω. Since $\nabla^2 f(x_k) + \lambda_k I$ is positive semidefinite, $\nabla^2 f(x^*)$ is also positive semidefinite, and $\nabla f(x^*) = 0$ follows by noting that

$$\|R_k p_k\|^2 \geqslant \frac{\|\nabla f(x_k)\|^2}{\|\nabla^2 f(x_k)\| + \lambda_k}$$

and that (5.7) implies that $\{\|R_k p_k\|\}$ converges to zero.

We can show that $\{\lambda_k\}$ is not bounded away from zero by contradiction. If $\lambda_k \geqslant \varepsilon > 0$ then (5.4) and (5.6) yield that

$$-\psi_k(s_k) \geqslant \tfrac{1}{2}\beta_1 \lambda_k \Delta_k^2 \geqslant \frac{1}{2}\left(\frac{\beta_1}{\beta_2^2}\right)\varepsilon\|s_k\|^2.$$

Now, a standard estimate is that

$$|f(x_k + s_k) - f(x_k) - \psi_k(s_k)|$$

$$\leqslant \tfrac{1}{2}\|s_k\|^2 \max_{0 \leqslant \xi \leqslant 1} \|\nabla^2 f(x_k + \xi s_k) - \nabla^2 f(x_k)\|, \qquad (5.9)$$

and thus the last two inequalities show that

$$|\rho_k - 1| \leqslant \left(\frac{\beta_2^2}{\beta_1 \varepsilon}\right) \max_{0 \leqslant \xi \leqslant 1} \|\nabla^2 f(x_k + \xi s_k) - \nabla^2 f(x_k)\|.$$

$$(5.10)$$

Inequality (5.7) implies that $\{\Delta_k\}$ converges to zero and hence $\{\|s_k\|\}$ also converges to zero. Thus the uniform continuity of $\nabla^2 f$

on Ω together with (5.10) implies that $\rho_k > \eta$ for all k sufficiently large and then the updating rules for Δ_k yield that $\{\Delta_k\}$ is bounded away from zero. This is in contradiction of the fact that $\{\Delta_k\}$ converges to zero. ∎

The results we have just established are only a sample of the available convergence results for Algorithm (5.2) under assumption (5.4) for s_k. This theorem extends results of Fletcher [21] and Sorensen [48]. The following additional results are known.

a) The sequence $\{\nabla f(x_k)\}$ converges to zero.
b) If x^* is an isolated limit point of $\{x_k\}$ then $\nabla^2 f(x^*)$ is positive semidefinite.
c) If $\nabla^2 f(x^*)$ is nonsingular for some limit point x^* of $\{x_k\}$ then $\{x_k\}$ converges to x^*.

Thomas [52] proved a), while Moré and Sorensen [37] established b) and c) as extensions of results due to Sorensen [48]. Of these results, b) is characteristic of the trust region approach, and is the only result that does not hold for Newton's method with a line search. This difference between the two approaches is of theoretical importance. From a practical viewpoint, however, it can be argued that a more important difference is that with a line search approach the search for a lower function value occurs in a 1-dimensional subspace, while with a trust region approach the search is not restricted to a lower dimensional subspace.

An additional result which is helpful in establishing rate of convergence results is that if $\{x_k\}$ converges to x^* and $\nabla^2 f(x^*)$ is positive definite, then the sequence $\{\Delta_k\}$ is bounded away from zero. To prove this first note that if $\varepsilon_0 > 0$ is a lower bound on the eigenvalues of $\nabla^2 f(x_k)$ then (5.5) shows that

$$|\psi_k^*| \geqslant \tfrac{1}{2}\varepsilon_0 \min\{\Delta_k^2, \|s_k^N\|^2\},$$

where

$$s_k^N \equiv -\nabla^2 f(x_k)^{-1} \nabla f(x_k). \tag{5.11}$$

Now, since $\psi_k(s_k) \leqslant 0$, we have that

$$\tfrac{1}{2}\|s_k\| \leqslant \|\nabla^2 f(x_k)^{-1}\| \|\nabla f(x_k)\|,$$

and thus $\frac{1}{2}\|s_k\| \leq \kappa\|s_k^N\|$ where κ is an upper bound on the condition number of $\nabla^2 f(x_k)$. Hence, assumption (5.4) shows that there is a constant $\varepsilon_1 > 0$ with

$$-\psi_k(s_k) \geq \tfrac{1}{2}\varepsilon_1\|s_k\|^2.$$

This estimate and (5.9) then yield that

$$|\rho_k - 1| \leq \left(\frac{1}{\varepsilon_1}\right) \max_{0 \leq \xi \leq 1} \|\nabla^2 f(x_k + \xi s_k) - \nabla^2 f(x_k)\|,$$

and thus $\rho_k > \eta$ for all k sufficiently large. It follows that $\{\Delta_k\}$ is bounded away from zero as desired.

Rate of convergence results can be obtained with the additional —but mild—assumption that there is a constant $\beta_3 > 0$ such that if $\|s_k\| \leq \beta_3 \Delta_k$ then $\nabla^2 f(x_k)$ is positive definite and $s_k = s_k^N$ where s_k^N is the Newton step (5.11). With this assumption in mind, suppose that $\{x_k\}$ converges to x^* and that $\nabla^2 f(x^*)$ is positive definite. Then $\{s_k\}$ converges to zero, and hence $\|s_k\| \leq \beta_3 \Delta_k$ for all k sufficiently large. Thus there is a $k_0 \geq 0$ such that $s_k = s_k^N$ for $k \geq k_0$, and then the rate of convergence results are provided by Theorem 2.8.

6. APPROXIMATIONS TO THE HESSIAN MATRIX

The methods we have described in the previous sections all require the computation of the Hessian matrix. This can be a difficult and error prone task, and in some cases analytic expressions for the entries of the Hessian matrix may not even be available. What can be done in these cases?

An obvious way to overcome these difficulties is to approximate the Hessian matrix with differences of gradients. However, there are several things to consider. Which difference approximation should be used? How large should the difference parameter be? How is the performance of the minimization method affected when difference approximations are used?

The two most common types of difference approximations use the forward difference and central difference formulas. The forward difference approximation is based on the Taylor's series expansion

$$\left(\frac{1}{\alpha}\right)[\nabla f(x+\alpha p)-\nabla f(x)] = \nabla^2 f(x)p + O(\alpha), \quad (6.1)$$

while the central difference approximation is based on

$$\left(\frac{1}{2\alpha}\right)[\nabla f(x+\alpha p)-2\nabla f(x)+\nabla f(x-\alpha p)] = \nabla^2 f(x)p + O(\alpha^2).$$

In optimization work, forward differences are quite common because they require fewer gradient evaluations and usually provide the necessary accuracy. If forward differences are used, an approximation $A(x)$ to the Hessian matrix at some $x \in R^n$ can be obtained by setting

$$A(x)e_j = \left(\frac{1}{\alpha_j}\right)[\nabla f(x+\alpha e_j)-\nabla f(x)], \quad 1 \leqslant j \leqslant n.$$

for some *difference parameter* $\alpha_j \neq 0$. Unfortunately, this approximation does not necessarily provide a symmetric matrix. This important feature of the Hessian can be obtained by using the symmetric matrix

$$\tfrac{1}{2}\left[A(x)+A(x)^T\right]$$

as the approximation to the Hessian matrix at x.

The choice of difference parameter presents a dilemma. In order to preserve the superlinear rate of convergence enjoyed by Newton's method it is necessary to force the difference parameter to zero. However, as the difference parameter α_j becomes small, the differences lose significance due to cancellation. To prevent this loss of significance, the difference parameter must stay above a certain threshold value. This dilemma can usually be resolved in practice because it is not necessary to provide a Hessian approximation of high accuracy. If the Hessian approximation has an accuracy comparable to the desired accuracy in the solution to the optimization problem, then convergence usually takes place at practically a quadratic rate. Less accurate Hessian approximations decrease the rate of convergence but do not prevent convergence. These remarks assume that the gradient is evaluated accurately; if this is not the case, we may not even be able to compute a descent direction.

Techniques for choosing the difference parameter in (6.1) require information about ∇f in a neighborhood of x which is obtained by evaluating ∇f at several points near x. For many practical problems it would be too expensive to acquire this information at each iterate. A sensible strategy for an optimization algorithm is to choose the difference parameter at a typical x (possibly the starting point x_0), and to use this choice until it is deemed unsuitable. The difference parameter is only recomputed when the quality of the difference approximation starts to degrade.

There are several algorithms for choosing the difference parameter at a point. Discussing these algorithms in detail is not within the scope of this paper, but we want to mention some of the ideas behind these algorithms. The main ideas are clear in the 1-dimensional case, so consider a differentiable function $\varphi: R \to R$, let $\varphi_c(\alpha)$ denote the computed value of $\varphi(\alpha)$, and let

$$\varepsilon(\alpha) = \varphi_c(\alpha) - \varphi(\alpha)$$

be the (absolute) error in the computed value. The smoothness of φ_c depends on the method used to evaluate φ on the computer, but in all cases φ_c is a step function. A reason for this is that a computer with l decimal digits of accuracy does not distinguish between numbers with the same first l digits. We mention this fact because it implies that φ_c is not differentiable. With these remarks in mind, note that our problem is to determine an α such that

$$\left(\frac{1}{\alpha}\right)\left[\varphi_c(\alpha) - \varphi_c(0)\right] \tag{6.2}$$

is close to $\varphi'(0)$. If we assume that we have an open neighborhood I of $\alpha = 0$, and a bound ε_0 such that

$$|\varepsilon(\alpha)| \leqslant \varepsilon_0, \qquad \alpha \in I,$$

then it is not difficult to determine the difference parameter. Note that a Taylor's expansion of φ shows that

$$\varphi_c(\alpha) - \varphi_c(0) - \alpha\varphi'(0) = \tfrac{1}{2}\varphi''(\xi)\alpha^2 + [\varepsilon(\alpha) - \varepsilon(0)]$$

for some ξ with $|\xi| < |\alpha|$, and hence

$$\left| \left(\frac{1}{\alpha} \right) [\varphi_c(\alpha) - \varphi_c(0)] - \varphi'(0) \right| \leqslant \tfrac{1}{2}\eta_0|\alpha| + \frac{2\varepsilon_0}{|\alpha|},$$

where η_0 is a bound for φ'' on I. This bound on the error between (6.2) and $\varphi'(0)$ has the correct qualitative behavior. If α is too small then the error is dominated by ε_0, while if α is too large then the error is determined by the curvature of φ. It is reasonable to choose α so that this bound is minimized, and this leads to a choice of

$$\alpha = 2 \left(\frac{\varepsilon_0}{\eta_0} \right)^{1/2}. \tag{6.3}$$

An algorithm for determining ε_0 and η_0 can be based on the work of Hamming [32], pages 163-173. The basic idea is that the 4th and 5th order differences of φ_c are a measure of ε_0 and that the 2nd order differences can be used to estimate η_0. It is necessary to take some precautions, but in general we have found that an algorithm based on these ideas and (6.3) is quite effective.

Another way to overcome the difficulties mentioned at the beginning of this section is to approximate the Hessian matrix directly. As an illustration, recall that in the example of Section 1, the (i, j) element of the Hessian $\nabla^2 f(x)$ is

$$\partial_{i,j}f(x) = \int_0^1 \left((\partial_{2,2}L)\varphi_i\varphi_j + (\partial_{2,3}L)(\varphi_i\dot{\varphi}_j + \dot{\varphi}_i\varphi_j) \right.$$

$$\left. + (\partial_{3,3}L)\dot{\varphi}_i\dot{\varphi}_j \right) d\tau,$$

where the partial derivatives of L are all evaluated at $(\tau, u(\tau), \dot{u}(\tau))$ and

$$u(\tau) \equiv \sum_{k=1}^n \xi_k\varphi_k(\tau). \tag{6.4}$$

In principle, these integrals can be evaluated with an appropriate quadrature, and the results used to define an approximation to the Hessian matrix. This requires the storage of a symmetric matrix of order n and the evaluation of $n(n+1)/2$ integrals over $[0,1]$. Since the dimension n must increase in order to refine the accuracy of u as an approximation to the continuous problem, it is usually necessary to solve large dimensional problems, and clearly, the cost of these requirements can then be prohibitive even for moderate values of n.

The above problems can be greatly reduced if we choose a basis $\{\varphi_i\}$ whose elements vanish on most of the interval $[0,1]$. For example, we could choose a B-spline basis. To illustrate this possibility, let $\tau_j = jh \equiv j/n$, and define

$$\varphi_j(\tau) = 1 - \frac{|\tau - \tau_j|}{h}, \qquad \tau \in [\tau_{j-1}, \tau_{j+1}], \qquad \varphi_j(\tau) = 0 \quad \text{otherwise.}$$

These functions are smooth B-splines of order 2. It is a simple matter to verify that

$$\varphi_i \varphi_j = \varphi_i \dot{\varphi}_j = \dot{\varphi}_i \dot{\varphi}_j \equiv 0, \qquad |i - j| > 1,$$

and therefore the Hessian matrix is tridiagonal. Thus the storage is now of order n, and it is only necessary to evaluate $2n$ integrals over intervals of length $1/n$. Similar remarks hold for the computation of the gradient. If a basis of smooth B-splines of order k is chosen then the Hessian has bandwidth $2k-1$. The computation of the gradient and Hessian is now more expensive, but there is an increase in the accuracy of (6.4) as an approximation to the continuous problem. For more information, see de Boor [13] on splines, and Gill and Murray [26] on the numerical solution of problems in the calculus of variations.

Large scale optimization problems frequently exhibit special structure such as sparsity in the Hessian matrix. Approximation of sparse Hessians by differences is attractive because the number of gradient differences required is often small compared to n. For example, if the Hessian matrix is tridiagonal (as in the above example), then 3 gradient differences suffice to approximate the Hessian.

A technique for estimating general sparse Hessian matrices is based on the work of Curtis, Powell, and Reid [9]. They pointed out that a group of columns of $\nabla^2 f(x)$ can be approximated with one gradient difference if columns in this group do not have a nonzero in the same row position. To see this, let I be the indices of a group of columns with this property, and let p be a vector with component $\rho_j \neq 0$ if j belongs to I and $\rho_j = 0$ otherwise. Then

$$\nabla^2 f(x)p = \sum_{j \in I} \rho_j \nabla^2 f(x) e_j,$$

and since the columns with indices in I do not have nonzeroes in the same row position, for each (i, j) element of $\nabla^2 f(x)$ with $j \in I$ we have that

$$\left(\nabla^2 f(x)p \right)_i = \rho_j e_i^T \nabla^2 f(x) e_j.$$

In view of (6.1), it follows that we can approximate all the columns with indices in I with just one gradient difference.

For a tridiagonal matrix, it is easy to see that columns with indices of the form $l \bmod 3$ can be placed in the lth group. Hence, as noted above, a tridiagonal matrix can be estimated with 3 gradient differences.

For general sparsity patterns it is not straightforward to partition the columns of the matrix into the least number of groups so that columns in a group do not have a nonzero in the same row position. Curtis, Powell, and Reid [9] suggested an algorithm but did not analyze the problem. Coleman and Moré [7] have approached this partitioning problem through its equivalence to certain graph coloring problems, and have used this point of view to analyze the partitioning problem and to suggest improved algorithms. Their numerical results show that these improved algorithms are nearly optimal on practical problems.

The partitioning technique that we have described for estimating sparse Hessians does not make any use of the symmetry of the matrix. Powell and Toint [43] have pointed out that it is often possible to use symmetry to reduce the number of required gradient differences. They proposed several ways of doing this, and with one of their methods it is possible to estimate a tridiagonal Hessian matrix with 2 gradient differences. It turns out that their methods can also be analyzed with graph theory techniques; a treatment from this point of view is given by Coleman and Moré [8].

7. QUASI-NEWTON METHODS

For some problems the objective function and its gradient are so expensive to calculate that we are not willing to compute a difference approximation to the Hessian matrix. These are not necessarily large dimensional problems. For example, the problem might be to minimize the L_2-norm of the solution to a differential equation that depends on a few parameters. In this case each function evaluation required by the optimization method actually involves a numerical solution of a differential equation.

In an effort to reduce the computational requirements of Newton's method, Davidon [11] introduced a revolutionary idea which provides a way to approximate the Hessian matrix using only the gradient information gathered at each iterate. This idea has led to a highly successful class of methods which today are usually called quasi-Newton methods. There is a huge literature on quasi-Newton methods; our purpose in this section is to provide a brief introduction to the two most powerful members of this class and to contrast quasi-Newton methods with methods in the Newton class. For a thorough discussion of various aspects of quasi-Newton methods, see the survey paper of Dennis and Moré [16].

In very simplistic terms a quasi-Newton method might be termed as an "earn while you learn" method. It is to be contrasted with methods in the Newton class through the manner of maintaining an approximate Hessian. In quasi-Newton methods, the approximate Hessian must satisfy the quasi-Newton equation. To derive this equation, suppose that we have a positive definite approximation B_k to the Hessian of f at x_k. We can then compute a descent direction p_k via

$$p_k = - B_k^{-1} \nabla f(x_k) \tag{7.1}$$

and a steplength α_k in $\mathrm{SR}(\mu, \eta)$ as in Section 4. This defines the step $s_k = \alpha_k p_k$ and the next iterate $x_{k+1} = x_k + s_k$. Since

$$\left(\int_0^1 \nabla^2 f(x_k + \xi s_k) d\xi \right) s_k = \nabla f(x_{k+1}) - \nabla f(x_k),$$

it might be reasonable to seek an update to the approximate Hessian which satisfies

$$B_{k+1} s_k = y_k \equiv \nabla f(x_{k+1}) - \nabla f(x_k). \tag{7.2}$$

This is the quasi-Newton equation and a method for generating B_{k+1} from B_k so that (7.2) holds is a quasi-Newton update. The quasi-Newton equation is essentially a gradient difference along the most recent search direction. Thus, quasi-Newton methods only use the search direction to obtain curvature information, while methods in the Newton class use n directions.

Various specific formulas exist for updating the matrix B_k, and we shall concentrate on those updates which guarantee that B_{k+1} is symmetric and positive definite. Note that if B_{k+1} is positive definite then (7.2) implies that $y_k^T s_k$ is positive. This condition is satisfied whenever α_k is in $\mathrm{SR}(\mu, \eta)$ because then

$$y_k^T s_k = \nabla f(x_{k+1})^T s_k - \nabla f(x_k)^T s_k \geqslant (1 - \eta) | \nabla f(x_k)^T s_k |.$$

We shall show below that if $y_k^T s_k$ is positive, then there are symmetric and positive definite matrices which satisfy the quasi-Newton equation.

In a discussion of quasi-Newton updates, it is customary and convenient to drop subscripts. Assume, therefore, that we have an approximate Hessian B, and vectors s and y with $y^T s$ positive. We then want to obtain update formulas that produce matrices B_+ according to the quasi-Newton equation

$$B_+ s = y. \tag{7.3}$$

The simplest derivation of such a formula is to ask for the nearest matrix to B which satisfies (7.3). If $B_+ = B + E$, then our problem is to find a solution to

$$\min\{\|E\|:(B+E)s=y\}, \tag{7.4}$$

where $\|\cdot\|$ is a suitable matrix norm. It is natural to choose the Frobenius norm defined by

$$\|A\|_F^2 = \sum_{i=1}^{n} \|Av_i\|^2 = \mathrm{trace}(A^TA),$$

for any set $v_1,\ldots,v_n$ of orthonormal vectors, because this is the Euclidean norm in the space of matrices. With the Frobenius norm it is a simple matter to verify that

$$E = \frac{(y-Bs)s^T}{s^Ts}$$

solves problem (7.4). Just note that E satisfies $(B+E)s=y$ and that if $\hat{E}$ is any other matrix that satisfies this equation then

$$\|E\|_F = \left\|\frac{(y-Bs)s^T}{s^Ts}\right\|_F = \left\|\frac{\hat{E}ss^T}{s^Ts}\right\|_F \leqslant \|\hat{E}\|_F\left\|\frac{ss^T}{s^Ts}\right\|_F = \|\hat{E}\|_F.$$

This E is the unique solution to (7.4) since $\|\cdot\|_F$ is an inner product norm and the constraint in (7.4) is linear. The explicit updating formula for B is therefore given by

$$B_+ = B + \frac{(y-Bs)s^T}{s^Ts}.$$

This is Broyden's [2] rank-1 update formula; it is a rank-1 update because $rank(E)\leqslant 1$. Broyden's update is the most powerful quasi-Newton update for the solution of systems of nonlinear equations. For minimization, however, there are more suitable updates. A reason for this is that B_+ is usually neither symmetric nor positive definite even though B might possess these properties. In Section 4 we saw the importance of these properties in obtaining convergence for descent methods. Therefore, we are very interested in updating formulas which maintain symmetry and positive definiteness in the matrices B. One way to obtain these properties is to require that $B = R^T R$ and obtain $R_+ = R + E$ such that $B_+ = R_+^T R_+$ satisfies (7.3). It would be quite natural to seek the correction E as a solution to the problem

$$\min\left\{\|E\|_F : (R + E)^T(R + E)s = y\right\}. \qquad (7.5)$$

Variational techniques can be used to show that the correction E which solves this problem is a rank-2 update to R. To our knowledge this approach has not been tried in practice because there is a rank-1 correction E which meets our requirements and has proven to be extremely successful. We might motivate a derivation of this rank-1 update by considering the implications of the quasi-Newton equation on the factors of B_+. If $B_+ = R_+^T R_+$ satisfies (7.3) then E must satisfy

$$(R + E)^T v = y, \qquad (R + E)s = v, \qquad v^T v = y^T s > 0. \quad (7.6)$$

Conversely, if E satisfies (7.6) for a given vector v, then $B_+ = R_+^T R_+$ satisfies (7.3). Thus a reasonable alternative to solving problem (7.5) is to specify a vector v of norm $(y^T s)^{1/2}$ and then obtain a correction matrix E which satisfies (7.6) as well as

$$\min\left\{\|E\|_F : (R + E)^T v = y\right\}. \qquad (7.7)$$

The solution to (7.7) for a given v is

$$E = \frac{v(y - R^T v)^T}{v^T v},$$

and it follows that this E satisfies $(R + E)s = v$ if and only if $v = \tau R s$ for some τ. Since we must have $v^T v = y^T s$, this condition determines τ. We have thus shown that

$$R_+ = R + \frac{v(y - R^T v)^T}{v^T v}, \qquad v = (y^T s)^{1/2}\frac{Rs}{\|Rs\|} \qquad (7.8)$$

induces a symmetric positive definite quasi-Newton update.

The updating formula we have just derived was discovered independently by Broyden [3], [4], Fletcher [20], Goldfarb [30], Shanno [46], and is often referred to as the BFGS formula. We have concentrated upon this particular update because it appears to work best in practice. The derivation of the BFGS update which we have presented is due to Dennis and Schnabel [17]. Davidon and Sorensen [12] have provided another derivation of the BFGS update by obtaining the quasi-Newton equation (7.6) as interpolation conditions on the gradient of the local quadratic model, and then showing that $v = \tau R s$ is a consistent choice. However, while numerous derivations of this update formula have been given, the superior performance of the BFGS update has not yet received a satisfactory explanation.

A quasi-Newton method based on the BFGS update would generate a sequence $\{x_k\}$ defined by $x_{k+1} = x_k + \alpha_k p_k$ where p_k is the direction (7.1) and $\{B_k\}$ is chosen by the BFGS formula. If B_0 is symmetric and positive definite, and if α_k satisfies the stopping rules $SR(\mu, \eta)$, then $y_k^T s_k$ is positive and thus B_k is well defined and positive definite for $k > 0$. For this method Powell [42] has shown that if $f: R^n \to R$ satisfies the assumptions of Theorem (4.15), and if f is convex on the level set Ω, then $\{\nabla f(x_k)\}$ converges to zero. Moreover, if $\{x_k\}$ converges to x^* and $\nabla^2 f(x^*)$ is positive definite, then $\{x_k\}$ converges superlinearly to x^*. In practice this method converges for general functions f, so there is a wide gap between this result and what is observed in practice.

At first sight, it would seem that Powell's result is a rather straightforward extension of the analysis of Newton's method. However, this would be the case only if we could show that the condition numbers of $\{B_k\}$ are uniformly bounded. Interestingly enough, Powell shows this, but only after convergence has been established.

The form of the update (7.8) is quite amenable to stable numerical computation. In particular, it is possible to maintain the matrices R in triangular form. This facilitates the solution of the system (7.1) and reduces the storage. The reduction to triangular form can be accomplished by standard (see, for example, Gill, Golub, Murray, and Saunders [25]) matrix updating techniques. If R is upper triangular, then a product of elementary rotations $Q = Q_1 Q_2 \cdots Q_{2n}$

can be constructed in such a way that

$$\hat{R}_+ = QR_+ = Q\left(R + \frac{v\left(y - R^T v\right)^T}{v^T v} \right) \qquad (7.9)$$

is also upper triangular. Since Q is orthogonal,

$$\hat{R}_+^T \hat{R}_+ = R_+^T Q^T Q R_+ = R_+^T R_+,$$

and thus $\hat{R}_+$ is the required factor of B_+. Since Q is the product of $2n$ elementary rotations, the arithmetic required in (7.9) is on the order of n^2 floating point operations. This is to be compared to the order of n^3 operations required to form B_+ and then factor. Another advantage of keeping R in triangular form is that the condition number of triangular matrices can easily be monitored. This provides the opportunity to alter these matrices when extreme ill-conditioning occurs.

8. CURRENT RESEARCH

It is fairly safe to say that Newton's method and quasi-Newton methods are understood well enough to provide reliable software for general small to medium size unconstrained minimization problems. Several subroutines are available through software libraries and others are under development.

Currently, researchers are focusing much of their attention upon large scale problems. The ground rules for what constitutes an effective algorithm can change drastically when the number of variables becomes large. We have tacitly assumed that the solution of a linear system of order n is, at worst, comparable in cost to the evaluation of the gradient and Hessian. This assumption may not be valid in large scale problems, and then it is necessary to take advantage of the special structure of the problem. With suitable modifications, Newton's method can still be an effective tool for large scale problems. We have already mentioned, in Section 6, one possible modification in connection with the estimation of sparse Hessian matrices by gradient differences. Modifications can also be made to the algorithms for determining the Newton direction. For example, since the Newton methods of this paper only require the Cholesky decomposition of a symmetric matrix, for sparse problems it is possible to reduce the amount of work and storage required by this decomposition. This is a well understood problem; see, for example, George and Liu [24]. Another possibility is to only determine an approximation to the Newton direction. A local analysis of this possibility is provided by Dembo, Eisenstat, and Steihaug [14]. Dembo and Steihaug [15] have proposed a global algorithm in which a conjugate gradient method is used to determine the modified Newton direction.

So far modifications of quasi-Newton methods to account for sparsity have not had the resounding success that these methods have had in the dense case. This is despite intensive effort in this area. There may be fundamental reasons for this as noted by Sorensen [49]. However, it would seem that this subject is just not fully understood at present, and thus this is still a very active research area. The interested reader should consult Steihaug [50] and Toint [53] for information and additional references.

The situation in greatest need of research at present arises when the Hessian matrix cannot be stored in fast memory. Currently the method of choice for this situation is a conjugate direction method. It would take a full article to describe these methods. Fletcher [21] has a nice introduction to the basic ideas behind conjugate direction methods, while Buckley [5], Buckley and Le Nir [6] and Gill and Murray [29] describe some of the recent work in this area.

The development of methods for these very difficult and highly practical situations hinges upon a thorough understanding of Newton's method. It is our hope that this paper will provide a basis for work in these areas.

Acknowledgments. We would like to thank Philip Gill and Michael Overton for their constructive comments on the paper.

REFERENCES

1. R. G. Bartle, *The Elements of Real Analysis*, John Wiley & Sons, NY, 1976.
2. C. G. Broyden, "A class of methods for solving nonlinear simultaneous equations," *Math. Comp.*, **19** (1965), 577–593.
3. _____, "A new double-rank minimization algorithm," *Notices Amer. Math. Soc.*, **16** (1969), 670.

4. C. G. Broyden, "The convergence of single-rank quasi-Newton methods," *Math. Comp.*, **24** (1970), 365–382.

5. A. Buckley, "A combined conjugate-gradient quasi-Newton minimization algorithm," *Math. Programming*, **15** (1978), 200–210.

6. A. Buckley and A. Le Nir, "QN-like variable storage conjugate gradients", *Math. Programming*, **27** (1983), 155–175.

7. T. F. Coleman, and J. J. Moré, "Estimation of sparse Jacobian matrices and graph coloring problems," *SIAM J. Numer. Anal.*, **20** (1983), 187–209.

8. _____, "Estimation of sparse Hessian matrices and graph coloring problems," *Math. Programming*, **28** (1984), 243–270.

9. A. R. Curtis, M. J. D. Powell, and J. K. Reid, "On the estimation of sparse Jacobian matrices," *J. Inst. Math. Appl.*, **13** (1974), 117–119.

10. J. W. Daniel, *The Approximate Minimization of Functionals*, Prentice-Hall, 1971.

11. W. C. Davidon, "Variable metric method for minimization," Argonne National Laboratory *Report ANL-5990 (Rev)*, (1959), Argonne, Illinois.

12. W. C. Davidon, and D. C. Sorensen, "Numerical methods for unconstrained minimization," *Proceedings COMPCON 80*, IEEE Computer Society, (1980), Long Beach, California.

13. C. De Boor, *A Practical Guide to Splines*, Springer-Verlag, 1978.

14. R. S. Dembo, S. C. Eisenstat, and T. Steihaug, "Inexact Newton methods," *SIAM J. Numer. Anal.*, **19** (1982), 400–408.

15. R. S. Dembo, and T. Steihaug, "Truncated Newton algorithms for large scale unconstrained optimization," *Math. Programming*, **26** (1983), 190–212.

16. J. E. Dennis and J. J. Moré, "Quasi-Newton methods, motivation and theory," *SIAM Rev.*, **19** (1977), 46–89.

17. J. E. Dennis, and R. B. Schnabel, "A new derivation of symmetric positive definite secant updates," *Nonlinear Programming*, *4*, O. L. Mangasarian, R. R. Meyer, and S. M. Robinson (eds.) Academic Press, 1981.

18. L. Eldén, "Algorithms for the regularization of ill-conditioned least squares problems," *BIT*, **17** (1977), 134–145.

19. W. H. Fleming, and R. W. Rishel, *Deterministic and Stochastic Optimal Control*, Springer-Verlag, 1975.

20. R. Fletcher, "A new approach to variable metric algorithms," *Comput. J.*, **13** (1970), 317–322.

21. _____, *Practical Methods of Optimation, Volume 1: Unconstrained Optimization*, John Wiley & Sons, 1980.

22. W. Gander, "On the linear least squares problem with a quadratic constraint," Computer Science Department, *Report 78-697* (1978), Stanford University, Stanford, California.

23. D. M. Gay, "Computing optimal locally constrained steps," *SIAM J. Sci. Stat. Comput.*, **2** (1981), 186–197.

24. A. George, and J. Liu, *Computer Solution of Large Positive Definite Systems*, Prentice-Hall, 1981.

25. P. E. Gill, G. H. Golub, W. Murray, and M. A. Saunders, "Methods for modifying matrix factorizations," *Math. Comp.*, **28** (1974), 505–536.

26. P. E. Gill and W. Murray, "The numerical solution of a problem in the calculus of variation," *Recent Mathematical Developments in Control*, D. J. Bell, (ed.), Academic Press, 1973.

27. _____, "Safeguarded steplength algorithms for optimization using descent methods," National Physical Laboratory, *Report NAC 37*, (1974), Teddington, England.

28. _____, "Newton-type methods for unconstrained and linearly constrained optimization," *Math. Programming*, 7 (1974), 311–350.

29. _____, "Conjugate-gradient methods for large-scale nonlinear optimization," Stanford University, Systems Optimization Laboratory, *Report SOL 79-15*, (1979), Stanford, California.

30. D. Goldfarb, "A family of variable-metric methods derived by variational means," *Math. Comp.*, 24 (1970), 23–26.

31. _____, "Curvilinear path steplength algorithms for minimization which use directions of negative curvature," *Math Programming*, 18 (1980), 31–40.

32. R. W. Hamming, *Introduction to Applied Numerical Analysis*, McGraw-Hill, 1971.

33. M. D. Hebden, "An algorithm for minimization using exact second derivatives, Atomic Energy Research Establishment," *Report T.P. 515*, (1973), Harwell, England.

34. H. B. Keller, "Global homotopies and Newton methods," *Recent Advances in Numerical Analysis*, C. de Boor and G. H. Golub (eds.), Academic Press, 1978.

35. G. P. McCormick, "A modification of Armijo's step-size rule for negative curvature," *Math. Programming*, 13 (1977), 111–115.

36. E. J. McShane, "The Lagrange multiplier rule," *Amer. Math. Monthly*, 80 (1973), 922–925.

37. J. J. Moré, and D. C. Sorensen, "On the use of directions of negative curvature in a modified Newton method," *Math. Programming*, 16, (1979), 1–20.

38. _____, "Computing a trust region step," *SIAM J. Sci. Stat. Comput.*, 4 (1983), 553–572.

39. J. M. Ortega, and W. C. Rheinboldt, *Iterative solution of nonlinear equations in several variables*, Academic Press, 1970.

40. A. Ostrowski, *Solutions of Equations and Systems of Equations*, Academic Press, 1966.

41. B. H. Pourciau, "Modern multiplier rules," *Amer. Math. Monthly*, 87 (1980) 433–452.

42. M. J. D. Powell, "Some global convergence properties of a variable metric algorithm for minimization without exact line searches", *Nonlinear Programming, SIAM-AMS Proceedings*, 9, American Mathematical Society, 1976.

43. M. J. D. Powell, and Ph. L. Toint, "On the estimation of sparse Hessian matrices," *SIAM J. Numer. Anal.*, 16 (1979), 1060–1074.

44. C. H. Reinsch, "Smoothing by spline functions," *Numer. Math.*, 10 (1967), 177–183.

45. _____, "Smoothing by spline functions II," *Numer. Math.*, 16 (1971), 451–454.

46. D. F. Shanno, "Conditioning of quasi-Newton methods for function minimization," *Math. Comp.*, **24**, (1970), 647–656.

47. D. R. Smith, *Variational Methods in Optimization*, Prentice-Hall, 1974.

48. D. C. Sorensen, "Newton's method with a model trust-region modification," Argonne National Laboratory, *Report ANL-80-106*, (1980), Argonne, Illinois. *SIAM J. Numer. Anal.*

49. _____ , "An example concerning quasi-Newton estimation of a sparse Hessian," *SIGNUM Newsletter*, **16** (1981), 8–10.

50. T. Steihaug, "Quasi-Newton methods for large scale nonlinear problems," Yale University, School of Organization and Management, *Report 49*, (1980), New Haven, Connecticut.

51. G. W. Stewart, *Introduction to Matrix Computations*, Academic Press, 1973.

52. S. W. Thomas, "Sequential estimation techniques for quasi-Newton algorithms," Ph. D. dissertation, (1975), Cornell University, Ithaca, New York.

53. Ph. L. Toint, "Towards an efficient sparsity exploiting Newton method for minimization," *Sparse Matrices and their Uses*, I. S. Duff, (ed.), Academic Press, 1981.

54. J. M. Varah, "A practical examination of some numerical methods for linear discrete ill-posed problems," *SIAM Rev.*, **21** (1979), 100–111.

55. P. Wolfe, "Convergence conditions for ascent methods," *SIAM Rev.*, **11** (1969), 226–235.

RESEARCH DIRECTIONS IN SPARSE MATRIX COMPUTATIONS

Iain S. Duff

1. INTRODUCTION

Our aim in this chapter is to convey a flavor of some of the present techniques used in the solution of sparse linear equations, stressing the areas holding most promise for future research. The beauty of this part of numerical analysis lies not so much in its mathematical elegance or axiomatic structure but rather in its unruly ferment, its practicality, and in its use of concepts and techniques from many areas of mathematics and computer science. Indeed, a quick survey of university science faculties reveals that the multidiscipline of numerical analysis has feet in both mathematics and computer science departments and nowhere is this dichotomy more marked than in the area of sparse matrix research. In fact, a slightly deeper survey of the literature (for example, Duff [10]) indicates that many valuable research contributions have been made by workers in various disciplines including chemists, physicists and engineers.

Sparse matrix research has a relatively short history since it was only in the late 1960s that attempts were made to design algorithms to solve general sparse problems. Admittedly, for about the previous twenty years, sparse linear equations arising from the discretization of simple partial differential equations were solved by various iterative techniques, and sparseness was being exploited in the simplex algorithm for linear programming. However, more general techniques were not developed until the work of Tinney and his colleagues at the Bonneville Power Administration, and the whole subject could be said to have had its coming-of-age party at Yorktown Heights in 1968 when IBM hosted the first conference specifically on sparse matrix research. Since that time, the sparse matrix theme has become firmly established on the conference circuit with major meetings in 1971 (Oxford), 1972 (Yorktown Heights), 1975 (Argonne), 1978 (Knoxville) and 1980 (Reading) (see references [1] to [7]). The 604 references in my survey paper (Duff [10]) included most of the publications in sparse matrix research up until January 1976 and from that time the growth in interest and papers has soared with currently upwards of 500 publications annually. Clearly, it is impossible to give a comprehensive survey in this chapter, the whole volume does not contain sufficient pages for that! I have instead chosen to restrict my comments to those areas of sparse matrix research with which I am most familiar. This, however, should be sufficient to give a flavor of the field and an indication of where current research trends are heading. For readers who are unfamiliar with the concept of sparsity or where it arises, the sparse world of this chapter is introduced in Section 2.

We concentrate on the solution of linear equations by direct methods based on Gaussian elimination. We do this because we can illustrate most of the points we wish to make by confining ourselves to such techniques and also because it is an area with which many of you may already have some familiarity in the full case. We review Gaussian elimination in Section 3 and discuss basic methods for adapting it for efficiency when the matrix is sparse.

We look further at pivoting for sparsity in Gaussian elimination in Section 4, in particular illustrating the vast gains made by recent algorithms in the case when the matrix is symmetric and positive definite. In Section 5, we take a more global view of organizing the direct solution of equations to take advantage of sparsity. This leads naturally to Section 6 where we reflect on the little that is understood about the structure of sparse systems and discuss approaches to the solution of sparse equations which take advantage of structure.

Although we concentrate on direct methods for the solution of sparse linear equations, we have already mentioned that systems arising in the discretization of simple elliptic partial differential equations were being solved for over thirty years by nondirect methods. The iterative methods used in the early days had severe limitations in the class of matrices they could handle and exhibited slow convergence on all but the best conditioned problems. However, very recently there has been a resurgence in new accelerated iterative methods often combining classical iterative methods with the new direct methods. Since such techniques could well dominate in the 1980s we study them a little in Section 7.

We then briefly discuss some aspects of sparse matrix software in Section 8 before, in our final section, giving a checklist of the important points that this chapter has been trying to convey together with suggestions for further reading. Indeed we have intentionally kept the text unencumbered by detailed references and suggest the references in the final section be used as a guide to further reading.

Throughout we illustrate many points by giving the results of runs on test examples. These problems have all arisen in practical computations and, unless we are using some peculiar feature of the problem (in which case its origins and features will be explicitly discussed), they have been chosen as representative of a much wider set of experiments.

Certainly, we do not wish to imply that exciting new advances are not being made elsewhere in sparse matrix research; perhaps some of the most dramatic recent developments have been in the solution of the sparse eigenproblem. Dramatic because, only a few years ago, there was no software and hardly any algorithms available for solving this problem. Now there are many good methods and several codes for solving sparse eigenproblems whether the whole spectrum is required or whether a group of end or interior eigenvalues and eigenvectors is needed. Many methods are now available for solving sparse least-squares problems where some problems in geodesy, for example, may have several million equations! There has also been much recent work in the utilization of sparsity in nonlinear optimization codes.

2. IT'S A SPARSE WORLD

The exploitation of sparsity increasingly plays a central role in large-scale scientific computing. In this section, we will discuss sparsity, illustrate the benefits of utilizing it, and indicate some of the major areas in which sparse matrix techniques are employed.

We call a system sparse if most equations involve only a small subset of the total number of variables. Otherwise the system is termed dense or full. Thus, if we consider the coefficient matrix of a linear or linearized sparse system, most of the entries in that matrix will be zero. This may seem a little imprecise but it is necessarily so because the actual percentage of non-zeros to total entries is often less important than their distribution. I prefer the following definition although it is somewhat circular. That is, a matrix (or system) is sparse if advantage can be taken of the percentage and/or distribution of zero entries.

Clearly we can, in general, represent the information content in a sparse matrix with much less storage than the n^2 words required for a full matrix of order n. We can, for example, hold an index list for each row indicating which variables are present in addition to the actual non-zero values themselves. We illustrate an index list storage scheme for sparse matrices in Figure 2.1. In the sparse storage mode, the information on the matrix is held in three arrays. A and ICN hold the non-zero values (ordered by rows) and their corresponding column indices, and IP holds pointers to indicate the position in A and ICN of the first non-zero in each row. The last entry in IP points to the position after the end of data in A and ICN. Thus the non-zero values in row i are in positions $A(IP(i))$ to $A(IP(i+1)-1)$, $i=1,\ldots,n$ where n is the order of the matrix. The storage needed for the sparse matrix storage scheme in

Figure 2.1 is m floating-point numbers and $m + n + 1$ integers if there are m non-zeros in the matrix.

Of course there are particular classes of sparse matrices for which we can economize storage even further. For band systems, where $a_{ij} = 0$ if $|i - j| > b$, we need only store the order of the system and the semibandwidth b in addition to the entries within the band. A generalization of the band storage scheme, variable-band storage, allows the first non-zero in each row to be a varying distance from the diagonal. The variable-band scheme need only store, for each row, the column index of the first non-zero in that row in addition to the non-zero values themselves. Both the band and variable-band schemes will store explicitly any zero entries within the band. When discussing any sparse storage scheme in the following text we will use the term overhead to signify any storage over and above that for the non-zero values themselves. Finally, it may be possible (for example, for the discretization of simple differential equations on regular grids) to avoid storing the matrix at all but instead have a formula for constructing all or part of it when it is needed.

Although savings in storage are very important, there is much more to the utilization of sparsity. In most practical applications we will want to perform some computation with the matrix, for example we may wish to calculate its eigenvalues or to solve a linear system with it as the coefficient matrix. We thus wish to retain as much sparsity as possible during the solution process. This partially accounts for the great popularity of iterative techniques, since often the coefficient matrix is only used to multiply a vector and so is unchanged during the computation. It is only with the development of clever data structure manipulation that methods which transform the matrix have become competitive. There are many different data

$$
\begin{bmatrix}
1.0 & 0.0 & 0.0 & 2.0 \\
0.0 & 5.0 & 0.0 & 4.0 \\
3.0 & 7.0 & 0.0 & 0.0 \\
0.0 & 0.0 & 8.0 & 0.0
\end{bmatrix}
$$

A:	1.0	2.0	5.0	4.0	3.0	7.0	8.0
ICN:	1	4	2	4	1	2	3
IP:	1	3	5	7	8		

Matrix   Index list storage scheme

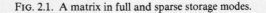

FIG. 2.1. A matrix in full and sparse storage modes.

structures and manipulation techniques used and some are quite complicated but, interesting though they are, we will not pursue them further here.

To sum up, we take advantage of sparsity by

(i)   only storing a small percentage of the entries in the matrix together with some form of indexing information;

(ii)   avoiding operations on zero entries; and

(iii)   organizing the computation to preserve as much of the initial sparsity as possible consonant with a fast and reliable solution.

We illustrate, in the table below, the gains that can be achieved by taking account of (i) to (iii) above. The computation chosen for this illustration is the direct solution of linear equations using Gaussian elimination. The matrix in this example arose in stress analysis calculations but we would like to emphasize that the magnitude of the savings is quite typical for problems in many areas.

Our sparse storage scheme in Figure 2.1 has illustrated that we can store a sparse matrix of order n with m non-zeros in $O(n)+$

TABLE 2.1

	Total storage required (in thousands of words)	Floating-point operations (in thousands)	Time (in seconds on an IBM 3033)
Treating the system as full	39.6	5254	3.382
Only storing and operating on non-zeros as in (i)	9.3	254	.211
Additionally organizing the computation to preserve much of initial sparsity	3.3	8	.019

An illustration of the gains when exploiting sparsity in the direct solution of a system of order 199 with 873 non-zeros.

$O(m)$ space*. Clearly we cannot store a general sparse matrix more economically and so a lower bound for the complexity of most sparse matrix algorithms is $O(n)+O(m)$. As we shall show later, there are some algorithms which attain this bound. Many algorithms have practical performances close to this bound although their worst case performance is poorer. Given this target, it is evident that $O(n^2)$ loops (for example, the scan of an array of length n at each stage of the elimination) should be avoided. This requirement imposes severe restrictions on the design of algorithms and software and causes sparse codes to be much more complicated than full codes for performing the same function.

We illustrate this in Figure 2.2, where we show a segment of Fortran code for performing the innermost loop in sparse Gaussian elimination (actually extracted directly from the Harwell Subroutine Library MA28 code) together with its full counterpart. The reason for such complicated codes is not that their authors like to exalt in the beauty and intricacy of the code (although, of course, they do) nor because they are paid for each line of code generated (would this were the case) but rather so that the run-time efficiency can be enhanced. The increase in efficiency can be very high as we illustrated in Table 2.1. Indeed it is often only by such savings that large problems become tractable.

Such complexities also create difficulties in assessing the merits of various competing codes. This is why band schemes can be competitive with their much more sophisticated counterparts. It also stresses the need for not just using fill-in or multiplication count as a yardstick for comparing codes since, in some instances, there can be more work in the bookkeeping concerned with the sparse data structures. Additionally, timing comparisons can be misleading unless we use a battery of test cases with widely varying structure. Even then, because some machines (for example, the IBM 3033) are very much more efficient than others (for example, the CDC 7600) at fixed-point calculations, any conclusions must necessarily be quite machine dependent and a rigid ranking, even for a single type of structure, may not be possible or meaningful.

---

*By $O(f(n))$ we mean $Kf(n)+g(n)$ where $g(n)/f(n) \to 0$ as $n \to \infty$, and K is independent of n.

```
C     ARRAY IQ IS AN INTEGER ARRAY ALL OF WHOSE ENTRIES
C     ARE POSITIVE EXCEPT THOSE ENTRIES IQ(I) FOR WHICH COLUMN I OF
C     THE PIVOT ROW IS NON-ZERO. THIS NON-ZERO WILL BE FOUND IN POSITION
C     IJPOS-IQ(I) OF ARRAY A. A PREVIOUS PASS THROUGH THE PIVOT ROW
C     HAS SO INITIALIZED IQ. THE COLUMN INDEX OF NON-ZERO A(I) IS
C     HELD IN ICN(I). FOR EACH NON-PIVOT ROW WITH A NON-ZERO IN THE
C     PIVOT COLUMN WE DO THE FOLLOWING
C
      DO 630 JJ=ROWI,IEND
      J=ICN(JJ)
      IF(IQ(J).GT.0) GO TO 630
      PIVROW=IJPOS-IQ(J)
      A(JJ)=A(JJ)+AU*A(PIVROW)
      ICN(PIVROW)=-ICN(PIVROW)
  630 CONTINUE
      DO 810 JJ=IJP1,PIVEND
      J=ICN(JJ)
      IF(J.LT.0) GO TO 800
C
C     FILL-IN PERFORMED. IN MA28 THERE IS EXTRA CODE HERE TO ENSURE
C     THAT POSITIONS IEND,... ARE FREE TO HOLD FILL-IN. THIS MAY HAVE
C     NECESSITATED MOVING THE NON-PIVOT ROW.
C
      A(IEND)=AU*A(JJ)
      ICN(IEND)=J
      IEND=IEND+1
C
C     IN MA28 WE NOW ALTER COLUMN ORIENTED REPRESENTATION TO ALLOW FOR
C     FILL-IN.
C
      GO TO 810
C
C     RESET ICN FOR NEXT PIVOT ROW
  800 ICN(JJ)=-J
C
  810 CONTINUE
```

Sparse code

```
      DO 630 J=1,N
      A(I,J)=A(I,J)+AU*A(IPIV,J)
  630    CONTINUE
```

Full code

FIG. 2.2. A comparison of sparse and full codes.

We finish this section by giving some idea of the diversity of areas where sparsity is exploited. Many problems in science and engineering yield sparse systems; examples arise in physics (fluid flow, atomic physics), chemistry (particularly chemical engineering), operations research, statistics (stochastic processes), civil engineering (structures problems), and electrical engineering (power systems, circuit design). The more complex models now used in the social sciences often give rise to sparse systems; for example, behavioral sciences (industrial relations), psychology (social dominance), politics and commerce (trade routes), demography (cultural interrelationships), economic modelling, business administration, and geography (surveying). In no way are these lists complete, they are intended only to give a flavor of subjects and subareas influenced by and influencing sparse matrix research.

Although we concentrate in the future sections on the solution of sparse linear equations, it should be appreciated that often other problems in numerical mathematics involve the solution of one or more linear systems. Among these are the solution of sets of initial-value ordinary differential equations using backward-difference methods, nonlinear systems, two-point boundary-value problems (using finite differences or finite elements), linear and nonlinear least-squares problems, eigenproblems, partial differential equations, and linear and nonlinear programming.

3. GAUSSIAN ELIMINATION

Gaussian elimination can be expressed as the LU decomposition of a matrix

$$A = LU \tag{3.1}$$

where L and U are lower and upper triangular matrices, respectively. The solution to the system

$$A\mathbf{x} = \mathbf{b} \tag{3.2}$$

is then accomplished by first solving the lower triangular set of equations

$$L\mathbf{y} = \mathbf{b} \tag{3.3}$$

(called the forward elimination or forward substitution) followed by back substitution to obtain x through the solution of the upper triangular set of equations

$$Ux = y. \tag{3.4}$$

Clearly, the work required to solve the two systems (3.3) and (3.4) for a full coefficient matrix of order n is n^2, while it can be readily shown that $n^3/3 + O(n^2)$ multiplications are required to produce the decomposition (3.1).

In the symmetric and positive definite case, we obtain the Cholesky decomposition

$$A = LL^T, \tag{3.5}$$

where the superscript T denotes matrix transpose.

The work required is $n^3/6 + O(n^2)$, although subsequent solutions still cost $n^2 + O(n)$ multiplications.

Of course, decompositions of the form (3.1) do not always exist even if A is nonsingular, the matrix

$$\begin{bmatrix} 0 & 1 \\ 1 & 1 \end{bmatrix}$$

being a simple example. In such cases, we must permute the matrix so that the diagonal throughout the decomposition is zero-free. This can be denoted by the equation

$$PAQ = LU \tag{3.6}$$

where P and Q are permutation matrices that can be generated as the elimination proceeds (note that the permutations are usually held in two arrays of length n with the integers 1 to n permuted as appropriate). We illustrate Gaussian elimination in Figure 3.1, where we show the situation partway through the decomposition (3.6). At this stage, the entries of L and U as indicated have been calculated and the reduced matrix A' (sometimes called the active matrix) has had its entries $a_{ij}(k < i, j < n)$ modified from those of the original matrix A. If entry a'_{kk} is zero, the elimination cannot

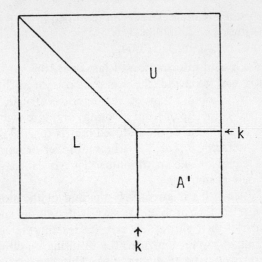

FIG. 3.1. Situation part-way through Gaussian elimination.

proceed until a non-zero entry from the reduced matrix has been permuted into that position. After this permutation, which will modify P and Q, the Gaussian elimination operation,

$$a'_{ij} \leftarrow a'_{ij} - a'_{ik} \left[a'_{kk} \right]^{-1} a'_{kj}, \qquad (3.7)$$

where a'_{kk} is called the pivot, can be performed on all the entries in the reduced matrix. In practice, such a permutation (called pivoting) will be performed if, otherwise, a'_{kk} would be small compared with a'_{ik} and/or a'_{kj}. We wish to avoid small pivots because the modulus of the triple product in (3.7) could then be much larger than $|a'_{ij}|$. Information might then be lost because of the finite word length of our computing machine. We illustrate this by observing that if the subtraction

$$.004468 - 5.124$$

is performed in four-figure floating-point arithmetic, the result would be

$$-5.120,$$

which would be the same as the result obtained if the first operand were replaced by

$$.004,$$

representing a loss of information of just over 10%.

A common way to avoid the worst effects of such numerical instability is to choose as pivot the largest entry in the whole reduced matrix (complete or total pivoting) or, more commonly, the largest entry in a column of the reduced matrix (partial pivoting) to ensure that the factor $a'_{ik}[a'_{kk}]^{-1}$ (the multiplier or entry in L) is less than or equal to one in modulus. This is called pivoting to preserve numerical stability.

For full systems this is effectively the end of the story, but for sparse matrices another problem arises. By definition, such systems will have most of their entries a'_{ij} zero (at least in the early stages) and if a'_{ik} and a'_{kj} were non-zero, the resulting modified entry in position (i, j) would be non-zero also. This occurrence is called fill-in. If it were repeated too many times, our reduced matrix would become so full that little advantage could be taken of sparsity. An extreme example of fill-in is shown in Figure 3.2 where the choice of the $(1, 1)$ entry as pivot would cause the reduced matrix to be completely full while pivoting on any other diagonal

$$
\begin{array}{cccccccc}
X & X & X & X & X & X & X & X \\
X & X & & & & & & \\
X & & X & & & & & \\
X & & & X & & & & \\
X & & & & X & & & \\
X & & & & & X & & \\
X & & & & & & X & \\
X & & & & & & & X \\
\end{array}
$$

FIG. 3.2. Extreme example of fill-in.

entry does not cause any fill-in, leaving the reduced matrix of the same structure as the original but with order one less. By induction, we can see that the whole LU decomposition of the example in Figure 3.2 can be effected without any fill-in.

In general, it is not so easy to forecast which entries make the best pivots on sparsity grounds, and several elaborate sparsity pivoting schemes have been proposed. As in many areas of useful mathematics, the earliest and one of the simplest criteria has really no serious rival as a general purpose scheme. This technique, usually named after Markowitz who proposed it in 1957, chooses as pivot that non-zero entry for which the product of the number of other non-zeros in its row and the number of other non-zeros in its column (both refer to the reduced matrix) is minimized. The Markowitz strategy clearly minimizes at each stage the number of nontrivial operations of the form (3.7) and also minimizes a bound on the fill-in at that stage. It is important to stress that the Markowitz strategy is purely heuristic and, although it has proven to produce good orderings for preserving sparsity over a wide range of problems, it is possible to produce examples on which it can perform badly. Indeed, it has been shown that the task of selecting the P and Q from the $(N!) \times (N!)$ possible permutation pairs to minimize the fill-in lies in the class of NP-complete problems.

Of course, such a pivoting strategy may conflict with the stability pivoting described earlier. For example, if the entry $(1,1)$ in the matrix in Figure 3.2 were the largest in the entire matrix, numerical stability considerations would require that it be selected as pivot. The Markowitz criterion, and indeed any reasonable sparsity pivoting strategy, would select as pivot any other diagonal entry.

It is possible to resolve this conflict by ordering the rows to reduce fill-in and performing partial pivoting within each row to maintain stability. Experience has shown that it is difficult to find a good row ordering and, for general systems, the fill-in is high.

Another way of combining numerical and sparsity pivoting is to relax the numerical test by using threshold pivoting. With threshold pivoting an entry a_{ij}, say, is acceptable for use as pivot if it satisfies the inequality

$$|a_{ij}| \geqslant u \cdot \max_{l} |a_{lj}|, \qquad (3.8)$$

where u is a preset scalar in the range 0 to 1 and the maximum is taken over all rows in the reduced matrix. Thus, by altering the value of u, we can strike a balance between the numerical stability of partial pivoting ($u = 1$) and ignoring the values of the non-zeros entirely ($u = 0$). The best value for u is problem dependent but, in most cases, good sparsity gains can be obtained with a value as high as .1 without too adversely affecting stability. We illustrate the effect of varying the threshold parameter on a range of problems in Table 3.1. The trend of the results are as one might expect although the higher number of non-zeros in the factors (and more arithmetic operations) when u is equal to one can cause the error to increase. The other apparently counter intuitive effect is that, at very low values of u, the fill-in can start increasing. This increase is because the growth in size of elements tends to be proportional to u^2 rather than u itself which may make non-zeros best suited on sparsity grounds unavailable as pivots.

When using threshold pivoting the combined stability and sparsity technique for the selection of pivots becomes: choose as pivot the non-zero in the reduced matrix with the lowest Markowitz count for which a threshold criterion of the form (3.8) is satisfied. It may be thought that we might have to test many entries to obtain those with lowest Markowitz count which satisfy (3.8) but the

TABLE 3.1

Order	147		199		822		541	
Non-zeros	2441		701		4790		4285	
u	Non-zeros in decomposition (m)	Error norm	m	Error	m	Error	m	Error
1(−10)	4881	1(2)	1350	4(−9)	6474	1(−8)	16553	5(15)
1(−4)	5028	3(−9)	1382	4(−9)	6474	1(−8)	16198	5(−2)
1(−2)	5067	4(−10)	1429	2(−11)	6495	2(−10)	15045	4(−6)
1(−1)	5095	2(−12)	1478	1(−12)	6653	1(−11)	13660	3(−9)
.25	6449	3(−12)	1598	8(−13)	6910	4(−12)	14249	8(−11)
.5	6381	2(−12)	1728	5(−13)	7231	1(−12)	14109	8(−11)
1.0	6772	2(−12)	1915	3(−13)	8716	6(−12)	16767	2(−10)

Varying the parameter in threshold pivoting. The notation $a(b)$ is used to represent the number $a \times 10^b$.

TABLE 3.2

Order of matrix	147	57	292
Number of non-zeros	2449	281	2208
Average length of search	3.2	2.8	3.1

Number of non-zeros accessed before finding pivot.

results of Table 3.2 indicate that with careful coding the amount of searching is not excessive. Of course, the efficiency of our pivot selection procedure depends crucially on how well we can access the non-zeros in order of increasing Markowitz count.

The use of threshold pivoting might alarm those schooled in classical numerical analysis since we have weakened a numerical pivoting strategy (partial pivoting) which is itself not foolproof. We cannot make any claims about the stability of our method except to note that the instability can be controlled by the choice of u, that we can monitor the instability, and that threshold pivoting has caused us few problems in practical situations. We now develop our comments on controlling and monitoring stability.

The calculated LU factors can be considered the exact factors of a perturbed matrix $A + E$ and our method is considered stable if the norm of E is small compared with that of A. If the (i, j)th entry of E is e_{ij} then

$$|e_{ij}| \leqslant 3.01 * \varepsilon * \text{ops} * a$$

where ε, ops and a are, respectively,

(i)   the machine precision;
(ii) the number of operations on each entry of the matrix; and
(iii) the size of the largest entry encountered during decomposition.

We can do little about (i) except to buy a new machine or work in a higher precision, and our sparsity pivoting will tend to keep (ii) low. It is thus (iii) that concerns us most.

With threshold pivoting as in (3.8), our growth in entries in the reduced matrix is bounded by $(1 + u^{-1})$ at each stage and so by changing u we can directly affect the value of a. Furthermore, the overhead in monitoring the growth need not be great. Additionally, it is possible to get a very cheap a posteriori bound on (iii) from the computed matrix factors. If the (i, j)th entries in L and U are l_{ij} and u_{ij}, respectively, this bound can be expressed as

$$a \leqslant \max_{i, j} |a_{ij}| + \max_i \|(l_{i1} \ldots l_{i, i-1})\|_p \max_j \|(u_{1j} \ldots u_{j-1, j})\|_q,$$

where $\frac{1}{p} + \frac{1}{q} = 1$ (the use of the max norm and 1-norm is common, i.e., $p = \infty$, $q = 1$).

If we discover that the growth in matrix entries is unacceptably large, we have two courses of action open. The first is to refactorize with an increased value for u, and the second is to use the inaccurate decomposition with iterative refinement or some more sophisticated scheme as discussed in Section 7.

In the case of symmetric positive definite systems, we can prove that pivoting down the diagonal is unconditionally stable but we will still wish to perform permutations, P, so that our factor, L, where

$$PAP^T = LL^T,$$

is as sparse as possible. The symmetric analogue to the Markowitz criterion is called the minimum degree criterion. Here we choose as pivot an entry from the diagonal whose row (in the reduced matrix) has the fewest number of non-zeros. The minimum degree strategy enjoys the same popularity and success as its unsymmetric counterpart and, additionally, great advantage can be taken of the fact that the non-zero values are not required when choosing pivots. As in the case of the Markowitz algorithm, the minimum degree strategy is a heuristic which keeps the fill-in and the number of floating-point operations low. The problem of obtaining a P for minimum fill-in in the subsequent factorization is NP-complete. We discuss the minimum degree ordering further in Section 4.

There is another important difference between Gaussian elimination on full and sparse matrices. Often, for example in the solution of nonlinear systems, we wish to solve a sequence of linear systems which have the same sparsity pattern although the actual values of the non-zeros may differ. If the matrices are full no worthwhile savings can be made but, for sparse systems, information from the first decomposition can be used to expedite subsequent factorizations and solutions. We call the first decomposition ANALYZE-FACTOR, because we need to analyze the structure as we perform the decomposition, and we call the subsequent decompositions FACTOR. SOLVE refers to the solution of the two triangular systems (3.3) and (3.4) after the decomposition has been effected. We illustrate the differences in times for these phases in Table 3.3, where the savings in utilizing structure information from the first decomposition are evident.

Of course, we are using the same pivotal sequence in FACTOR as in ANALYZE-FACTOR and so could suffer from numerical instability in subsequent factorizations. In many situations, however, the numerical values in the matrices of such a sequence do not differ widely and instability is not a problem. If stability is of concern we could monitor the FACTOR decomposition and perform another ANALYZE-FACTOR if necessary.

TABLE 3.3

Order of matrix Number of non-zeros	199 701	822 4841	900 4380
ANALYZE-FACTOR FACTOR SOLVE	.24 .05 .01	1.76 .26 .04	11.40 1.52 .09

Times for three phases of sparse Gaussian elimination (in seconds on an IBM 370/168).

TABLE 3.4

| Order of matrix | 147 | 1176 | 778 | 503 | 1005 |
Number of non-zeros	1298	9864	3025	3265	4813
ANALYZE	.033	.250	.149	.161	.416
SYMBOLIC-FACTOR	.014	.124	.060	.051	.090
NUMERIC-FACTOR	.046	.285	.302	.283	.693
SOLVE	.007	.032	.039	.030	.060

Times for different phases for Gaussian elimination on sparse symmetric positive definite systems (in seconds on an IBM 3033).

As we mentioned earlier and will discuss at greater length in Section 4, the situation for symmetric positive definite matrices is quite different. Here ANALYZE and FACTOR can be totally decoupled and, for reasons which will be explained in Section 5, the time for ANALYZE may be substantially less than the time for subsequent FACTORs. We illustrate this in Table 3.4. In this table we have divided the FACTOR phase into a SYMBOLIC-FACTOR, which depends only on the matrix pattern (and so is executed only once for each ANALYZE), and a NUMERIC-FACTOR where the actual numerical factorization is performed.

4. DIAGONAL PIVOTING IN SPARSE GAUSSIAN ELIMINATION

There is a large body of problems where we do not have to fear numerical instability, symmetric positive definite systems being the commonest and largest subclass. As we mentioned in Section 3, enormous savings can be obtained if we do not have to consider numerical values while pivoting. Not only do we not need to store or reference the non-zero values during the pivot selection stage but we do not necessarily have to access or update the full structure of subsequent reduced matrices. Additionally by restricting pivoting to the diagonal, we decrease greatly the complexity of the pivot selection process. For most of this section we will be concerned with symmetric matrices where diagonal pivoting has the added merit of preserving properties like symmetry and definiteness.

Elementary graph theory has been used as a powerful tool in the analysis and implementation of diagonal pivoting on symmetric matrices (see, for example, George and Liu [13]). We will now explore one use of graph theory a little further. For this illustration of the use of graph theory, we consider finite undirected graphs, $G(V, E)$, which consist of a finite set of distinct vertices V and an edge set E consisting of unordered pairs of vertices. An edge $e \in E$ will be denoted by (u, v) for some $u, v \in V$. The graph is labelled if the vertex set is in 1-1 correspondence with the integers $1, 2 \ldots |V|$, where $|V|$ is the number of vertices in V. In this application of graph theory, the set E, by convention, does not include self-loops (edges of the form (u, u), $u \in V$) or multiple edges. Also, since the graph is undirected, edge (u, v) is equal to edge (v, u). A subgraph $H(U, F)$ of $G(V, E)$ has vertex set $U \subseteq V$ and edge set $F \subseteq E$. A complete subgraph $H(U, F)$ is such that $(u, v) \in F$ for all $u, v \in U$. With any symmetric matrix, say, of order n, we can associate a labelled graph with n vertices such that there is an edge between vertex i and vertex j (edge (i, j)) if and only if entry a_{ij} (and, by symmetry, a_{ji}) of the matrix is non-zero. We give an example of a matrix and its associated graph in Figure 4.1. The main benefits of using such a correspondence can be summarized as:

(i) The structure of the graph is invariant under symmetric permutations of the matrix (they correspond merely to a relabelling of the vertices).

(ii) For mesh problems, there is usually an equivalence between the mesh and the graph associated with the resulting ma-

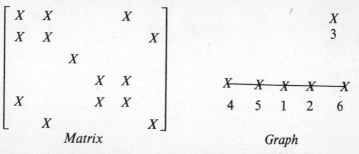

FIG. 4.1. Symmetric matrix and associated graph.

trix. We can thus work more directly with the underlying
structure.

(iii)   We can represent cliques (complete subgraphs) in a graph
by listing the vertices in a clique without storing all the
interconnection edges.

To illustrate the importance of the clique concept in Gaussian
elimination, we show in Figure 4.2 a matrix and its associated
graph (also the underlying mesh, if, for example, the matrix represents the five-point discretization of the Laplacian operator). If the
circled diagonal entry in the matrix were chosen as pivot (ad-

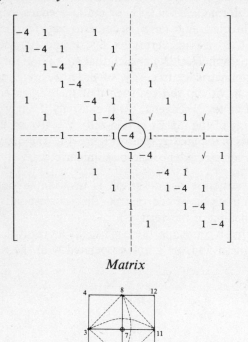

Matrix

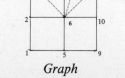

Graph

FIG. 4.2. Use of cliques in Gaussian elimination.

mittedly not a very sensible choice on sparsity grounds) then the resulting reduced matrix would have the dashed (pivot) row and column removed and have additional non-zeros (fill-ins) in the checked positions and their symmetric counterparts. The corresponding changes to the graph cause the removal of the circled vertex and its adjacent edges and the addition of all the dashed edges shown.

Thus the vertices $3, 6, 8, 11$ become a clique. It is an easily proven property of Gaussian elimination that all vertices of the graph of the reduced matrix which were connected to the vertex associated with the pivot will become pairwise connected after that step of the elimination process. Although this clique formation (and, later in the process, clique amalgamation) has been observed for some time it is only very recently that techniques making use of (iii) above have been used in ordering algorithms. For our small clique above, there are 4 vertices and 6 interconnecting edges but as the elimination progresses this difference will be more marked since a clique on n vertices has $n(n-1)/2$ edges corresponding to the off-diagonal non-zeros in the associated full submatrix. Since any subsequent clique amalgamation can be performed in a time proportional to the number of edges in the cliques concerned, both work and storage are linear rather than quadratic in the number of nodes in the cliques.

We illustrate clique amalgamation in Figure 4.3, where the circled element is being used as a pivot and the vertices in each rectangle are all pairwise connected. We do not show the edges internal to each rectangle in Figure 4.3 because we wish to reflect the storage and data manipulation scheme which will be used. There are two cliques (sometimes called elements or generalized elements by anal-

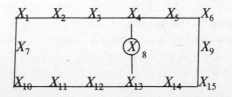

FIG. 4.3. Illustration of clique amalgamation.

ogy with elements in the finite element method) held only as lists of constituent vertices (1, 2, 3, 4, 7, 8,10, 11, 12, 13) and (4,5,6,8,9,13,14,15). After the elimination, the variables in both of these cliques will all be pairwise connected to form a new clique given by the list (1,2,3,4,5,6,7,9,10,11,12,13,14,15) and we can see that not only is a list merge the only operation required but additionally the storage after elimination (for the single clique) will be less than before the elimination (for the two cliques).

To summarize, there are two very important aspects of such an approach which make this development one of the most exciting in the last decade. First, because we do not have to mimic the Gaussian elimination operations during the ordering phase, the ordering can be significantly faster than the actual elimination (see, for example, Table 3.4). This is in sharp contrast to the situation for unsymmetric systems or indeed to the situation for symmetric systems ten years ago. The second, and in many ways more important, aspect is that of storage where the ordering can be implemented so that it requires only slightly more storage (a small multiple of n) than that for the matrix itself. This means that even for very large problems whose decomposition must necessarily be out-of-core, it may be possible to do an in-core ordering. Additionally, since the storage required for computing the ordering is independent of the fill-in and is known in advance, we can ensure sufficient storage is available for successful completion of this phase. When computing the ordering we can determine the storage required for symbolic and numerical factorization and the work for numerical factorization (the symbolic factorization can be performed in time proportional to the storage). This forecast could be crucially important if we want to know whether the in-core solution by a direct method is feasible or whether an out-of-core solver or iterative method must be used.

It is worth observing that the overhead storage (say $|R_L|$) for the factors produced by the minimum degree ordering is much less than that for the non-zero values themselves (say $|L|$). Indeed, for model problems on an $n \times n$ grid we have $|R_L| = o(|L|)$*. This reduction in overhead essentially occurs because neighbouring rows in the

---

*$f(n) = o(g(n))$ implies $\lim_{n \to \infty} (f(n)/g(n)) = 0$.

factorization have many columns in common and the column indexing information need not be repeated for each row. Recently, George has shown that the structure of the factors can be computed from the ordering in $O(|R_L|)$ time. The calculation of the ordering itself, if there are m non-zeros in the matrix, executes in $O(m)$ space and typically requires $O(m)$ time although there are no execution-time bounds known with the exception of some rather trivial cases. In the unsymmetric case, it can be shown that the evaluation of the structure of the factors takes $\min\{O(m),$ number of multiplications to compute $LU\}$ time and so is likely to be as expensive as actually computing the numerical factors.

To illustrate the remarkable power of current implementations of minimum degree we show, in Table 4.1, ordering times for various algorithms on three problems arising from successively finer finite element triangulations of an L-shaped region. Codes MA17A, and MA17E are from the Harwell Subroutine Library and do not use the clique amalgamation concept. The difference between MA17A and MA17E is due to the use of more efficient internal data structures by MA17E. YSMP (Yale Sparse Matrix Package), SPARSPAK (University of Waterloo), and MA27A (Harwell Subroutine Library) all use successive refinements of the generalized element model. When we realize that the MA17 codes were considered very powerful in the early 1970s, we see just how dramatic the figures in Table 4.1 are.

There are unsymmetric matrices for which pivoting down the diagonal in any order is stable and for which, in theory, ordering techniques could be developed to take advantage of the improvements illustrated in Table 4.1. The class of such problems is

TABLE 4.1

Order Non-zeros	265 1009	1009 3937	3466 13681
MA17A (1970)	1.56	29.9	> 250
MA17E (1973)	.68	6.86	62.4
YSMP (1978)	.24	1.27	6.05
SPARSPAK (1980)	.27	1.11	4.04
MA27A (1981)	.15	.58	2.05

Ordering times on graded-L (IBM 370/168 seconds).

unknown although there is both theoretical and practical evidence
to suggest that matrices arising from the discretization of second
order non-self-adjoint elliptic partial differential equations with
constant coefficients fall into this category.

There are two main dangers associated with diagonal pivoting on
unsymmetric systems. The most obvious one lies in determining
that such pivoting is numerically stable and the second one lies in
the approximation of the structure of an unsymmetric system by a
symmetric one. Common choices are to use the structure of $A + A^T$
or the symmetric pattern whose upper (or lower) triangle is equivalent to the upper (or lower) triangle of A. These choices could,
however, be very poor if A is far from symmetric. Additionally, the
information from the ordering on work and storage for subsequent
factorization may not be at all accurate. We illustrate the perfor-

TABLE 4.2

Order of matrix	541	822
Number of non-zeros	4285	5607
Non-zeros after decomposition		
Diagonal pivoting	14025	95292
Markowitz	13623	6653
TIME		
Analyze-Factor		
Diagonal pivoting	1.979	31.885
Markowitz	3.986	1.498
Factor		
Diagonal pivoting	.293	16.182
Markowitz	.560	.272
Solve		
Diagonal pivoting	.032	.213
Markowitz	.049	.035
Residual (l_2 norm)		
Diagonal pivoting	8×10^{-6}	8×10^{-9}
Markowitz	4×10^{-9}	1×10^{-11}

Diagonal pivoting on unsymmetric matrices. Time is in seconds on
an IBM 370/168.

mance of a diagonal pivoting algorithm on two sparse matrices in Table 4.2. The matrix of order 541 has a nearly symmetric structure while that of order 822 is very unsymmetric. These results illustrate the dangers of indiscriminate use of diagonal pivoting on unsymmetric matrices.

If it is desired to obtain some of the benefits indicated in Table 4.1 in the decomposition of unsymmetric systems, a possible approach is to perform an a priori ordering ignoring non-zero values and then to perturb this ordering to take account of asymmetry and numerical stability. We discuss this further in Section 6.

5. OTHER CONSIDERATIONS IN SPARSE GAUSSIAN ELIMINATION

There are several nonequivalent criteria for the choice of permutations for the decomposition

$$PAQ = LU. \tag{5.1}$$

A possible set of criteria for selecting P and Q are:

  (i)   to preserve sparsity;
  (ii)   to optimize data structures;
  (iii) to take advantage of underlying structure;
and
  (iv) to maintain stability.

In a sense (iii) includes (i) and (ii) but we choose to think of (iii) as a separate, more global ordering phase which we will discuss in Section 6. We have already considered (i) (in Sections 3 and 4) and its combination with (iv) in Section 3. To illustrate the difference between (i) and (ii), we compare the minimum degree criterion with the variable-band technique whose storage scheme was discussed in Section 2. When implementing the variable-band scheme (also called profile elimination), it is normal to order the non-zeros of the matrix so that the quantity

$$\sum_i \max_{\substack{a_{ij} \neq 0 \\ j < i}} |i - j|,$$

called the profile, is minimized or at least kept low.

Under a mild restriction (each row after the first has a non-zero to the left of the diagonal), pivoting down the diagonal of this permuted matrix will cause no fill-in outside the leftmost non-zero in each row but will fill in the profile (between that non-zero and the diagonal) completely. The defect of this strategy for general systems is that the fill-in can be very much higher than that obtained from a minimum degree ordering. The benefit of the variable-band approach is that the storage overhead can be greatly reduced. Apart from the non-zeros themselves, the only storage required is an array to indicate the length of each row. The problem of profile or bandwidth minimization can be shown to be NP-complete so, as was the case with the general ordering algorithms, the profile minimization algorithms will only be heuristic. They will find a low rather than a minimum profile.

We illustrate the use of these two ordering schemes diagrammatically in Figure 5.1, where the matrix arose in eigensystem calculations in atomic physics at the University of Lund. This matrix is a good candidate for the variable-band scheme pivoting down the diagonal in the natural order. This gives complete fill-in within the band as shown in Figure 5.1(b), whereas the minimum degree ordering disperses the non-zeros as in Figure 5.1(c) but surprisingly causes much less fill-in in the factors (Figure 5.1(d)). It is the dispersal of non-zeros by the minimum degree algorithm which increases the overhead for this method since integer pointers or indices are required to identify the position of every non-zero or every block of non-zeros. We quantify the effect of this overhead in Table 5.1 where it is apparent that for overall economy in storage, particularly if each floating-point number and each integer occupies the same amount of storage, the variable band scheme is to be preferred. However, to illustrate the power of the general minimum

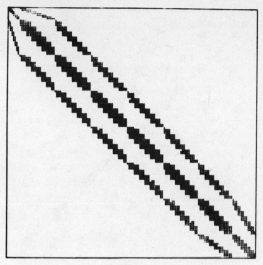

(a) Original matrix.

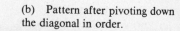

(b) Pattern after pivoting down
the diagonal in order.

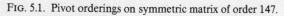

FIG. 5.1. Pivot orderings on symmetric matrix of order 147.

Iain S. Duff

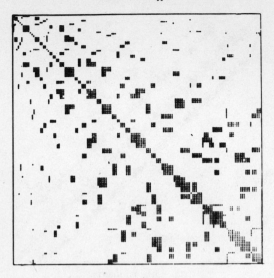

(c)   Reordered according to minimum degree strategy.

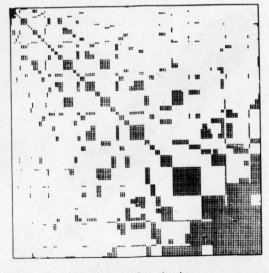

(d)   Pattern after pivoting in order down the diagonal of figure (c).

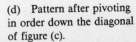

FIG. 5.1. Pivot orderings on symmetric matrix of order 147.

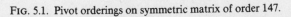

TABLE 5.1

	Non-zeros in factors	Overhead storage required
Minimum degree	2340	949
Variable band (natural ordering)	3017	147
Variable band (profile reduction ordering)	2450	147

Comparison of variable band and minimum degree algorithms on matrix from Figure 5.1 of order 147 with 1298 non-zeros.

TABLE 5.2

Order	443	1723	1009	918	778
Number of non-zeros	1033	4117	3937	4151	3025
Non-zeros in factors					
Minimum degree	1403	6313	19352	13900	13688
Variable band	11227	80983	26254	23105	17936
Total storage including overhead (in words)					
Minimum degree	5725	23729	33189	26634	24447
Variable band	14773	94769	34328	30451	24162
Multiplications in factorization (in thousands)					
Minimum degree	2	13	272	143	168
Variable band	159	2209	350	317	211
Time (in seconds on an IBM 3033) ANALYZE					
Minimum degree	.235	.733	.649	.794	.548
Variable band	.136	.263	.239	.235	.202
FACTOR					
Minimum degree	.054	.220	.761	.500	.503
Variable band	.267	2.957	.764	.694	.501

A comparison of variable band and minimum degree orderings on a selection of sparse symmetric matrices.

degree scheme, we present comparisons of these two ordering
strategies on a selection of sparse matrices in Table 5.2. In general,
variable-band techniques are only competitive on problems with a
regular sparsity structure such as those arising in structural analysis
and do very badly on more irregular problems such as those arising
in networks (the first two problems in Table 5.2).

In the general unsymmetric case, account must be taken of the
value of the non-zeros when computing the ordering, and the
combination of threshold pivoting with the Markowitz criterion, as
described in Section 3, is normally used. We illustrate in Figures
5.2(a) and 5.2(b) how the Markowitz ordering will tend to destroy
any structure originally present in the matrix, scattering the non-
zeros apparently randomly and necessitating an even higher integer
overhead than with the minimum degree ordering in the symmetric
case.

Another complication of a general scheme that scatters non-zeros
is that the innermost loop of most codes employing such techniques
is of the form

```
        DO 100 JJ = J1, J2
        J = ICN (JJ)
        W(J) = W(J) + AMULT*A(JJ)
100     CONTINUE
```

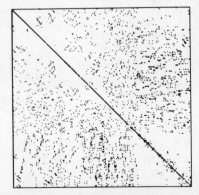

(a)   Original matrix.   (b)   Matrix reordered so that pivots
                                            are on the diagonal in order.

FIG. 5.2. Markowitz ordering on unsymmetric matrix of order 541.

TABLE 5.3

Order Non-zeros	147 2441	1176 18552	199 701	292 2208	130 1282	541 4285
Time for ANALYZE-FACTOR						
No switch	.427	2.585	.0613	.276	.1078	.749
1.0	.365	2.610	.0601	.278	.1047	.739
0.8	.277	1.541	.0569	.250	.1006	.723
0.6	.207	1.429	.0545	.228	.0992	.696
0.4	.153	1.055	.0523	.201	.0972	.657
0.2	.114	.806	.0489	.169	.0899	.587
0.05	.132	1.025	.0645	.252	.0900	.810
Time for FACTOR						
No switch	.073	.529	.0087	.0417	.0179	.105
1.0	.058	.526	.0083	.0408	.0150	.101
0.8	.047	.304	.0080	.0379	.0138	.098
0.6	.042	.292	.0079	.0366	.0134	.097
0.4	.042	.283	.0080	.0375	.0133	.097
0.2	.052	.331	.0094	.0501	.0157	.149
0.05	.093	.618	.0330	.1745	.0653	.514
Time for SOLVE						
No switch	.0050	.0230	.0024	.0057	.0018	.0127
1.0	.0044	.0230	.0024	.0057	.0018	.0124
0.8	.0038	.0193	.0023	.0055	.0017	.0122
0.6	.0035	.0189	.0023	.0053	.0017	.0121
0.4	.0032	.0178	.0023	.0052	.0017	.0117
0.2	.0030	.0170	.0024	.0051	.0018	.0111
0.05	.0027	.0186	.0031	.0061	.0023	.0129

Time (in seconds) for runs of Harwell code MA28 on the CRAY-1. The density of reduced matrix at which the switch to full code is made is shown in the left hand column.

TABLE 5.4

Order Non-zeros	147 2441	1176 18552	999 701	292 2208	130 1282	541 4285
Order of full matrix						
No switch	0	0	0	0	0	0
1.0	37	8	12	13	13	26
0.8	53	92	17	28	18	34
0.6	65	116	21	39	22	48
0.4	82	165	26	60	31	72
0.2	108	211	40	95	55	148
0.05	147	285	96	185	130	273
Number entries in **factors**						
No switch } 1.0 }	5669	21177	1464	5215	1331	12571
0.8	6055	22858	1499	5323	1371	12701
0.6	6734	26554	1576	5695	1481	13301
0.4	8454	37048	1733	7171	1869	15237
0.2	12497	51338	2492	11640	3709	28965
0.05	21609	86326	9674	35260	16900	78251
Integer storage for **factors**						
No switch	5669	21177	1464	5215	1331	12571
1.0	4303	21116	1323	5049	1165	11898
0.8	3249	14397	1213	4542	1050	11548
0.6	2512	13101	1138	4177	1000	11000
0.4	1733	9826	1060	3574	911	10056
0.2	836	6820	895	2618	687	7064
0.05	3	5104	461	1038	3	3725
Total storage						
No switch	11338	42354	2928	10430	2662	25142
1.0	9972	42293	2787	10264	2496	24469
0.8	9304	37255	2712	9865	2421	24249
0.6	9246	39655	2714	9872	2481	24301
0.4	10187	46874	2793	10745	2780	22301
0.2	13333	58158	3387	14258	4396	36029
0.05	21612	91430	10135	36298	16903	81976

Storage and order of full matrix after switch over to full code at densities of reduced matrix shown in the left hand column.

where the array W is accessed using indirect addressing. Some modern computers (for example, the CRAY-1 or the Cyber 205) can execute loops involving vectors very efficiently but such vectorization is inhibited when indirect addressing is used.

We would like to be able to use direct addressing of the form

```
          DO 100 JJ = J1, J2
          W(JJ) = W(JJ) + AMULT*A(JJ)
100       CONTINUE
```

in our innermost loop, both for better vectorization and because it would also be more efficient on non-vectorizing machines. We discuss general classes of techniques using direct addressing in sparse elimination in Section 6.

In this present section we observe that, because of fill-in, the reduced matrix will become increasingly dense. Towards the end of the elimination it may pay us to abandon our sparse data structures and treat the matrix as if it were full. We illustrate the benefits of allowing such a switch to direct addressing by showing, in Table 5.3, some runs of a sparse code for unsymmetric matrices on the CRAY-1. We see that significant gains can be achieved by switching to full code when the matrix is still quite sparse. Naturally, we cease to take advantage of any zeros when we switch to the full code and so the storage for the non-zero values will increase as the switch over density decreases. At the same time, however, we do not need any integer overhead for the reduced matrix which we are treating as full. We illustrate the effect of these two opposing trends in Table 5.4 where the figures are from the same runs as in Table 5.3. We see that the total storage first decreases a little when we switch to full code and it is only at quite low densities that the increase due to storing explicit zeros dominates.

6. THE IMPORTANCE OF STRUCTURE

One problem with the ordering techniques discussed earlier is that they take little account of the global structure of the underlying problem and a simple use of the variable-band ordering works well only on rather simple geometries. In this section, we study techniques which attempt to use the underlying structure of the problem. We believe this to be a very exciting development in sparse matrix research that is still very much in its infancy.

Many fortunes (and reputations) have been won and lost in the quest for better solution techniques to discretizations of elliptic partial differential equations and, in particular, to 5- or 9-point discretizations of the Laplacian operator. For some people, the only type of sparse matrix is of the form shown in Figure 4.2. However, most of us would agree that such matrices have a rather special structure and special properties and indeed it is by their very exploitation that efficient solution techniques have been developed.

It would be quite unfair to say that the structural exploitation story ends there since the profile or variable-band techniques knowingly or unknowingly make use of the peculiar connectivity found in, for example, structural engineering problems, and the preordering to block triangular form or bordered block triangular form (see later in this section) is very effective in many areas of linear programming and chemical engineering. To illustrate the relationship between physical structures and sparse matrices we show three structures in Figures 6.1(a), (c) and (e) and their corresponding matrices in Figures 6.1(b), (d) and (f). If we compare these examples with those in Figures 6.2 and 6.3, which are from completely different application areas, it is immediately evident that the underlying structure of these sparse matrices is quite different. It is not at all evident how we might quantify the differences, a necessary first step towards recognizing and thence capitalizing on underlying structure.

Process flow engineering, power systems and power networks, scheduling problems, partial differential equations, stochastic modelling, economic modelling, and linear programming all produce sparse matrices with considerably different characteristics. One might ask whether it is feasible or worthwhile to attempt to develop general-purpose methods, or whether it is preferable to concentrate on just one problem area (which may itself be very diverse) and develop highly efficient techniques which are almost inapplicable elsewhere. As an author of software for a multipurpose library, I believe the former to be difficult but feasible and, if one can use

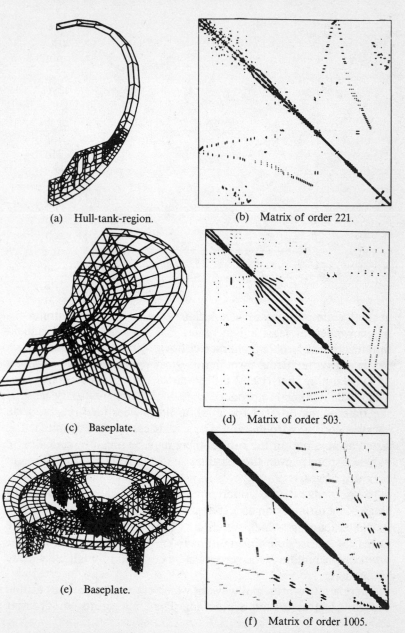

(a) Hull-tank-region.

(b) Matrix of order 221.

(c) Baseplate.

(d) Matrix of order 503.

(e) Baseplate.

(f) Matrix of order 1005.

FIG. 6.1. Physical structures and related sparse matrices.

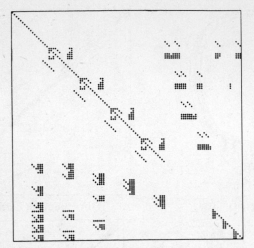

Fig. 6.2. A matrix of order 113 from statistical modelling.

general codes to obtain a solution, albeit without optimum efficiency, then at least a benchmark has been set for subsequent
optimization to problem characteristics.

Obviously the model formulation greatly influences the structure
of the resulting matrix and indeed historically the formulation has
often been designed to produce a smaller, denser matrix because at
the time of the original formulation there were few, if any, good
sparse algorithms around and sparse matrices were to be avoided as
much as possible. In the future, there may be much remodelling to
enhance sparsity even though the size of the resulting model might
be much larger.

A knowledge of the underlying structure of the problem at the
stage of its formulation as a sparse matrix is invaluable if dissection
or substructuring, which we discuss later in this section, is used.
Often, of course, the information is known a priori, for example
when one is building a larger model by combining smaller "solved"
submodels.

There are two principal ways in which current techniques exploit
the global structure of a problem. The first selects an ordering

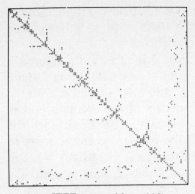

(a)   IEEE test problem (118).

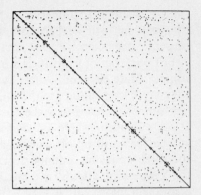

(b)   Equivalenced reduced representation of entire US power network (274).

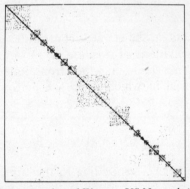

(c)   Representation of Western US Network (1454).

FIG. 6.3. Matrices from electric power networks. (Order in parentheses).

based on the structure of the underlying problem and then perturbs it locally to preserve numerical stability. The second preorders the system so that it can be solved as a concurrent or consecutive sequence of subproblems or as an update to the solution of a simpler problem. We consider each in turn.

We will discuss two related approaches which base their orderings on the underlying problem, namely the frontal methods and dissection techniques. Frontal schemes, long loved by engineers and finite element practitioners, need not be restricted to finite element problems but will tend to do better if there is an underlying mesh present rather than if one is created artificially.

For purposes of exposition we will describe the frontal approach using a finite element model. Two neighboring triangles in a finite element discretization are shown in Figure 6.4(a), together with the contributions they will make to the matrix of the overall problem. If there is one variable associated with each node shown, then the submatrices will be of order 6 and the overlap (corresponding to the three nodes in common to both triangles) of order 3. We call the process of adding a contribution from an element into the coefficient matrix an "assembly" and there will be a contribution to entry a_{ij} of the matrix of the form

$$a_{ij} = a_{ij} + e_{56}^{(l)} + e_{23}^{(l+1)}, \tag{6.1}$$

where the subscripts on the element contributions denote the local coordinate representations of position (i, j) in the assembled matrix. The crucial observation on which frontal methods are based is that it is immaterial in which order we perform the operations corresponding to (6.1) and (3.7), viz.,

$$a_{ij} = a_{ij} - a_{ik}[a_{kk}]^{-1}a_{kj}, \tag{6.2}$$

so long as the entries corresponding to the triple product in (6.2) are fully summed (i.e., have had all contributions of the form (6.1)) before the Gaussian elimination operation (6.2) is effected. In particular, a_{ij} may yet receive additional contributions. Thus we

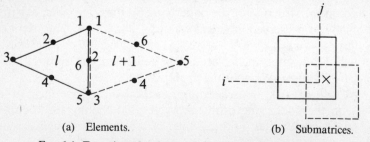

(a) Elements. (b) Submatrices.

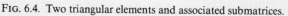

FIG. 6.4. Two triangular elements and associated submatrices.

can assemble the matrix as we eliminate, which, as we will now see, can lead to considerable organizational simplification.

Consider the triangulation in Figure 6.5, where there are 7 variables associated with each triangle (at the vertices, midpoints of the sides, and the centroid). We assemble our matrix in the order given in that figure. After the assembly of element 6 our matrix has the form shown in Figure 6.6. The important thing about this figure is the presence of the zero blocks at the top right and bottom left-hand corners. This means that, if we choose our pivots from anywhere within the top left-hand block (which we call the fully-assembled (or pivot) block), we do not affect any part of the matrix which is not yet assembled (bottom right region in figure). We will, of course, alter the values of the entries in the pivot block itself and the hatched regions (we call this whole region the assembled block) by the elimination equation (6.2). Although the doubly hatched area and the pivot block are completely assembled when this operation is performed, contributions to the singly hatched region have yet to come from elements 7–13. However, as we observed above, this does not prevent us from performing eliminations with pivots from the pivot block.

In the absence of any need for numerical pivoting (for example, when the assembled matrix is positive definite) the whole of the pivot block of Figure 6.6 could be eliminated and, since these rows and columns take no further part in the factorization of the matrix, they could be written to backing store. Since the lower right-hand

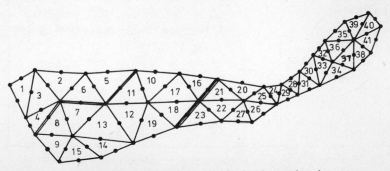

FIG. 6.5. Triangulated region with triangles numbered.

part of the matrix in Figure 6.6 is not yet even partially assembled, the only part that need be held in main storage at this time is the singly hatched submatrix in the figure.

The "front" at the stage shown in Figure 6.6 corresponds to the 7 variables in the double-lined triangle sides in Figure 6.5. The size of the front will of course vary as the assembly and elimination progresses; for example, after assembly of element 19 and consequent eliminations, the front will consist of variables on the triple-lined triangle sides in Figure 6.5. For any elements, geometry and ordering on the elements, there is clearly a maximum front size (for

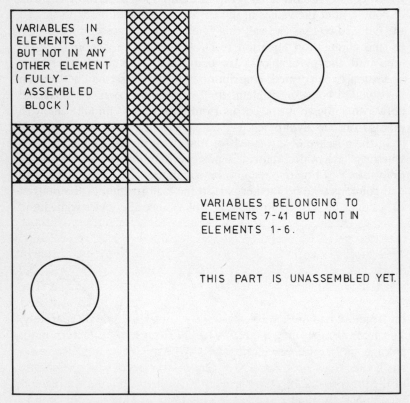

VARIABLES IN ELEMENTS 1-6 BUT NOT IN ANY OTHER ELEMENT (FULLY – ASSEMBLED BLOCK)

VARIABLES BELONGING TO ELEMENTS 7-41 BUT NOT IN ELEMENTS 1-6.

THIS PART IS UNASSEMBLED YET.

FIG. 6.6. Matrix of partially assembled problem (after assembly of element 6).

our example, after assembly of element 10) which will determine the space requirements and efficiency of the method. Often, an ordering of the elements to keep this maximum front size low will be evident from the geometry. When it is not, an ordering scheme rather like those used to reduce the profile of variable-band matrices is needed.

If the matrix is unsymmetric, then the entries in the pivot block of Figure 6.6 may not be suitable for pivoting on numerical grounds. Indeed we will only choose an entry from this block (not necessarily from the diagonal) as pivot if it satisfies a relative tolerance threshold of the form (3.8), viz.,

$$|a_{lk}| \geqslant u \cdot \max_i |a_{ik}|, \qquad 0 < u \leqslant 1,$$

where the maximum is taken over all rows in the assembled block. If we cannot eliminate all of the pivot block because its entries are too small, then we simply perform some more assemblies, eventually causing the large entries in the doubly hatched regions to move into the pivot block where they can then be used as pivots. The only penalty we pay is that our active front may become larger than the structure itself dictates but, in return for that, stability is as assured as in the unsymmetric codes discussed previously. All we are doing is making local perturbations to an ordering based globally on the structure.

In a similar way, a frontal scheme for symmetric indefinite matrices could be designed where diagonal pivoting on 1×1 or 2×2 blocks is used. The Harwell code MA27, which we used in Section 4, employs this kind of pivoting.

There are three great benefits in using the frontal approach. The first, with which we introduced this scheme, is that orderings based on structure and on numerical stability can be computed separately. The second is that only the active front matrix need be kept in main storage and so primary storage requirements can be kept low even when the whole problem is very large. The third lies in the coding of the innermost loop. We normally treat the frontal matrix (assembled block of Figure 6.6) as a full matrix. Thus we avoid any indirect addressing in the innermost loop, making it much simpler and more easily vectorizable, an important consideration which we already discussed in Section 5.

Of course, the simple frontal scheme can perform very badly on a multilobed structure like that of a hub with propellor blades and for this we need a more general multifrontal approach. Here the basic idea is similar to that described above but now the entire frontal matrix can be output to backing store and another front started in the structure. Later in the elimination process the output fronts are combined and further eliminations take place. The notion of working independently on different parts of the structure is what the engineers have for many years called substructuring. The Harwell code MA27, which was used in runs for Table 4.1, combines a general minimum degree ordering with a multifrontal approach and uses direct addressing in the innermost loop of the numerical factorization.

Another route to methods based on the global structure of the underlying problem is to extend the techniques of dissection and nested dissection. With dissection techniques we choose a set of variables which divide the problem into two or more subproblems and put these variables at the end of the pivot sequence. An example is shown in Figure 6.7 where variables corresponding to the cross are placed at the end of the ordering and the resulting

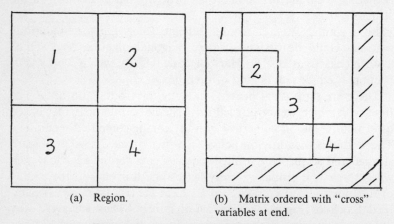

(a)   Region.   (b)   Matrix ordered with "cross"
                                        variables at end.

Fig. 6.7. Illustration of dissection technique.

problem splits into the four subproblems corresponding to the submatrices shown in Figure 6.7(b). The dissections can be nested so that, in our example, each of the subproblems 1 to 4 would themselves be split into 4 and so on. If the nested dissection ordering is applied to a regular grid problem it can be shown theoretically that the work and storage for the solution of the whole problem are $\mathcal{O}(q^3)$ and $O(q^2 \log q)$, respectively, if the grid has q points in each direction. This dependence on q can be shown to be

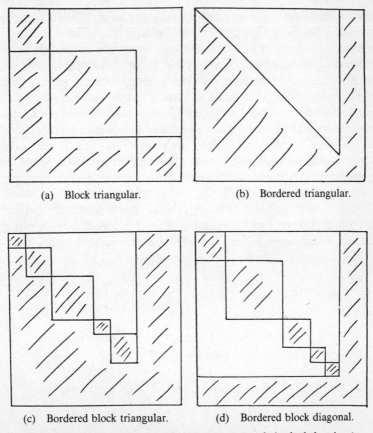

(a)   Block triangular.   (b)   Bordered triangular.

(c)   Bordered block triangular.   (d)   Bordered block diagonal.

FIG. 6.8. Block forms for preordering (non-zeros only in shaded regions).

asymptotically optimal for all solution techniques based on Gaussian elimination because we must always solve a full $q \times q$ subsystem at some stage.

Recently there has been much work on automating nested dissection algorithms. For general problems there are no bounds on the complexity of these orderings and the algorithms used are heuristic. For two-dimensional finite element problems, automatic nested dissection algorithms have been designed which, if there are m non-zeros in the matrix, execute in $O(m \log m)$ time and $O(m)$ space and yield factors with $O(m \log m)$ non-zeros. Furthermore, these bounds can be shown to be optimal. There is, however, no implementation of these optimal algorithms readily available and so we rely on heuristic algorithms in this planar case also.

Global orderings of this type are not, of course, applicable to all problems and there are many systems in linear programming, network analysis, economic modelling and chemical engineering, for example, whose underlying structure is too unsymmetric to be amenable to such techniques. Here again we can often preorder the matrix to a suitable form either automatically or by a knowledge of the structure of the underlying problem. Tewarson [9] suggests several so-called desirable forms. We show four such forms in Figure 6.8.

When we have preordered our matrix to one of the forms shown in Figure 6.8 the solution to our problem can be greatly simplified. For example, if our block triangular form has 3 blocks as shown in Figure 6.9 and the solution and right-hand-side vectors are partitioned conformally, then the solution

$$\begin{bmatrix} A_{11} & & \\ A_{21} & A_{22} & \\ A_{31} & A_{32} & A_{33} \end{bmatrix} \begin{bmatrix} x_1 \\ x_2 \\ x_3 \end{bmatrix} = \begin{bmatrix} b_1 \\ b_2 \\ b_3 \end{bmatrix}$$

FIG. 6.9. Block triangular partitioning.

to the complete problem can be effected in three stages, viz,

$$A_{11}x_1 = b_1$$
$$A_{22}x_2 = b_2 - A_{21}x_1$$
$$A_{33}x_3 = b_3 - A_{31}x_1 - A_{32}x_2,$$

where we observe that we need never solve a system larger than the largest block on the diagonal. Additionally no Gaussian elimination operations are performed on the off-diagonal blocks, which are used only for forward substitution and do not suffer from any fill-in.

There are good algorithms available for permuting a general matrix to block triangular form and it can be shown that this form is essentially unique. The block triangularization is usually performed in two stages, the first employs row (or column) permutations to make the diagonal zero-free, and the second stage uses a symmetric ordering to partition the matrix into block triangular form. The best known algorithm for the first stage has a worst case time bound of $O(n^{1/2}m)$, but we have found that a simpler algorithm with a worst case time bound of $O(n \cdot m)$ typically executes in only $O(n) + O(m)$ time. There are algorithms available for the second stage, which employ a depth first search of the graph of the matrix, and which have a worst case bound of only $O(n) + O(m)$.

For a large class of problems the bordered triangular form is only attainable at the cost of a relatively wide border, but the removal of only a few columns can make it possible to order the rest into block triangular form. These bordered block triangular forms occur commonly in chemical process engineering, operations research problems and econometrics. Although complicated heuristics have been suggested for finding such a structure (see references Rose [3] or Himmelblau [8]), there are few very good algorithms and little software, and indeed one could surmise that the trial removal of successive columns followed by an attempt to block triangularize using the highly efficient algorithms mentioned earlier might prove as effective as any more sophisticated technique. One way of obtaining a bordered block diagonal form is by the simple dissection strategy discussed earlier in this section (see Figure 6.7).

With the bordered patterns there are basically two modes of solution. We can perform a matrix splitting of the form

$$A = M - N, \tag{6.3}$$

where the matrix N corresponds to the border (sometimes called the torn columns) and then solve using the iteration

$$M\mathbf{x}^{(k+1)} = N\mathbf{x}^{(k)} + \mathbf{b}, \tag{6.4}$$

although we have no guarantee of convergence. We notice that in each of the cases illustrated in Figure 6.8, the matrix M has a particularly simple form, thus facilitating the solution of equation (6.4) at each step. The other approach is to use a purely direct method based on partitions.

If our bordered matrix is of the form shown in Figure 6.10, a solution method based on partitioning

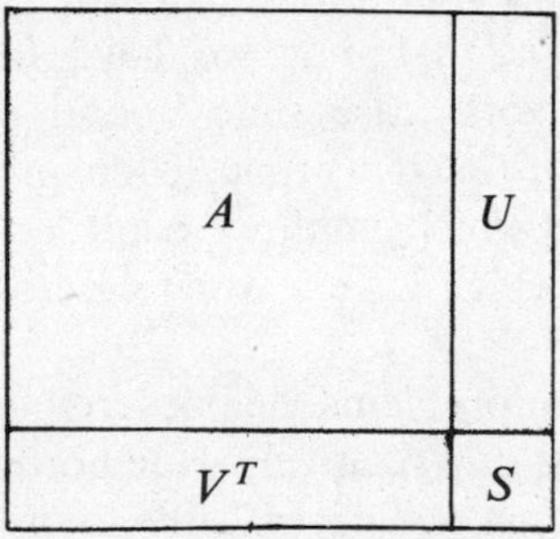

FIG. 6.10. Bordered matrix.

solves the system

$$\begin{bmatrix} A & U \\ V^T & S \end{bmatrix} \begin{bmatrix} \mathbf{x}_1 \\ \mathbf{x}_2 \end{bmatrix} = \begin{bmatrix} \mathbf{b}_1 \\ \mathbf{b}_2 \end{bmatrix} \tag{6.5}$$

by solving

$$(S - V^T A^{-1} U)\mathbf{x}_2 = \mathbf{b}_2 - V^T A^{-1} \mathbf{b}_1, \tag{6.6}$$

followed by

$$A\mathbf{x}_1 = \mathbf{b}_1 - U\mathbf{x}_2, \tag{6.7}$$

where the inverse operator A^{-1} in (6.6) is not held explicitly.

Another problem, which is related to the solution of a partitioned system, is that of solving a system where the coefficient matrix differs by only a low rank change from that of a system solved previously. If the low rank change to A, of order n, can be expressed as WTZ^T where W, T and Z^T are of order $n \times k$, $k \times k$, and $k \times n$, respectively ($k \ll n$), then the modification formula

$$(A + WTZ^T)^{-1} = A^{-1} - A^{-1}W(T^{-1} + Z^TA^{-1}W)^{-1}Z^TA^{-1}$$

$$(6.8)$$

expresses the inverse of the modified matrix in terms of the inverse of the original matrix and the inverse of a $k \times k$ matrix. As in the use of the partitioned formulae (6.6) and (6.7) we would never compute the inverses explicitly but would hold them as a sparse factorization.

The relationship of the modification formula (6.8) to the solution to the partitioned problem (6.5) is evident if we set $S = -T^{-1}$, $V = Z$, $U = W$ and $\mathbf{b}_2 = \mathbf{O}$ in (6.5). It is possible to express a general partitioned problem (6.5) as a matrix modification of order $2k$. This is not, however, an attractive route algorithmically and it is better to solve (6.5) as a partitioned system using (6.6) and (6.7).

7. ITERATIVE METHODS

A great source of entertainment in the early sparse matrix conferences was the battle between those supporting the new wave of direct solvers and those whose family fortunes were based on iterative methods. It is becoming clear, retrospectively, why nobody won this battle. Many successful techniques can be classed in neither camp and are semi-iterative or semidirect depending on which side of the fence you stand. Indeed there is a whole spectrum of techniques from straight Gaussian elimination through to the classical iterative methods like SOR. We will examine a few such approaches to give a flavor of this very fruitful area which looks likely to yield good solution techniques for very large systems from three-dimensional partial differential equations. We include this section only to show the interplay between direct and iterative methods. For details on iterative methods themselves the chapter in this volume by Widlund should be consulted.

Even in the case of full systems, an iteration is often used to improve a solution obtained by direct methods. A common technique is iterative refinement where an initial estimate, $\mathbf{x}^{(0)}$, of the solution to the linear system

$$A\mathbf{x} = \mathbf{b} \tag{7.1}$$

is improved by the correction

$$\mathbf{x}^{(k+1)} = \mathbf{x}^{(k)} + \boldsymbol{\delta}\mathbf{x}^{(k)}, \qquad k = 0,1,2\ldots$$

where $\boldsymbol{\delta}\mathbf{x}^{(k)}$ is the solution of

$$A\,\boldsymbol{\delta}\mathbf{x} = \mathbf{r}^{(k)} \tag{7.2}$$

with

$$\mathbf{r}^{(k)} = \mathbf{b} - A\mathbf{x}^{(k)}$$

the residual corresponding to the vector $\mathbf{x}^{(k)}$.

If the true solution to (7.1) is $\mathbf{x}^*$ we can define the error $\mathbf{e}^{(k)}$ by

$$\mathbf{e}^{(k)} = \mathbf{x}^{(k)} - \mathbf{x}^*.$$

Further, if B is some estimate of the inverse of A used to solve (7.2) (for example, B might be held as approximate LU factors of A), then

$$\mathbf{e}^{(k+1)} = (I - BA)\mathbf{e}^{(k)}.$$

Therefore, for our method to converge, it is not necessary that B be a good estimate to the inverse of A, but only that the spectral radius of $(I - BA)$ be less than one. This observation leads to the possibility of choosing L and U in (5.1) so that, instead of being accurate factors of A, they retain much of the sparsity of A but are only a factorization of A in the sense that the iteration above converges. Evidence suggests that for some classes of problems a partial factorization of A which retains non-zeros only in positions which are non-zero in the original A is suitable. Thus, we totally avoid fill-in and the data organization and storage problems concomitant with it.

However, even if the method outlined above converges, convergence may be very slow. Several ways of accelerating convergence have been suggested including the use of Chebyshev acceleration, generalized conjugate gradients, or Lanczos recurrences.

A highly successful method when the matrix A is symmetric definite is to use a partial Cholesky decomposition

$$A = LL^T + E, \tag{7.3}$$

where the entries in E are not necessarily relatively small, coupled with conjugate gradients. In the conjugate gradient iteration the operator

$$L^{-1}A(L^T)^{-1} \tag{7.4}$$

is used instead of A. The idea of the preconditioning (7.4) is to alter the spectrum of the iteration matrix so that it is favorable for the conjugate gradient algorithm. The art of choosing a good preconditioning is still very much in its infancy and clearly the preconditioning matrix can range from being a representation of the exact inverse of A (resulting in a direct method) to the identity matrix (ordinary conjugate gradients). For the 5- or 9-point discretization of Laplace's equation a suitable preconditioning is obtained by performing Cholesky decomposition allowing only one or two diagonal bands of fill-in as shown in the figure below. This class of methods are called the ICCG methods—Incomplete Cholesky Conjugate Gradient methods.

Although the conjugate gradient algorithm works only for symmetric positive definite systems, we could transform any system to one with a positive definite coefficient matrix by solving the normal equations

$$A^TA\mathbf{x} = A^T\mathbf{b}. \tag{7.5}$$

However, this method is not favored because the conditioning of

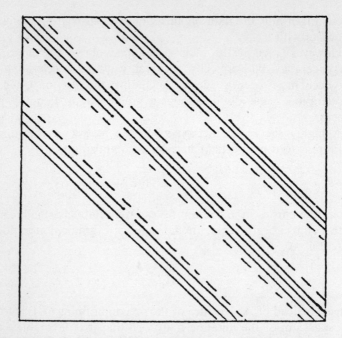

FIG. 7.1. Dashed diagonals illustrate positions where fill-ins are allowed during a partial Cholesky factorization for an ICCG algorithm.

the problem (7.5) is usually much worse than the original system (7.1) and because the work in solving (7.5) is about twice that for (7.1).

A common way of extending solution techniques to more general systems is to perform the matrix splitting of (6.3), viz.,

$$A = M - N \tag{7.6}$$

where M is a matrix on which the unmodified technique is applicable. The solution to (7.1) is then found by using the iteration (6.4), viz.,

$$M\mathbf{x}^{(k+1)} = N\mathbf{x}^{(k)} + \mathbf{b}. \tag{7.7}$$

Examples of commonly used splittings are to choose M to be positive definite, or, for elliptic partial differential equations, to choose a lower order finite difference approximation to the operator which A approximates or a discretization of a restriction of the operator to a subregion of the problem domain. In the latter case, we might, for example, wish to solve over a nearly rectangular region. M could be chosen to be a discretization over the rectangular region (for which fast and efficient solution techniques might be available), while N would account for the deviation of the region from this rectangle. At worst N might be of the form

$$\begin{bmatrix} I & 0 \\ 0 & N' \end{bmatrix}$$

where N' is a small full matrix.

Another very recent class of methods which blurs the distinction between iterative and direct methods are the multigrid techniques. Here a simple iterative technique (for example, SOR) is used on a fine grid chosen so that the truncation error in approximating the differential equation is sufficiently low. An approximation to the solution x_1 and its residual

$$r_1 = b - Ax_1$$

are calculated, and the residual is then interpolated onto a coarser grid to form r_2, whence the equations

$$A_2 \delta x_2 = r_2, \tag{7.8}$$

where A_2 is a discretization of the operator on the coarser grid, are solved for δx_2. δx_2 is then interpolated back onto the fine grid to give a correction to x_1. The coarse grid correction may be repeated after a few more SOR smoothing iterations on the fine grid, starting from this adjusted solution. Of course, this procedure can be nested inasmuch as system (7.8) can be solved by this two-level approach, and so on. It is common to use a direct solver at the innermost level (coarsest grid), thus giving a method which is both iterative and direct. The choice of interpolation used as well as the various smoothers and coarse grid corrections employed give a whole battery of methods, some of which have been successfully used to solve some three-dimensional partial differential equations.

There are many specialized techniques for solving linear systems arising from discretizations of partial differential equations, particularly separable equations in two dimensions on a rectangular region. We will not describe these so-called fast methods here. They are discussed in the chapter by Swarztrauber. Most of these direct methods are based on the use of Fourier transforms (using the fast Fourier transform, FFT) and/or cyclic reduction and can solve a system on a $q \times q$ grid in $O(q^2 \log \log q)$ time. Techniques akin to matrix splitting can be used to extend the domain of these methods through an approach called the capacitance-matrix method.

8. SOFTWARE CONSIDERATIONS

In this section on software considerations, we discuss the general philosophy of mathematical software design as it affects sparse matrix research. We do not give details on software available for the methods we have discussed. Readers interested in this aspect should consult the survey on sparse matrix software by Duff [15].

For earlier generations of algorithm designers, the work stopped as soon as the method was proven to be efficient on some class of problems and the paper (or two) was published. It was left to some first year graduate student to extend its domain to cover some genuinely interesting problems and then to release it to the remarkably uncomplaining user.

Only recently has the art of mathematical software design been appreciated as a worthwhile pursuit for qualified researchers. There has been a growing awareness among the numerical analysis community that much of the real usefulness of their research, in all but the most esoteric reaches of the subject, is lost unless it can be packaged in a form suitable for the customer, perhaps a physicist or engineer who may be very naive about the inner-workings of numerical computation. As algorithm designers, we have to decide whether we merely wish to receive the adulation of our peers, or whether we wish to benefit a much larger scientific community.

These general comments on numerical software are even more true in the present context. Most of sparse matrix research is motivated by the need to solve practical problems or to provide tools so that others may do so. Yet even here, it is not fully appreciated how much work is involved in, or the benefits derived from, building a convenient, versatile and robust user interface.

One of the main benefits of dressing our algorithmic servant in good clothes is that it enhances the confidence the user will have in the software and often determines whether it will be used at all. If some problem in the presentation of data to the software arises, it will normally be detected and diagnosed by the software package and some suggestion for corrective action returned. This robustness is extremely important for two reasons. First, as we indicated in Section 2, the solution of the sparse linear equations could well be in an inner loop of a more complex calculation and so the "user" is quite likely to be another program. Second, such robustness increases a user's confidence in the code. Assuming the code justifies this confidence, any errors in the user's program are more easily isolated and detected since the "black box" is known to be error free. We might even dare to suggest that the user, perhaps dangerously, is attracted by the gloss on the outside irrespective of what evils that gloss may be covering. Certainly some well-used and well-trusted packages in statistics and structural analysis have had software of dubious quality embedded within.

There are costs, of course, involved in providing such an interface. Not only the manpower required to create it but also the overhead in run-time and the length of the code. However, applications where 10% savings in time are critical are few, and the savings in human effort when good robust software is readily available far outweighs these expenses.

It is particularly important in sparse matrix software to shield the user from the complexity of the data structures involved and in which he is very unlikely to have any interest. The screening in SPARSPAK (from the University of Waterloo) is more or less complete since the user must merely declare a single array of storage, entry of data being by one of several subroutines. There are different subroutines for inputting an entry or a row at a time or by full submatrices, as is common in finite element applications. Input to the Harwell subroutine MA28 is through three arrays. One array, A, say, holds the non-zero values and the other two, IRN and ICN, say, hold row and column indices for the corresponding non-zero entry. That is to say, the value $A(i)$ is in row $IRN(i)$ and column $ICN(i)$, $i = 1,\ldots, m$, where m is the number of entries input. These entries allow duplicates (i.e., $i \neq j$ but $IRN(i) = IRN(j)$ and $ICN(i) = ICN(j)$) whose values are summed and can be in any order. The Yale Sparse Matrix Package (YSMP) accepts the structure shown in Figure 2.1 as input. That is, the non-zero values are input by rows (or columns) together with an accompanying array of column (row) indices and pointers to the position in the arrays of the start of each row (column). This interface is less flexible than the previous two but allows a much simpler and shorter sort prior to the analysis and factorization.

Another important aspect of a good user interface is to provide useful diagnostics in the event of failure and in some instances take further action to provide the user with more information on the error. In addition to providing error diagnostics, SPARSPAK has a subroutine to enable the user to extract useful information about the run and a parameter to allow various levels of detail on the progress of the solution to be output. In MA32, the Harwell unsymmetric out-of-core frontal solver, computation can continue symbolically even if insufficient space has been allocated to the in-core frontal matrix, the user finally receiving a lower bound on the size required for subsequent runs. Additionally, if insufficient space has been allocated for the factors (either in-core or out-of-core), then the amount required for subsequent success will be returned at the end of the run.

Indeed these facilities and the ability to perform orderings in a small predetermined amount of storage from which an accurate forecast of time and space requirements for subsequent solution can be obtained (see Section 5) open the door to an exciting new prospect: that of interactive computing with sparse matrices. This could well be an important growth area in the next decade.

We commented earlier on the reasons for the long time-lag between algorithms and good software. As a postscript to this section, when we observe the large gap between algorithms and software and the almost total dearth of good software in some areas, we can clearly see where many advances will be made in the near future.

9. SUMMARY AND FURTHER READING

In this concluding section we present a checklist of areas covered in the preceding sections which we believe will be growth areas in the future. Thus this section is a lazy reader's guide to a sparse future. We have intentionally avoided encumbering the text with references but since we realize that you are now champing at the bit for more sparse matrix information, we include a small bibliography at the end of this section which will act as a guide to a second-level introduction to this exciting and expanding field.

Our checklist summarizes many of the points made earlier and should not be thought of as being comprehensive. For example, we do not list any of the advances being made in the solution of sparse eigenproblems, sparse least-squares problems, or more general sparse minimization.

1. Understanding structure.
   (a) Collection of test examples.
   (b) Discovery of common characteristics between matrices from different applications.
   (c) Further development of preordering and partitioning schemes and use of substructuring.
   (d) Work on global decomposition schemes where an overlying or underlying ordering based on structural considerations is perturbed locally to ensure stability, economy, or efficiency.
2. Gaussian elimination.
   (a) Recognition of which classes of problems do not require numerical pivoting.
   (b) Extension of clique algorithms to unsymmetric systems where stability is handled by local perturbations to pivot order.
3. Solutions of linear equations.
   (a) A deeper practical and theoretical understanding of preconditioning.
   (b) Extension of semidirect techniques to more general systems.
   (c) More exploitation of "tearing" (see also 1(c)).

4. Adaptation to new computer architectures.
    Little tuning will be needed if we come to grips with struc-
    ture and get our algorithm design correct initially.
    (a) Use of full matrix techniques.
    (b) Use of decomposition and substructuring for minicom-
         puters and multiprocessor environments.
5. Software.
    (a) Continued development of high-quality, robust numeri-
         cal software.
    (b) Incorporation of this new software in existing packages.
    (c) Design of large sparse matrix packages with a simple
         user interface and the capability for interactive work.

We divide our references into two groups. We first list proceedings of conferences on sparse matrices and then list some books
together with comments on their relative strengths. Additionally,
although it is now somewhat dated, the reader may find the survey
article useful, particularly the bibliography which fairly comprehen-
sively covers publications prior to January 1976. Also included is a
reference to a recent survey on sparse matrix software.

ACKNOWLEDGEMENT

I would like to thank the anonymous reviewers of this chapter for
their helpful comments.

REFERENCES

Sparse Matrix Proceedings.

1. R. A. Willoughby, *Proceedings of the Symposium on Sparse Matrices and their
    Applications* (Conference at IBM, Yorktown Heights), IBM report RA1 ‡11707,
    1969.
2. J. K. Reid, *Large Sparse Sets of Linear Equations.* (Conference at Oxford),
    Academic Press, New York, 1971.
3. D. J. Rose and R. A. Willoughby, *Sparse Matrices and their Applications.*
    (Conference at IBM, Yorktown Heights), Plenum Press, New York, 1972.
4. V. A. Barker, *Sparse Matrix Techniques.* (Summer School at Copenhagen),
    Springer-Verlag, 1976.

5. J. R. Bunch and D. J. Rose, *Sparse Matrix Computations.* (Conference at Argonne, Illinois), Academic Press, New York, 1976.
6. I. S. Duff and G. W. Stewart, *Sparse Matrix Proceedings 1978.* (Conference at Knoxville), SIAM Press, 1979.
7. I. S. Duff, *Sparse Matrices and their Uses* (Conference at Reading), Academic Press, New York, 1981.

Other Publications.

8. D. M. Himmelblau, *Decompositions of Large Scale Problems* (Conference in Cambridge, England), North Holland, 1973. Many application papers with most from chemical engineering. Several contributions on partitioning and tearing.
9. R. P. Tewarson, *Sparse Matrices*, Academic Press, New York, 1973. The first book solely on sparse matrices. Now somewhat dated and fairly elementary. Principally on the solution of linear equations. Discusses preordering techniques.
10. I. S. Duff, "A survey of sparse matrix research," *Proc. IEEE* **65** (1977), pp. 500–535. Survey paper arranged by subject area with an extensive bibliography.
11. A. Brameller, R. N. Allan, Y. M. Harman, *Sparsity*, Pitman, 1976. Extremely elementary. Has electrical network emphasis.
12. A. Jennings, *Matrix Computation for Engineers and Scientists*, John Wiley, New York, 1977. A good proportion of this book is concerned with sparsity. Direct and iterative methods for solving linear equations and techniques for the solution of eigenproblems are discussed.
13. A. George and J. W. H. Liu, *Computer Solution of Large Sparse Positive Definite Systems*, Prentice-Hall, 1981. Discusses graphical interpretation of Gaussian elimination, nested dissection and extensions, and much recent work on the solution of symmetric problems.
14. Å. Björck, R. J. Plemmons, H. Schneider, *Large Scale Matrix Problems*, North Holland, 1981. Contains the papers which appeared in volume 34 of Linear Algebra and its Applications. Includes current research papers on least squares, linear equations, eigenvalue problems and sparse optimization.
15. I. S. Duff, "A survey of sparse matrix software," Harwell Report AERE R10512, 1982. To appear in *Sources and Development of Mathematical Software*, W. R. Cowell (Ed.), Prentice-Hall. Surveys sparse software, giving details on the availability of codes for linear equations (by direct, iterative and semidirect methods), for least squares, for eigenvalue calculations and for linear and nonlinear programming. Particular attention is given to sparse matrix software applicable in the solution of partial differential equations.

QUESTIONS OF NUMERICAL CONDITION RELATED TO POLYNOMIALS[†]

*Walter Gautschi**

1. INTRODUCTION

Polynomials (in one variable) permeate much of classical numerical analysis, either in the role of approximators, or as gauge functions for a variety of numerical methods, or in the role of characteristic polynomials of one kind or another. It seems appropriate, therefore, to study some of their basic properties as they relate to computation. In the following we wish to consider one particular aspect of polynomials, namely, the extent to which they, or quantities related to them, are sensitive to small perturbations. In other words, we are interested in the numerical condition of polynomials. We shall examine from this angle three particular

[†]Revised and in part reprinted (with permission of the publisher) from *Recent Advances in Numerical Analysis* (C. de Boor and G. H. Golub, eds.), pp. 45–72, Academic Press, New York, 1978.

*Supported in part by the National Science Foundation under Grant MCS 7927158A01.

problem areas: (1) The representation of polynomials (polynomial bases); (2) Algebraic equations; (3) The problem of orthogonalization. Before embarking on these topics, however, we must briefly consider ways of measuring the condition of problems. We do this in the framework of maps from one normed space into another, for which we define appropriate condition numbers.

2. THE CONDITION OF MAPS

2.1. Nonlinear maps. Let X, Y be normed linear spaces, and let $y = f(x)$ define a map $M: \mathscr{D} \subset X \to Y$, with $\mathscr{D}$ an open domain. Let $\mathring{x} \in \mathscr{D}$ be fixed, and $\mathring{y} = f(\mathring{x})$, and assume that neither $\mathring{x}$ nor $\mathring{y}$ is the zero element in the respective space. The sensitivity of the map M at $\mathring{x}$, with respect to small relative changes in $\mathring{x}$, will be measured by the (*asymptotic*) *condition number* (see Rice [27])

$$\text{cond}(M; \mathring{x}) = \lim_{\delta \to 0} \sup_{\|h\| = \delta} \left\{ \frac{\|f(\mathring{x} + h) - f(\mathring{x})\|}{\|f(\mathring{x})\|} \Big/ \frac{\|h\|}{\|\mathring{x}\|} \right\},$$

$$(2.1)$$

provided the limit exists. The number in (2.1) measures the maximum amount by which a relative perturbation of $\mathring{x}$ (given by $\delta/\|\mathring{x}\|$) is magnified under the map M, in the limit of infinitesimal perturbations. Maps with large condition numbers are called *ill-conditioned*.

If M has a Fréchet derivative $[\partial f / \partial x]_0$ at $\mathring{x}$, then

$$\text{cond}(M; \mathring{x}) = \frac{\|\mathring{x}\|}{\|\mathring{y}\|} \left\| \left[\frac{\partial f}{\partial x} \right]_0 \right\| \qquad (\mathring{y} = f(\mathring{x})). \qquad (2.2)$$

In the important case of finite-dimensional spaces, $X = \mathbb{R}^n$, $Y = \mathbb{R}^m$, the Fréchet derivative, as is well known, is the linear map defined by the Jacobian matrix of f. We may then use in (2.2) any family of vector norms and subordinate family of matrix norms (see Stewart [31], p. 177).

For composite maps $K \circ M$, the chain rule for Fréchet derivatives (see Ortega & Rheinboldt [25], p. 62) can be used to show that

$$\text{cond}(K \circ M; \mathring{x}) \leqslant \text{cond}(K; \mathring{y})\text{cond}(M; \mathring{x}). \qquad (2.3)$$

If the composite map is known to be ill-conditioned, the inequality (2.3) permits us to infer the ill-conditioning of (at least) one of the component maps.

2.2. Linear maps. If $M: y = f(x)$ is a linear (bounded) map, then

$$\sup_{\|h\| = \delta} \frac{\|f(\mathring{x} + h) - f(\mathring{x})\|}{\|h\|} = \sup_{\|h\| = \delta} \frac{\|f(h)\|}{\|h\|}$$

is independent of $\mathring{x}$ and δ and equal to the norm of M. Equation (2.1) then reduces to

$$\text{cond}(M; \mathring{x}) = \frac{\|\mathring{x}\|}{\|\mathring{y}\|}\|M\| \qquad (M \text{ linear, } \mathring{y} = M\mathring{x}). \qquad (2.4)$$

If in addition M is invertible, we can ask for the supremum of (2.4) as $\mathring{x}$ varies in X (or, equivalently, $\mathring{y}$ varies in MX), and we find, since $\mathring{x} = M^{-1}\mathring{y}$, that

$$\sup_{x \in X} \text{cond}(M; x) = \|M^{-1}\|\,\|M\|. \qquad (2.5)$$

The number on the right, usually referred to as the *condition number of* M, will be denoted by

$$\text{cond } M = \|M^{-1}\|\,\|M\|. \qquad (2.6)$$

We have, alternatively,

$$\text{cond } M = \frac{\displaystyle\sup_{x \in X} \left(\|Mx\|/\|x\|\right)}{\displaystyle\inf_{x \in X} \left(\|Mx\|/\|x\|\right)}. \qquad (2.7)$$

Condition numbers such as those proposed cannot be expected to do more than convey general guidelines as to the susceptibility of the respective maps to small changes in the elements of their domains. By their very definition they reflect "worst case" situations and therefore are inherently conservative measures.

3. THE CONDITION OF POLYNOMIAL BASES

Let $\mathbb{P}_{n-1}$ denote the class of (real) polynomials of degree $\leqslant n-1$, and let $p_1, p_2, \ldots, p_n$ be a basis in $\mathbb{P}_{n-1}$. For any $p \in \mathbb{P}_{n-1}$, we denote by $u_1, u_2, \ldots, u_n$ the coefficients of p with respect to this basis,

$$p(x) = \sum_{k=1}^{n} u_k p_k(x). \qquad (3.1)$$

We wish to determine how strongly the values of p on some given finite interval $[a, b]$ react to small perturbations in the coefficients u_k and, vice versa, how the coefficients of p are affected by small changes in p.

The question may be formalized as one concerning the condition of the linear map $M_n : \mathbb{R}^n \to \mathbb{P}_{n-1}[a, b]$, which associates to each vector $u^T = [u_1, u_2, \ldots, u_n] \in \mathbb{R}^n$ the polynomial p in (3.1), restricted to $[a, b]$,

$$(M_n u)(x) = \sum_{k=1}^{n} u_k p_k(x), \qquad a \leqslant x \leqslant b. \qquad (3.2)$$

We are thus interested in the condition number of M_n, see (2.6),

$$\text{cond } M_n = \|M_n^{-1}\| \|M_n\|, \qquad (3.3)$$

in particular, how fast it grows as $n \to \infty$, and how this growth depends on the particular interval chosen.

For definiteness, we consider only uniform norms, i.e., $\|u\| = \max_{1 \leqslant k \leqslant n} |u_k|$ in $\mathbb{R}^n$, and $\|p\| = \max_{a \leqslant x \leqslant b} |p(x)|$ in $\mathbb{P}_{n-1}[a, b]$, although there are circumstances in which other norms may be preferable (see, e.g., Geurts [19], Gautschi [16], Sections 3.1, 3.2).

We shall use the notation u_p to denote the coefficient vector of p,

$$u_p = M_n^{-1} p, \qquad p \in \mathbb{P}_{n-1}. \qquad (3.4)$$

3.1. Power basis.

For the power basis

$$p_k(x) = x^{k-1}, \qquad k = 1, 2, \ldots, n, \tag{3.5}$$

it is natural to assume an interval $[a, b]$ that contains the origin. We shall do so in the following, but other intervals could also be treated (in fact more easily). For definiteness, we assume further that $[a, b]$ is centered to the right of the origin, i.e., $0 \leqslant |a| \leqslant b$. It then follows immediately that

$$\|M_n\| = \sup_{\|u\| = 1} \max_{a \leqslant x \leqslant b} \left| \sum_{k=1}^{n} u_k x^{k-1} \right| = \sum_{k=1}^{n} b^{k-1},$$

hence

$$\|M_n\| = \frac{b^n - 1}{b - 1}. \tag{3.6}$$

(It is understood, here and below, that the value of the function on the right equals n if $b = 1$.)

For the inverse map M_n^{-1} we have

$$\|M_n^{-1}\| = \sup_{\|p\| = 1} \max_{1 \leqslant k \leqslant n} \frac{|p^{(k-1)}(0)|}{(k-1)!} = \max_{1 \leqslant k \leqslant n} \sup_{\|p\| = 1} \frac{|p^{(k-1)}(0)|}{(k-1)!}.$$

Therefore, in terms of the linear functionals $\lambda_k : \mathbb{P}_{n-1}[a, b] \to \mathbb{R}$ defined by $\lambda_k p = p^{(k-1)}(0)/(k-1)!$,

$$\|M_n^{-1}\| = \max_{1 \leqslant k \leqslant n} \|\lambda_k\|. \tag{3.7}$$

Our problem thus reduces to determining the norm of λ_k. This is related to the problem of best uniform approximation of binomials $f_{n,\tau}(x) = (1 - |\tau|)x^n + \tau x^{n-1}$, where $-1 \leqslant \tau \leqslant 1$, by polynomials g of degree $\leqslant n-2$, which, in turn, gives rise to the Zolotarev polynomials

$$z_{n,\tau}(x) = \frac{1}{E_{n,\tau}} \left(f_{n,\tau}(x) - g_{n,\tau}^*(x) \right), \qquad -1 \leqslant \tau \leqslant 1,$$

where

$$E_{n,\tau} = \inf_{g \in \mathbb{P}_{n-2}} \|f_{n,\tau} - g\| = \|f_{n,\tau} - g^*_{n,\tau}\|.$$

The extremal for the functional λ_k, indeed, is a Zolotarev polynomial (of degree $n-1$, since we are working with $\mathbb{P}_{n-1}[a, b]$), that is, for $2 \leqslant k \leqslant n$,

$$\|\lambda_k\| = \sup_{\|p\|=1} |\lambda_k p| = |\lambda_k z_{n-1,\tau}| \text{ for some } \tau \in [-1,1] \quad (3.8)$$

(see, e.g., Schönhage [30], Satz 6.11). Unfortunately, if the interval $[a, b]$ is arbitrary, the value of the parameter τ in (3.8) is not easily expressible, and may be different for different values of k. The exact determination of $\|M_n^{-1}\|$ in (3.7) indeed is cumbersome (see Voronovskaja [33], Ch. III and the appendix by V. A. Gusev). Upper bounds for $\|M_n^{-1}\|$ are obtained in Gautschi ([14], Theorem 4.1).

For the power basis (3.5), the most natural interval, however, is an interval symmetric about the origin, $[-\omega, \omega]$, $\omega > 0$. In this case, the Zolotarev polynomials reduce to Chebyshev polynomials of the first kind (Schönhage [30], p. 167). Making use of this, it then follows readily from (3.6) and (3.7) that

$$\text{cond } M_n = \frac{\omega^n - 1}{\omega - 1} \max\left\{ \|u_{T_{n-1}(x/\omega)}\|, \|u_{T_{n-2}(x/\omega)}\| \right\}, \quad (3.9)$$

where T_m denotes the Chebyshev polynomial of degree m, and $u_{T_m(x/\omega)}$ the coefficient vector of $T_m(x/\omega)$; see (3.4).

Using asymptotic estimates of $\|u_{T_m(x/\omega)}\|$ as $m \to \infty$, Gautschi ([14], Eq. (2.2)), it can be deduced from (3.9) that

$$(\text{cond } M_n)^{1/n} \sim \begin{cases} 1 + \sqrt{1 + \omega^2}, \ \omega \geqslant 1, \\ \\ \dfrac{1 + \sqrt{1 + \omega^2}}{\omega}, \ \omega < 1, \end{cases} \quad \text{as } n \to \infty. \quad (3.10)$$

TABLE 3.1

	$(\text{cond } M_n)^{1/n}$				
n	$\omega = .1$	$\omega = .2$	$\omega = 1$	$\omega = 5$	$\omega = 10$
5	9.767	5.743	2.091	3.789	6.444
10	13.977	7.579	2.377	4.616	8.027
20	16.719	8.706	2.437	5.210	9.252
40	18.286	9.330	2.447	5.588	10.023
$\vdots$	$\vdots$	$\vdots$	$\vdots$	$\vdots$	$\vdots$
∞	20.050	10.099	2.414	6.099	11.050

The condition of M_n on $[-\omega, \omega]$.

The condition of M_n on $[-\omega, \omega]$ thus grows exponentially with n, the asymptotic growth rate being smallest (equal to $1 + \sqrt{2}$) when $\omega = 1$. Some numerical values are shown in Table 3.1.[†]

Similar results hold for intervals $[0, \omega]$, $\omega > 0$, in which case (Gautschi [14])

$$(\text{cond } M_n)^{1/n} \sim \begin{cases} (1 + \sqrt{1 + \omega})^2, \omega \geqslant 1, \\[2mm] \dfrac{(1 + \sqrt{1 + \omega})^2}{\omega}, \omega < 1, \end{cases} \quad \text{as } n \to \infty.$$

(3.11)

Again, the minimum growth rate ($= (1 + \sqrt{2})^2$) occurs when $\omega = 1$.

It is interesting to note that exponential growth of the condition is also observed for piecewise polynomial functions if represented

---

[†] The information in Table 3.1 might suggest that the asymptotic growth rate is approached monotonically. The reader, however, will have noticed that the limit rate in the case $\omega = 1$ is smaller than the seemingly increasing approach rates! In reality, the approach is indeed monotone, if $\omega < 1$, but changes from increasing to decreasing, if $\omega = 1$, and from decreasing to increasing, if $\omega > 1$. The changeover occurs near $n = 35$, if $\omega = 1$ (hence is not visible in Table 3.1), and near $(e/2\pi)\omega$, if $\omega \gg 1$. The latter would begin to be visible in Table 3.1 if $(e/2\pi)\omega \doteq 10$, i.e., $\omega \doteq 23$. The reason for this behavior can be found in the more precise relations $\text{cond } M_n \sim (\gamma^2 n)^{1/2} \rho^n$ if $\omega = 1$, and $\text{cond } M_n \sim (\gamma^2/n)^{1/2} \rho^n$, if $\omega \neq 1$, where ρ is the limit rate and $\gamma = \gamma(\omega)$ can be explicitly computed; Gautschi [14].

in terms of normalized B-splines. In fact, for splines of degree $k-1$, the condition of the B-spline basis is known to lie between $(1-1/k)2^{k-3/2}$ and $2k \cdot 9^k$; see de Boor [2], Lyche [23]. Empirical evidence seems to suggest that the conditon is indeed $O(2^k)$; de Boor [3].

3.2. Bases of orthogonal polynomials.

We now consider the case of an orthogonal basis, i.e.,

$$p_k(x) = \pi_{k-1}(x), \qquad k=1,2,\ldots,n, \qquad (3.12)$$

where $\pi_0, \pi_1, \ldots, \pi_{n-1}$ are the first n of a sequence of polynomials orthogonal on the (finite) interval $[a, b]$ with respect to a nonnegative measure $d\sigma(x)$. We consider the condition of this basis on the interval of orthogonality, $[a, b]$. Since the coefficients u_k in (3.1) of any polynomial $p \in \mathbb{P}_{n-1}$ are now representable as Fourier coefficients of p, it is easy to estimate the condition of M_n with the aid of Schwarz' inequality. One finds (Gautschi [12])

$$\text{cond } M_n \leqslant \max_{1 \leqslant k \leqslant n} \left(\frac{\mu_0}{h_{k-1}} \right)^{1/2} \cdot \max_{a \leqslant x \leqslant b} \sum_{k=1}^{n} |\pi_{k-1}(x)|,$$

$$(3.13)$$

where

$$\mu_0 = \int_a^b d\sigma(x), \qquad h_k = \int_a^b \pi_k^2(x)\, d\sigma(x), \qquad k=0,1,\ldots.$$

$$(3.14)$$

The first maximum in (3.13) is a bound for $\|M_n^{-1}\|$, the second an obvious bound for $\|M_n\|$. It should be noted that neither cond M_n nor the bound in (3.13) is invariant under different normalizations of the orthogonal polynomials $\{\pi_{k-1}\}$, and the bound indeed is minimized in the case of an orthonormal system.

It follows from (3.13) that the condition of an orthogonal basis, typically, exhibits only polynomial growth in n. For Chebyshev polynomials $\pi_r = T_r$ on $[-1,1]$, for example, one finds

$$\operatorname{cond} M_n \leqslant 2^{1/2} n \qquad (\pi_r = T_r),$$

while for Legendre polynomials $\pi_r = P_r$ on $[-1,1]$,

$$\operatorname{cond} M_n \leqslant n(2n-1)^{1/2} \qquad (\pi_r = P_r).$$

The improvement over the power basis is substantial.

3.3. Lagrangian bases. All bases $\{p_k\}$ considered previously have the property that $\deg p_k = k-1$, $k = 1, 2, 3, \ldots$. We now consider an example of a basis in which each p_k is a polynomial of degree $n-1$, namely the familiar Lagrange polynomials

$$p_k(x) = l_k(x), \qquad l_k(x) = \prod_{\substack{\nu=1 \\ \nu \neq k}}^{n} \frac{x - s_\nu}{s_k - s_\nu}, \qquad k = 1, 2, \ldots, n,$$

corresponding to a set of distinct nodes $s_1, s_2, \ldots, s_n$ in $[a, b]$. Lagrange's interpolation formula

$$p(x) = \sum_{k=1}^{n} p(s_k) l_k(x)$$

shows immediately that $u_k = p(s_k)$ in (3.1). By standard arguments in approximation theory, one finds $\|M_n\| = L_n$, $\|M_n^{-1}\| = 1$, where

$$L_n = \max_{a \leqslant x \leqslant b} \sum_{k=1}^{n} |l_k(x)|$$

is the Lebesgue constant for the nodes s_ν. Consequently (see also de Boor [5], p. 19),

$$\operatorname{cond} M_n = L_n. \tag{3.15}$$

By a result of Faber and Bernstein, Natanson ([24], p. 24), one has

$$L_n > \frac{\ln n}{8\sqrt{\pi}}$$

for arbitrary (distinct) nodes s_ν, while for Chebyshev nodes, on the other hand,

$$L_n \sim \frac{2}{\pi}\ln n, \qquad n \to \infty$$

(see, e.g., Rivlin [28], p. 18). The basis consisting of the Lagrange polynomials $\{l_k\}$ for Chebyshev nodes, therefore, is optimally conditioned among all Lagrangian bases, and indeed among all polynomial bases (de Boor [4]), in the sense of attaining the optimal growth rate $O(\ln n)$.

4. THE CONDITION OF ALGEBRAIC EQUATIONS

We now turn our attention to roots of algebraic equations and their sensitivity to small changes in the coefficients. (We assume that the equation is expressed linearly in terms of basis polynomials.) An interesting, though largely unexplored, aspect of this question is the manner in which this sensitivity depends on the choice of polynomial basis. By far best understood is the case of equations expressed in the usual power form.

In order to give a formal statement of the problem, we assume, first of all, that the basis polynomials p_k have $\deg p_k = k - 1$, $k = 1, 2, \ldots$, so that an algebraic equation of exact degree n can be written in normalized form

$$p(x) = 0, \qquad p(x) = p_{n+1}(x) + \sum_{k=1}^{n} u_k p_k(x), \qquad (4.1)$$

with leading coefficient 1. (To enhance clarity, we sometimes write $p(u; x)$ instead of $p(x)$, where $u = [u_1, u_2, \ldots, u_n]^T$.)

In general, one might be interested in just one, or in several, or collectively in all the roots of the equation, and again, there may be one single coefficient, or several, or all of them that are subject to perturbation. We treat all these cases in one, by considering q (simple) roots $\boldsymbol{\xi} = [\xi_1, \xi_2, \ldots, \xi_q]^T$, $1 \leqslant q \leqslant n$, of (4.1), corresponding to $u = \mathring{u}$, and by introducing a multi-index $\mathbf{k} = (k_1, k_2, \ldots, k_p)$,

$1 \leqslant k_1 < k_2 < \cdots < k_p \leqslant n$, to indicate which of the coefficients in $\mathring{u}$ are to undergo changes. We write $\mathbf{k}^c$ for the multi-index complementary to $\mathbf{k}$, and denote by $\mathbf{u} \in \mathbb{R}^n$ the vector whose kth component is $\mathring{u}_k$, if $k \in \mathbf{k}^c$, and u_k, if $k \in \mathbf{k}$. There will be a neighborhood $\mathscr{D} = N(\mathring{u}_{\mathbf{k}}) \subset \mathbb{R}^p$ such that the equation (4.1) with $u = \mathbf{u}$, $u_{\mathbf{k}} \in \mathscr{D}$, continues to have q simple zeros $\boldsymbol{\xi} = [\xi_1, \xi_2, \ldots, \xi_q]^T$ and $\boldsymbol{\xi} \to \mathring{\boldsymbol{\xi}}$ as $\mathbf{u} \to \mathring{u}$. We assume that neither $\mathring{\boldsymbol{\xi}}$, nor $\mathring{u}_{\mathbf{k}}$, is the zero vector in the space $\mathbb{C}^q$ and $\mathbb{R}^p$, respectively. (Clearly, $\mathring{\boldsymbol{\xi}} \neq \mathbf{0}$ if $q > 1$, since each $\mathring{\xi}_j$ is simple.) Our interest, then, is in the condition of the map $M_{\mathbf{k}, q} : \mathscr{D} \subset \mathbb{R}^p \to \mathbb{C}^q$ defined by

$$M_{\mathbf{k}, q} : \boldsymbol{\xi} = \mathbf{f}(u_{\mathbf{k}}), \qquad u_{\mathbf{k}} \in \mathscr{D} \subset \mathbb{R}^p,$$

where $p(\mathbf{u}; f_j(u_{\mathbf{k}})) \equiv 0$ on $\mathscr{D}$, for each $j = 1, 2, \ldots, q$, and $\mathbf{f}(u_{\mathbf{k}}) \to \mathring{\boldsymbol{\xi}}$ as $u_{\mathbf{k}} \to \mathring{u}_{\mathbf{k}}$.

It is now a straightforward matter to use (2.2) to calculate the condition number of the map $M_{\mathbf{k}, q}$ at $\mathring{u}_{\mathbf{k}}$. If we denote by

$$V_{\mathbf{k}, q}(\boldsymbol{\xi}) = \begin{bmatrix} p_{k_1}(\xi_1) & p_{k_1}(\xi_2) & \cdots & p_{k_1}(\xi_q) \\ p_{k_2}(\xi_1) & p_{k_2}(\xi_2) & \cdots & p_{k_2}(\xi_q) \\ \cdots\cdots\cdots\cdots\cdots\cdots\cdots\cdots\cdots \\ p_{k_p}(\xi_1) & p_{k_p}(\xi_2) & \cdots & p_{k_p}(\xi_q) \end{bmatrix} \in \mathbb{C}^{p \times q}$$

the "generalized Vandermonde matrix", and by $D(\mathbf{u}; \boldsymbol{\xi})$ the diagonal matrix

$$D(\mathbf{u}; \boldsymbol{\xi}) = \operatorname{diag}\left[p'(\mathbf{u}; \xi_1), p'(\mathbf{u}; \xi_2), \ldots, p'(\mathbf{u}; \xi_q) \right] \in \mathbb{C}^{q \times q}$$

(where the prime in $p'(\mathbf{u}; x)$ indicates differentiation with respect to x), we find that

$$\operatorname{cond}(M_{\mathbf{k}, q}; \mathring{u}_{\mathbf{k}}) = \frac{\|\mathring{u}_{\mathbf{k}}\|}{\|\mathring{\boldsymbol{\xi}}\|} \left\| D^{-1}(\mathring{u}; \mathring{\boldsymbol{\xi}}) V_{\mathbf{k}, q}^T(\mathring{\boldsymbol{\xi}}) \right\|. \qquad (4.2)$$

Specifically, if $\mathbf{k} = (1, 2, \ldots, n)$ and $q = n$, which is the extreme case of all roots being considered simultaneously and all coefficients undergoing changes, and if we choose the l_1-vector norm and subordinate matrix norm, we get

$$\text{cond}_1(M_{\mathbf{k},n}; \mathring{u}) = \frac{\sum\limits_{k=1}^{n} |\mathring{u}_k|}{\sum\limits_{j=1}^{n} |\mathring{\xi}_j|} \cdot \max_{1 \leqslant k \leqslant n} \sum\limits_{j=1}^{n} \left| \frac{p_k(\mathring{\xi}_j)}{p'(\mathring{\xi}_j)} \right|,$$

(4.2a)

where $p'(\mathring{\xi}_j) = p'(\mathring{u}; \mathring{\xi}_j)$. This provides the most overall description of the condition of the algebraic equation (4.1), assuming all roots are simple. The other extreme is $p = q = 1$, in which case we write $\mathbf{k} = k$, $\xi_1 = \xi$, and we find

$$\text{cond}(M_{k,1}; \mathring{u}) = \left| \frac{\mathring{u}_k p_k(\mathring{\xi})}{\mathring{\xi} p'(\mathring{\xi})} \right|, \qquad k = 1, 2, \ldots, n. \quad (4.2b)$$

Each condition number in (4.2b) measures the sensitivity of the root $\mathring{\xi}$ to perturbations in one single (nonvanishing) coefficient, $\mathring{u}_k$, and provides the most detailed description of the condition of the root $\mathring{\xi}$. Note, in (4.2b), that only $\mathring{\xi}$ is assumed to be simple; some or all of the other roots may well be multiple.

A compromise between (4.2a) and (4.2b) for characterizing the condition of a single root, $\mathring{\xi}$, is cond $\mathring{\xi} = \sum_{k=1}^{n} \text{cond}(M_{k,1}; \mathring{u})$, that is (we drop superscripts from now on),

$$\text{cond } \xi = \frac{1}{|\xi p'(\xi)|} \sum\limits_{k=1}^{n} |u_k p_k(\xi)|. \quad (4.3)$$

In the following, we adopt (4.3) as the condition number of the root ξ of (4.1). (Alternatively, we could use (4.2) with $q = 1$ and $\mathbf{k} = (1, 2, \ldots, n)$.)

4.1. Equations in power form. Here, $p_k(x) = x^{k-1}$, and (4.3) assumes the form

$$\text{cond } \xi = \frac{1}{|\xi p'(\xi)|} \sum\limits_{k=1}^{n} |u_k \xi^{k-1}|. \quad (4.4)$$

The condition number (4.4) is easily seen to be invariant under scaling of the independent variable by an arbitrary complex number. Denoting the zeros of $p(x)$ by $\xi_1, \xi_2, \ldots, \xi_n$, additional insight may be provided by estimating the condition of one of these, ξ_μ, in terms of all of them. A result in this vein is the inequality (Gautschi [13])

$$
\operatorname{cond} \xi_\mu \leqslant \frac{2 \prod_{\substack{\nu=1 \\ \nu \neq \mu}}^{n} \left(1 + \left|\frac{\xi_\nu}{\xi_\mu}\right|\right) - 1}{\prod_{\substack{\nu=1 \\ \nu \neq \mu}}^{n} \left|1 - \frac{\xi_\nu}{\xi_\mu}\right|},
\tag{4.5}
$$

in which equality holds precisely when all zeros ξ_ν are located on a half-ray through the origin. A similar inequality, resp. equality, holds if the zeros are pairwise symmetric with respect to the origin. Note that the bound in (4.5), like the condition number itself, is invariant with respect to scaling.

We illustrate (4.5) by several examples, beginning with the well-known example due to Wilkinson of a severely ill-conditioned equation.

EXAMPLE 4.1 (Wilkinson [38], p. 41ff): $\xi_\nu = \nu$, $\nu = 1, 2, \ldots, n$.

This is a root configuration for which (4.5) holds with equality sign. There follows, by a simple computation,

$$
\operatorname{cond} \xi_\mu = \frac{(\mu + n)! - \mu^n \mu!}{\mu!^2 (n - \mu)!}, \qquad \mu = 1, 2, \ldots, n.
$$

An asymptotic analysis for large n will show (Gautschi [13]) that the worst conditioned root is the one near $n/\sqrt{2} = .7071 \ldots n$. (For $n = 20$, the case considered by Wilkinson, the distinction goes to $\xi_{14} = 14$.) Its condition number grows exponentially,

$$
\max_{1 \leqslant \mu \leqslant n} \operatorname{cond} \xi_\mu \sim \frac{1}{\pi(2 - \sqrt{2})n} \left(\frac{\sqrt{2} + 1}{\sqrt{2} - 1}\right)^n, \qquad n \to \infty.
\tag{4.6}
$$

The best conditioned root is $\xi_1 = 1$, with a condition number that grows very slowly,

$$\text{cond } \xi_1 \sim n^2, \qquad n \to \infty. \tag{4.7}$$

It is instructive to observe what happens if one of the coefficients u_k in

$$\prod_{\nu=1}^{n} (x - \nu) = x^n + u_n x^{n-1} + \cdots + u_1,$$

say the coefficient u_{k_0}, $k_0 = [(n + 3)/2]$, is continuously perturbed,

$$u_{k_0}(t) = (1 + t) u_{k_0}, \qquad 0 \leqslant t \leqslant \varepsilon,$$

all other coefficients being held constant. The resulting motion of the roots[†] is shown in Figure 4.1 for $n = 10$, $\varepsilon = 8 \times 10^{-7}$, and in Figure 4.2 for the ten "most active" roots in the case $n = 20$, $\varepsilon = 7 \times 10^{-14}$(!). Initially, of course, the zeros are all confined to move along the real axis. Not before long, however, a number of them will collide, each time branching off into pairs of conjugate complex roots. When $n = 10$, there are three collisions within $0 \leqslant t \leqslant \varepsilon$, occurring at $t_1 = 1.02567 \times 10^{-7}$, $t_2 = 1.53420 \times 10^{-7}$, and $t_3 = 7.21568 \times 10^{-7}$, and one further collision (not shown in Figure 4.1) at $t_4 = 1.17328 \times 10^{-4}$. These are the only collisions in $0 \leqslant t \leqslant 1$, as far as we could determine. For $n = 20$, there are five collisions within $0 \leqslant t \leqslant \varepsilon$, at approximately $t_1 = 4.0 \times 10^{-15}$, $t_2 = 5.4 \times 10^{-15}$, $t_3 = 8.9 \times 10^{-15}$, $t_4 = 2.75 \times 10^{-14}$, $t_5 = 6.25 \times 10^{-14}$, and several more later on (e.g., at $t_6 = 1.01 \times 10^{-12}$, $t_7 = 1.67 \times 10^{-12}$).

The behavior in Figures 4.1 and 4.2 may be viewed as an elementary example of a bifurcation phenomenon (catastrophe theory, if you will), the special feature here being the almost infinitesimal time scale on which the phenomenon takes place.

[†] The graphs were obtained by numerical integration of the differential equations satisfied by $\xi_\nu(t)$, $\nu = 1, 2, \ldots, n$. The exact instances of collision were determined by finding the t-zeros of the resultant of p and p'. The graphs after each collision represent the absolute values of the conjugate complex roots produced by the collision.

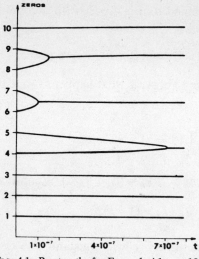

FIG. 4.1. Root paths for Example 4.1, $n = 10$.

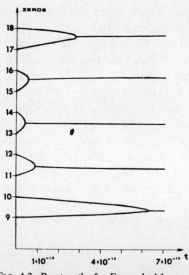

FIG. 4.2. Root paths for Example 4.1, $n = 20$.

EXAMPLE 4.2 (Wilkinson [38], p. 44ff): $\xi_\nu = 2^{-\nu}$, $\nu = 1, 2, \ldots, n$.

The roots ξ_ν accumulate rapidly near the origin, which at first might suggest that they become more and more ill-conditioned. In reality, however, they are all quite well-conditioned. This can be seen from the inequality

$$\text{cond}\, \xi_\mu < 2 \prod_{\substack{\nu = 1 \\ \nu \neq \mu}}^{n} \frac{1 + \left| \dfrac{\xi_\nu}{\xi_\mu} \right|}{\left| 1 - \dfrac{\xi_\nu}{\xi_\mu} \right|}, \qquad (4.8)$$

which follows at once from (4.5), and which in the case at hand yields

$$\text{cond}\, \xi_\mu < 2 \prod_{\nu = 1}^{\infty} \left(\frac{1 + 2^{-\nu}}{1 - 2^{-\nu}} \right)^2 = 136.32\ldots. \qquad (4.9)$$

The condition is thus bounded by a relatively small number (as condition numbers go), uniformly in n.

Since the bound in (4.8) is invariant with respect to reciprocation, the same result holds for the roots $\xi_\nu = 2^\nu$, $\nu = 1, 2, \ldots, n$.

EXAMPLE 4.3 (Roots of unity): $\xi_\nu = e^{2\pi i \nu / n}$, $\nu = 1, 2, \ldots, n$.

Since $p(x) = x^n - 1$, Equation (4.4) gives at once

$$\text{cond}\, \xi_\mu = \frac{1}{n}, \qquad \mu = 1, 2, \ldots, n. \qquad (4.10)$$

All roots are equally well-conditioned, the condition in fact getting better with increasing degree! The example, of course, is quite trivial, and (4.10) is just another way of saying that $(1 + \varepsilon)^{1/n} - 1 \sim \varepsilon / n$ as $\varepsilon \to 0$.

4.2. Equations expressed in terms of orthogonal polynomials.
We now assume that the equation is written in the form

$$p(x) = 0, \qquad p(x) = \pi_n(x) + \sum_{k=1}^{n} u_k \pi_{k-1}(x), \qquad (4.11)$$

where $\{\pi_r\}$ is a set of orthogonal polynomials. It is to be noted that
the normalization in (4.11) is such that $p(x)$ and $\pi_n(x)$ have the
same leading coefficient, if expressed in powers of x. The condition
number of any (simple) root ξ of (4.11) is

$$\operatorname{cond} \xi = \frac{1}{|\xi p'(\xi)|} \sum_{k=1}^{n} |u_k \pi_{k-1}(\xi)|. \qquad (4.12)$$

It is easily seen that $\operatorname{cond} \xi$ does not depend on the particular
way the orthogonal polynomials π_r are normalized. Note, however,
again, that changing the normalization of π_n also changes p,
according to the normalization of the equation adopted in (4.11).
An easy lower bound can be had by noting that

$$\sum_{k=1}^{n} |u_k \pi_{k-1}(\xi)| \geqslant \left| \sum_{k=1}^{n} u_k \pi_{k-1}(\xi) \right| = |\pi_n(\xi)|.$$

Thus,

$$\operatorname{cond} \xi \geqslant \left| \frac{\pi_n(\xi)}{\xi p'(\xi)} \right|. \qquad (4.13)$$

For an upper bound we could apply Schwarz' inequality to the sum
in (4.12), but the result is not particularly revealing. It appears
difficult, indeed, to extract from (4.12) much detailed information
concerning the qualitative behavior of the condition of ξ. We may
note, however, that there are three factors which influence its
magnitude: (i) the magnitude of the Fourier coefficients u_k of p;
(ii) the magnitude of the orthogonal polynomials π_k evaluated at
the root ξ; (iii) the magnitude of $\xi p'(\xi)$. Since orthogonal polynomials grow rapidly outside their interval of orthogonality, it

seems imperative, in view of (i) and (ii), that the interval of orthogonality be selected so as to contain ξ, if ξ is real.

It is quite possible that equations that are ill-conditioned in power form become well-conditioned when expanded in orthogonal polynomials, and vice versa. We can see this already by reexamining the examples discussed previously. We begin with Wilkinson's example, whose roots we now scale to be enclosed in the interval $[0,1]$. As we have noted earlier, such a scaling does not affect the condition of the roots, if the equation is in power form.

EXAMPLE 4.1′: $\xi_\nu = \nu/n$, $\nu = 1, 2, \ldots, n$.

The condition number (4.12) can be computed for various (classical) polynomials $\{\pi_r\}$ orthogonal on $[0,1]$. It turns out that the Chebyshev polynomials of the second kind perform best (in the sense of making $\max_\mu \text{cond}\, \xi_\mu$ smallest). Some numerical results are shown in the second column of Table 4.1. They are contrasted in the third column with the analogous condition numbers for the equation in power form (see Example 4.1). The improvement is clearly significant. We remark, nevertheless, that the condition still grows exponentially with n (though at a moderate rate), as can be deduced from the inequality (4.13); if n is even, e.g., one finds

$$\max_\mu \text{cond}\, \xi_\mu \geqslant \frac{(n/4)^n}{(n/2)!^2} \sim \frac{1}{\pi n}(e/2)^n, \qquad n \to \infty.$$

TABLE 4.1

	$\max_\mu \text{cond}\, \xi_\mu$	
n	Example 4.1′	Example 4.1
5	1.85	5.87×10^2
10	2.64×10^1	2.32×10^6
15	6.35×20^2	1.05×10^{10}
20	1.40×10^4	5.40×10^{13}

The condition of the roots in Examples 4.1′ and 4.1.

EXAMPLE 4.2′: $\xi_\nu = 2 \cdot 2^{-\nu}$, $\nu = 1, 2, \ldots, n$.

All orthogonal (on $[0,1]$) polynomials $\{\pi_r\}$ tried on this example led to condition numbers that grow extremely rapidly with n. For the "best" of these, the Chebyshev polynomials of the first kind, the results are shown in the second column of Table 4.2. The third column again contains the condition numbers of the same roots for the equation in power form. The contrast is striking!

It is not difficult to identify the culprit in Example 4.2′: it is the derivative of p at ξ_ν, which becomes extremely small as ν approaches n. Indeed, if p is normalized to have leading coefficient one, $p(x) = x^n + \cdots$, one finds for $\nu = n$ that

$$\left| \xi_n p'(\xi_n) \right| \sim \frac{.28879}{2^{n(n-1)/2}}, \qquad n \to \infty.$$

The Fourier coefficients u_k in (4.12), although reasonably small (of order 10^{-3}), are no match for this kind of decay! In the case of the power basis, the small denominator $|\xi_n p'(\xi_n)|$ in (4.4) is neutralized by an equally small numerator,

$$\sum_{k=1}^{n} |u_k \xi_n^{k-1}| \sim \frac{2.3842}{2^{(n+1)(n-2)/2}}, \qquad n \to \infty.$$

EXAMPLE 4.3′: $\xi_\nu = e^{2\pi i\nu/n}$, $\nu = 1, 2, \ldots, n$.

TABLE 4.2

n	max cond ξ_μ	
	Example 4.2′	Example 4.2
5	5.03×10^1	4.91×10^1
10	1.44×10^{12}	1.13×10^2
15	1.19×10^{30}	1.32×10^2
20	3.58×10^{55}	1.36×10^2

The condition of the roots in Examples 4.2′ and 4.2.

The roots of unity, extremely well-conditioned in the power basis (Example 4.3), continue to be quite well-conditioned in orthogonal bases, provided the interval of orthogonality is reasonably chosen. The most natural choice is $[-1,1]$, and all (classical) polynomials orthogonal on this interval do quite well, yielding condition numbers ranging from about .5 for $n = 5$ to about 35 for $n = 20$.

Just to show how an unreasonable choice of orthogonality interval may turn even the roots of unity into poorly conditioned roots, consider the case of Laguerre polynomials (orthogonal on $(0, \infty)$). Here,

$$\frac{(-1)^n}{n!}(x^n - 1) = (-1)^n\left(1 - \frac{1}{n!}\right) + \sum_{r=1}^{n} (-1)^{n-r}\binom{n}{r}L_r(x),$$

(4.14)

as can be derived from an integral formula for Laguerre polynomials (see Buchholz [7], p. 120, Eq. (4β)). Therefore, from (4.12), one gets

$$\text{cond}\,\xi_\mu = (n-1)!\left(1 - \frac{1}{n!} + \sum_{r=1}^{n-1}\binom{n}{r}|L_r(\xi_\mu)|\right).$$

Some numerical values are shown in Table 4.3; they speak for themselves!

TABLE 4.3

n	$\max\limits_{\mu}\text{cond}\,\xi_\mu$
5	3.17×10^3
10	5.45×10^9
20	9.71×10^{24}
40	1.92×10^{61}

The condition of the roots of unity in the Laguerre polynomial basis.

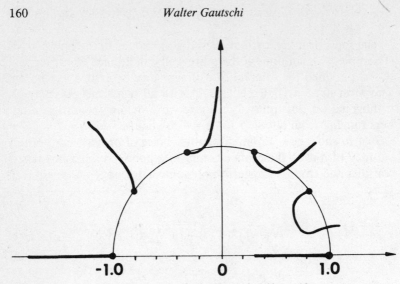

FIG. 4.3. Root paths for Example 4.3′, $n = 10$.

The case $n = 10$ is further illustrated in Figure 4.3, which shows the motion of the roots ξ_μ induced by a multiplication of the single coefficient

$$(-1)^{n-r_0}\binom{n}{r_0}, \qquad r_0 = \left[\frac{n+1}{2}\right],$$

in (4.14) by $1 + t$ and variation of t from 0 to 10^{-8}.

5. GENERATION OF ORTHOGONAL POLYNOMIALS.

Generating orthogonal polynomials is fairly straightforward once the three-term recurrence relation, which they are known to satisfy, is explicitly available. Such is the case for all classical orthogonal polynomials. We are interested here in the more difficult task of generating the recurrence relation in cases where it is not explicitly known. A related problem is the construction of the Gaussian quadrature formulae; we use this connection to discuss the condition of the problem.

5.1. Statement of the problem. Let the desired (monic) polynomials $\{\pi_r\}$ be orthogonal with respect to a nonnegative measure $d\sigma(x)$ on the real line $\mathbb{R}$, where $\sigma(x)$ has at least $n+1$ points of increase. We assume that the first $2n$ moments of $d\sigma$ exist,

$$\mu_k = \int_{\mathbb{R}} x^k \, d\sigma(x), \qquad k = 0, 1, 2, \ldots, 2n-1. \tag{5.1}$$

It is well known that the orthogonal polynomials π_r then exist for $r = 0, 1, \ldots, n$, and satisfy a recurrence relation of the form

$$\pi_{-1}(x) = 0, \qquad \pi_0(x) = 1, \tag{5.2}$$

$$\pi_{k+1}(x) = (x - \alpha_k)\pi_k(x) - \beta_k \pi_{k-1}(x), \qquad k = 0, 1, 2, \ldots, n-1,$$

with real α_k and $\beta_k > 0$. (β_0 is arbitrary, but is conveniently defined as $\beta_0 = \int_{\mathbb{R}} d\sigma(x)$.) Associated with (5.2) is the symmetric tridiagonal matrix of order n,

$$J_n = \begin{bmatrix} \alpha_0 & \sqrt{\beta_1} & & & 0 \\ \sqrt{\beta_1} & \alpha_1 & \sqrt{\beta_2} & & \\ & \ddots & \ddots & \ddots & \\ & & & & \sqrt{\beta_{n-1}} \\ 0 & & & \sqrt{\beta_{n-1}} & \alpha_{n-1} \end{bmatrix}, \tag{5.3}$$

called *Jacobi matrix*, whose rth leading principal submatrix has the characteristic polynomial π_r, $r = 1, 2, \ldots, n$. Let $\xi_1, \xi_2, \ldots, \xi_n$ be the zeros of π_n, hence the eigenvalues of J_n (they are all real and simple, as is well known). Let $v_j = [v_{1j}, v_{2j}, \ldots, v_{nj}]^T$ be the normalized eigenvector of J_n belonging to the eigenvalue ξ_j,

$$J_n v_j = \xi_j v_j, \qquad v_j^T v_j = 1.$$

Then (Wilf [37], Ch. 2, Exercise 9, Golub & Welsch [20])

$$\int_{\mathbb{R}} f(x) \, d\sigma(x) = \sum_{j=1}^{n} \lambda_j f(\xi_j) + R_n(f),$$

$$\lambda_j = \mu_0 v_{1j}^2, \qquad j = 1, 2, \ldots, n, \tag{5.4}$$

is the *Gaussian quadrature formula* associated with the measure $d\sigma$, i.e., $R_n(f) = 0$ for each $f \in \mathbb{P}_{2n-1}$.

There are classical procedures for generating the Jacobi matrix J_n, hence the Gaussian quadrature formula (5.4), from the given moments μ_k in (5.1). Unfortunately, as will be seen in the next subsection, the underlying map is likely to be severely ill-conditioned. The moments μ_k, indeed, are a poor way of "codifying" the measure $d\sigma$. An alternative way is through *modified moments*,

$$m_k = \int_{\mathbb{R}} p_k(x)\, d\sigma(x), \qquad k = 0,1,2,\dots,2n-1, \qquad (5.5)$$

where $\{p_r\}$ is a suitably selected set of polynomials (see Sack & Donovan [29]). We shall assume that the p_r, like the π_r, satisfy a recurrence relation of the type

$$\begin{cases} p_{-1}(x) = 0, \qquad p_0(x) = 1, \\ p_{k+1}(x) = (x - a_k)p_k(x) - b_k p_{k-1}(x), \end{cases} \qquad (5.6)$$

$$k = 0,1,2,\dots,2n-2,$$

but now with coefficients a_k, b_k that are known. For example, $\{p_r\}$ may consist of a set of known (classical) orthogonal polynomials. If all a_k, b_k are zero, then $p_k(x) = x^k$, $k = 0,1,2,\dots$, and the modified moments reduce to ordinary moments, $m_k = \mu_k$, $k = 0,1,2,\dots$.

The problem we wish to consider is the following: Given the modified moments m_k in (5.5), determine the Jacobi matrix (5.3). In particular, this will also determine the orthogonal polynomials $\{\pi_r\}_{r=0}^n$, by virtue of (5.2), and the associated Gaussian quadrature formula, by virtue of (5.4).

The map in question thus is $K_n : \mathbb{R}^{2n} \to \mathbb{R}^{2n}$ which associates to the first $2n$ modified moments m_r the $2n$ recursion coefficients α_k, β_k, $k = 0,1,\dots,n-1$, for the respective orthogonal polynomials:

$$K_n : m \to \rho$$

$$m^T = [m_0, m_1, \dots, m_{2n-1}], \qquad \rho^T = [\alpha_0, \dots, \alpha_{n-1}, \beta_0, \dots, \beta_{n-1}].$$

$$(5.7)$$

(Recall that $\beta_0 = \int_{\mathbb{R}} d\sigma(x) = m_0$.) For the purpose of analyzing the condition of the map K_n it is convenient to think of K_n as the composition of two maps,

$$K_n = H_n \circ G_n, \tag{5.8}$$

where $G_n : \mathbb{R}^{2n} \to \mathbb{R}^{2n}$ takes us from the modified moments m_r to the Gaussian quadrature rule,

$$G_n : m \to \gamma, \qquad \gamma^T = [\lambda_1, \ldots, \lambda_n, \xi_1, \ldots, \xi_n], \tag{5.9}$$

and $H_n : \mathbb{R}^{2n} \to \mathbb{R}^{2n}$ from the Gaussian quadrature rule to the recursion coefficients,

$$H_n : \gamma \to \rho.$$

The map H_n is usually (but not always) quite well-conditioned, as can be inferred from the discussion in Gautschi [16], Section 3.1. Here we look only at the map G_n which is by far the more critical one. We first demonstrate the ill-conditioned character of this map when $m = \mu$ are ordinary moments (5.1), and then see what can be gained from taking m to be modified moments with respect to a system of (classical) orthogonal polynomials.

5.2 Condition of the map G_n in the case of ordinary moments. Here $m = \mu$, or $a_k = b_k = 0$ in (5.6). For definiteness we assume $d\sigma(x)$ supported on the interval $[0, 1]$ and normalized such that $\mu_0 = 1$. The Jacobian matrix of the map G_n in (5.9) is easily seen to be the inverse of a confluent Vandermonde matrix in the nodes ξ_j, multiplied (from the left) by the inverse of the diagonal matrix $\mathrm{diag}[1, \ldots, 1, \lambda_1, \ldots, \lambda_n]$. Using two-sided estimates of the (uniform) norm of inverses of confluent Vandermonde matrices, it is possible to prove (Gautschi [10]) that

$$\mathrm{cond}_{\infty}(G_n; \mu) > \frac{1}{2} \max_{1 \leqslant j \leqslant n} \left[\frac{\pi_n(-1)}{\pi_n'(\xi_j)} \right]^2, \tag{5.10}$$

where π_n is the (desired) orthogonal polynomial of degree n with respect to the measure $d\sigma$. Since the point -1 is well outside the interval of orthogonality $[0,1]$, it is evident from (5.10) that the condition of G_n grows at least exponentially with n. An idea as to the numerical value of the growth rate can be had by considering the (representative) example $\pi_n = T_n^*$, the "shifted" Chebyshev polynomial of the first kind. In this case, (5.10) yields

$$\text{cond}_\infty(G_n;\mu) > \frac{(3+\sqrt{8})^{2n}}{64n^2} \qquad (\pi_n = T_n^*). \qquad (5.11)$$

The lower bound in (5.11) happens to grow at the same exponential rate as the (Turing-) condition number of the $n \times n$-Hilbert matrix!

5.3 Condition of the map G_n in the case of modified moments.
We assume now that the vector m consists of modified moments

$$m^T = [m_0, m_1, \ldots, m_{2n-1}], \qquad m_k = \int_{\mathbb{R}} p_k(x)\sigma(x),$$

where $\{p_k\}$ is a system of (monic) polynomials orthogonal with respect to some measure $ds(x)$,

$$\int_{\mathbb{R}} p_k(x)p_l(x)\, ds(x) = 0, \qquad k \neq l.$$

The support of ds normally coincides with that of $d\sigma$, but need not do so necessarily. The analysis of the condition of the map $G_n: m \to \gamma$ in (5.9) is somewhat simplified if, instead of G_n, one considers the map

$$\tilde{G}_n: \tilde{m} \to \gamma, \qquad \gamma = [\lambda_1, \ldots, \lambda_n, \xi_1, \ldots, \xi_n], \qquad (5.12)$$

where $\tilde{m}$ is the vector of *normalized* modified moments,

$$\tilde{m}_k = d_k^{-1/2} m_k, \qquad d_k = \int_{\mathbb{R}} p_k^2(x)\, ds(x),$$

$$k = 0, 1, \ldots, 2n-1. \qquad (5.13)$$

The additional diagonal map, $D_n : m \to \tilde{m}$, of course, is quite harmless as far as the numerical condition is concerned, but makes the modified moments independent of the normalization of the polynomials p_k.

The condition of $\tilde{G}_n$ can be estimated rather realistically in terms of the fundamental Hermite interpolation polynomials associated with the Gaussian nodes $\xi_1, \ldots, \xi_n$. These are the polynomials of degree $2n - 1$,

$$
\begin{aligned}
h_i(x) &= l_i^2(x)\big[1 - 2l_i'(\xi_i)(x - \xi_i)\big], \\
k_i(x) &= l_i^2(x)(x - \xi_i),
\end{aligned}
\qquad i = 1, 2, \ldots, n, \quad (5.14)
$$

satisfying

$$
\begin{aligned}
h_i(\xi_j) &= \delta_{ij}, & h_i'(\xi_j) &= 0, \\
k_i(\xi_j) &= 0, & k_i'(\xi_j) &= \delta_{ij},
\end{aligned}
\qquad i, j = 1, 2, \ldots, n. \quad (5.15)
$$

Here, δ_{ij} is the Kronecker symbol and l_i in (5.14) are the fundamental Lagrange interpolation polynomials. Using the Euclidean norm $\|\cdot\|_2$, one can show (Gautschi [16], Section 3.3) that

$$
\mathrm{cond}_2(\tilde{G}_n; \tilde{m}) \leqslant \frac{\|\tilde{m}\|_2}{\|\gamma\|_2} \left\{ \int_{\mathbb{R}} \sum_{i=1}^{n} \left(h_i^2(x) + \frac{1}{\lambda_i^2} k_i^2(x) \right) ds(x) \right\}^{1/2}.
$$

$$(5.16)$$

It is interesting to note the interplay between the two measures $d\sigma$ and ds in this formula. Both the integrand and the vector γ depend solely on the target measure $d\sigma$. Integration, on the other hand, is with respect to the given measure ds, while the vector $\tilde{m}$ depends on both measures. To evaluate the integral in (5.16), once ξ_j and λ_j are known, one can use the $2n$-point Gaussian quadrature rule associated with ds, which produces the integral exactly (up to rounding errors). Since ds is usually one of the classical measures, the Gauss formula in question is readily available.

The critical quantity in (5.16) is the square root of the integral, which in fact represents the Frobenius norm of the Fréchet derivative of $\tilde{G}_n$; see Gautschi [16], Section 3.3. This quantity in turn depends critically on the behavior of the function

$$g_n(x) = \sum_{i=1}^{n} \left(h_i^2(x) + \frac{1}{\lambda_i^2} k_i^2(x) \right). \tag{5.17}$$

This is a nonnegative polynomial of degree $4n-2$ satisfying, in particular,

$$g_n(\xi_j) = 1, \qquad g_n'(\xi_j) = 0, \qquad j = 1, 2, \ldots, n. \tag{5.18}$$

It is useful to classify the nodes ξ_j into *weak* nodes and *strong* nodes according as $g_n''(\xi_j) < 0$ or $g_n''(\xi_j) > 0$. Near a weak node, the function g_n is less than 1 and is likely (but not necessarily so; see Example 5.1) to remain below 1 between consecutive weak nodes. In contrast, g_n is larger than 1 near a strong node and must peak (possibly at very large values) on either side of it. We expect the map $\tilde{G}_n$ to be well-conditioned if the majority of the nodes are weak, and the strong nodes, if any, accumulate closely. This is often the case when the support of $d\sigma$ is a finite interval (see Examples 5.2 and 5.3). On the other hand, if there are nodes (strong or weak ones) which are separated by relatively large gaps, then there is a potential danger of g_n shooting up to considerable heights on the gaps, giving rise to ill-conditioning. This kind of predicament is likely to occur if the support of $d\sigma$ is an infinite interval or if it consists of separate intervals. In the former case, the gaps arise because of the relatively wide spacing of the absolutely larger nodes ξ_j, whereas in the latter case the gaps are the holes between the separate support intervals. Note that this all depends solely on the measure $d\sigma$. There are other factors depending on the measure ds which may also significantly influence the magnitude of cond $\tilde{G}_n$; see, in particular, Examples 5.4 and 5.5.

A simple computation based on (5.14) and (5.17) shows that

$$\frac{1}{2} g_n''(\xi_j) = 2l_j''(\xi_j) - 6\left[l_j'(\xi_j)\right]^2 + \lambda_j^{-2}$$

$$= 2 \sum_{k \neq j} \sum_{l \neq j; \, l \neq k} \frac{1}{(\xi_j - \xi_k)(\xi_j - \xi_l)} \tag{5.19}$$

$$-6\left(\sum_{k \neq j} \frac{1}{\xi_j - \xi_k} \right)^2 + \frac{1}{\lambda_j^2}.$$

The node ξ_j is therefore strong or weak depending on whether the quantity on the right of (5.19) is positive or negative. Unfortunately, little is known concerning this matter, and a detailed analysis would seem to provide an interesting and rewarding area of research.

It is known, however, that all Chebyshev nodes $\xi_j = \cos\vartheta_j$, $\vartheta_j = (2j-1)\pi/(2n)$, $j=1,2,\ldots,n$ (corresponding to the Chebyshev measure $d\sigma(x) = (1-x^2)^{-\frac{1}{2}} dx$ on $[-1,1]$) are indeed weak; Gautschi [18]. For measures $d\sigma$ that behave similarly to the Chebyshev measure, one should expect, therefore, that most, if not all, nodes ξ_j are weak and hence that the map $\tilde{G}_n$ is quite well-conditioned; see Example 5.2.

5.4. An algorithm. A number of algorithms are known for carrying out the map K_n in (5.7) from the modified moments m_j to the recursion coefficients α_k, β_k. We describe a particularly simple one due, in the form given below, to Wheeler [34],[†] and in a different form, to Sack & Donovan [29]. The algorithm actually goes back to Chebyshev [8] who proposed it in the special case of ordinary moments ($a_k = b_k = 0$) and discrete measures $d\sigma$.

We introduce the "mixed moments"

$$\sigma_{kl} = \int_{\mathbb{R}} \pi_k(x) p_l(x) \, d\sigma(x), \qquad k,l \geqslant -1, \qquad (5.20)$$

and note that by orthogonality, $\sigma_{kl} = 0$ for $k > l$, and

$$\int_{\mathbb{R}} \pi_k^2(x) \, d\sigma(x) = \int_{\mathbb{R}} \pi_k(x) x p_{k-1}(x) \, d\sigma(x) = \sigma_{kk}, \qquad k \geqslant 1.$$

The relation $\sigma_{k+1, k-1} = 0$, therefore, together with (5.2), yields immediately $\sigma_{kk} - \beta_k \sigma_{k-1, k-1} = 0$, hence

$$\beta_k = \frac{\sigma_{kk}}{\sigma_{k-1, k-1}}, \qquad k = 1, 2, 3, \ldots. \qquad (5.21)$$

(Recall that β_0 is set equal to m_0.) Similarly, $\sigma_{k+1, k} = 0$ gives

$$\int_{\mathbb{R}} \pi_k(x) x p_k(x) \, d\sigma(x) - \alpha_k \sigma_{kk} - \beta_k \sigma_{k-1, k} = 0,$$

[†]Equation (3.4) in Wheeler [34] is misprinted; a_k, b_k should read a_l and b_l, respectively.

and using (5.6) in the form

$$xp_k(x) = p_{k+1}(x) + a_k p_k(x) + b_k p_{k-1}(x), \qquad (5.22)$$

yields $\sigma_{k,k+1} + (a_k - \alpha_k)\sigma_{kk} - \beta_k \sigma_{k-1,k} = 0$, hence, together with (5.21),

$$\begin{cases} \alpha_0 = a_0 + \dfrac{\sigma_{01}}{\sigma_{00}}, \\[2mm] \alpha_k = a_k - \dfrac{\sigma_{k-1,k}}{\sigma_{k-1,k-1}} + \dfrac{\sigma_{k,k+1}}{\sigma_{kk}}, \qquad k = 1,2,3,\ldots. \end{cases} \qquad (5.23)$$

The σ's in turn satisfy the recursion

$$\sigma_{kl} = \sigma_{k-1,l+1} - (\alpha_{k-1} - a_l)\sigma_{k-1,l}$$
$$- \beta_{k-1}\sigma_{k-2,l} + b_l \sigma_{k-1,l-1}, \qquad (5.24)$$

as follows from (5.2) and (5.22) (where k is replaced by l). To construct orthogonal polynomials π_r of degrees $r \leqslant n$, we thus have the following algorithm.

Initialization:

$$\begin{cases} \sigma_{-1,l} = 0, & l = 1,2,\ldots,2n-2, \\[1mm] \sigma_{0,l} = m_l, & l = 0,1,\ldots,2n-1, \\[1mm] \alpha_0 = a_0 + \dfrac{m_1}{m_0}, \\[2mm] \beta_0 = m_0. \end{cases}$$

Continuation: For $k = 1,2,\ldots,n-1$ (5.25)

$$\begin{cases} \sigma_{kl} = \sigma_{k-1,l+1} - (\alpha_{k-1} - a_l)\sigma_{k-1,l} - \beta_{k-1}\sigma_{k-2,l} \\[1mm] \qquad + b_l \sigma_{k-1,l-1}, \qquad l = k, k+1,\ldots,2n-k-1, \\[2mm] \alpha_k = a_k - \dfrac{\sigma_{k-1,k}}{\sigma_{k-1,k-1}} + \dfrac{\sigma_{k,k+1}}{\sigma_{kk}}, \\[2mm] \beta_k = \dfrac{\sigma_{kk}}{\sigma_{k-1,k-1}}. \end{cases}$$

The algorithm requires as input $\{m_l\}_{l=0}^{2n-1}$ and $\{a_k, b_k\}_{k=0}^{2n-2}$; it furnishes $\{\alpha_k, \beta_k\}_{k=0}^{n-1}$, hence the orthogonal polynomials $\{\pi_r\}_{r=0}^{n}$, and also, incidentally, the normalizing factors $\sigma_{kk} = \int_{\mathbf{R}} \pi_k^2(x)\,d\sigma(x)$, $k \leqslant n-1$. The number of multiplications and divisions required is $3n^2 - n - 1$, the number of additions, $4n^2 - 3n$; the algorithm thus involves $O(n^2)$ operations altogether.

The success of the algorithm (5.25), of course, depends on the ability to compute all required modified moments m_l accurately and reliably. Most frequently, these moments are obtained from recurrence relations, judiciously employed, as for example in the case of Chebyshev or Gegenbauer moments (Piessens & Branders [26], Branders [6], Luke [22], Lewanowicz [21]). Sometimes they can be computed directly in terms of special functions, or in integer form (Gautschi [11], Examples (ii), (iii), Wheeler & Blumstein [36], Blue [1], Gautschi [15], Gatteschi [9]). Still another possibility is to use a suitable discretization process (Gautschi [16], Section 2.5).

5.5 Examples. We begin with a measure of discrete type, already considered by Chebyshev [8].

EXAMPLE 5.1: $d\sigma(x) = (1/N)\sum_{k=0}^{N-1}\delta(x - k/N)$, where $\delta(\cdot)$ is the Dirac delta function.

The associated N orthogonal polynomials $\pi_0, \pi_1, \ldots, \pi_{N-1}$ are explicitly known. There is no need, therefore, to carry out maps such as G_n in (5.9). Nevertheless, we briefly consider the condition of this map in order to explain an interesting phenomenon observed previously in Gautschi [16], Example 4.1, namely the gradual worsening of the condition of $\tilde{G}_n$ as n approaches N. The underlying moments are those corresponding to the shifted (monic) Legendre polynomials $p_k(x) = (k!^2/(2k)!)P_k(2x - 1)$, $0 \leqslant x \leqslant 1$.

Computation reveals that all zeros ξ_j of π_n, $n \leqslant N$, are weak nodes. Yet the condition of $\tilde{G}_n$ deteriorates significantly as n approaches N; see Table 5.1. The reason for this can be found in the following peculiar behavior of the function $g_n(x)$ of (5.17). In the central zone of the interval $[0,1]$, g_n wiggles rapidly, always remaining $\leqslant 1$. In both end zones, however, g_n becomes $\geqslant 1$ and exhibits spikes of increasing magnitudes as one moves toward the endpoints. On the last interval $[\xi_n, 1]$, soon after g_n'' becomes positive, g_n increases rapidly to a global maximum at $x = 1$. The behavior is illustrated in Table 5.1, where we show the approximate magnitude of the largest spike attained on $[\xi_{n-1}, \xi_n]$ (and of a similar spike on $[\xi_1, \xi_2]$), as well as $\max_{0 \leqslant x \leqslant 1} g_n(x) = g_n(1)$. The neighboring spikes to the left of $[\xi_{n-1}, \xi_n]$ (or to the right of $[\xi_1, \xi_2]$) are typically several orders of magnitude smaller. When n is relatively small, there may be no spikes at all, which is indicated in Table 5.1 by a dash. Also shown are the values of the estimate (5.16) of the condition of $\tilde{G}_n$. It can be seen that the main contribution to cond $\tilde{G}_n$ comes from the two spikes of g_n on $[\xi_1, \xi_2]$ and $[\xi_{n-1}, \xi_n]$ and from the final upward surge of g_n on $[\xi_n, 1]$. The formation of these spikes and their increasing magnitudes as n approaches N is undoubtedly caused by the fact that the nodes ξ_j approach more and more a uniform distribution; they are exactly uniformly distributed when $n = N$. It is well known that interpolation polynomials (and those of Hermite are no exception!) are prone to violent oscillations in such cases.

EXAMPLE 5.2: $d\sigma(x) = [(1 - k^2 x^2)(1 - x^2)]^{-1/2} dx$
on $[-1, 1]$, $0 < k < 1$.

This example also has been considered previously in Gautschi [16], Example 4.4, where the use of Chebyshev moments was found to work extremely well. Here we point out that all nodes ξ_j of the

TABLE 5.1

N	n	$\max g_n$ $[\xi_{n-1}, \xi_n]$	$g_n(1)$	cond $\tilde{G}_n$	N	n	$\max g_n$ $[\xi_{n-1}, \xi_n]$	$g_n(1)$	cond $\tilde{G}_n$
10	5	—	4×10^2	2.5×10^0	40	15	1.4×10^0	5×10^5	2.0×10^1
	10	8×10^3	3×10^{11}	6.3×10^4		25	9×10^6	3×10^{15}	1.3×10^6
20	5	—	3×10^1	7.9×10^{-1}		35	3×10^{22}	4×10^{32}	5.0×10^{14}
	10	—	1×10^5	1.9×10^1	80	10	—	5×10^1	4.9×10^{-1}
	15	6×10^3	3×10^{11}	3.0×10^4		20	1.2×10^0	2×10^5	6.5×10^0
	20	3×10^{14}	4×10^{23}	3.3×10^{10}		30	2×10^3	1×10^{11}	3.9×10^3
40	5	—	7×10^0	6.5×10^{-1}		40	1×10^{10}	2×10^{19}	4.8×10^7
	10	—	6×10^2	1.0×10^0		50	2×10^{20}	3×10^{30}	1.7×10^{13}

The behavior of $g_n(x)$ in Example 5.1 and the condition of $\tilde{G}_n$

n-point Gauss formula for $d\sigma$ appear to be weak nodes. This was verified numerically for various values of k^2 as close to 1 as $k^2 = .99$, and for values of n as large as $n = 80$. In all cases computed, moreover, the function g_n was found never to exceed 1 on $[-1,1]$. No wonder, therefore, that the map $\tilde{G}_n$ is extremely well-conditioned!

EXAMPLE 5.3: $d\sigma(x) = x^\alpha \ln(1/x)\, dx$ on $[0,1]$, $\alpha > -1$.

Here, the modified moments with respect to the shifted Legendre polynomials $p_k(x) = (k!^2/(2k)!) P_k(2x-1)$ can be obtained explicitly. For example, if α is not an integer, then

$$\frac{(2l)!}{l!^2} m_l = \frac{1}{\alpha+1}\left\{ \frac{1}{\alpha+1} + \sum_{k=1}^{l}\left(\frac{1}{\alpha+1+k} - \frac{1}{\alpha+1-k}\right)\right\}$$

$$\times \prod_{k=1}^{l} \frac{\alpha+1-k}{\alpha+1+k}, \qquad l = 0,1,2,\dots . \qquad (5.26)$$

(Similar formulas hold for integral α; see Blue [1] for $\alpha = 0$, Gautschi [15] for $\alpha > 0$, and Gatteschi [9] for still more general cases.) The appropriate recursion coefficients for $\{p_r\}$ are

$$a_k = \frac{1}{2}, \qquad\qquad k = 0,1,2,\dots,$$

$$b_k = \frac{1}{4(4-k^{-2})}, \qquad k = 1,2,3,\dots . \qquad (5.27)$$

With the quantities in (5.26) and (5.27) as input, algorithm (5.25) now easily furnishes the recursion coefficients $\alpha_k, \beta_k, 0 \le k \le n-1$, for the orthogonal polynomials with respect to $d\sigma(x) = x^\alpha \ln(1/x)\, dx$. For $\alpha = -\frac{1}{2}$, and $n = 2,4,8,\dots,80$, and single-precision computation on the CDC 6500 computer (approx. 14 decimal digit accuracy), the mean square errors $\varepsilon_n(\alpha,\beta) = (\sum_{k=0}^{n-1}[\varepsilon^2(\alpha_k) + \varepsilon^2(\beta_k)])^{1/2}$, where $\varepsilon(\alpha_k)$, $\varepsilon(\beta_k)$ are the relative errors in the coefficients α_k, β_k, are shown in the left half of Table 5.2. The right half displays the analogous results for the power moments $\mu_l = (\alpha+1+l)^{-2}$, and $a_k = b_k = 0$, all k. In the first case, all coefficients are obtained close to machine precision, attesting not only to the extremely well-conditioned nature of the problem, but also to the

TABLE 5.2

	Legendre moments	power moments
n	$\varepsilon_n(\alpha,\beta)$	$\varepsilon_n(\alpha,\beta)$
2	9.22×10^{-14}	9.92×10^{-15}
4	2.42×10^{-13}	3.25×10^{-12}
8	6.53×10^{-13}	6.29×10^{-7}
12	1.09×10^{-12}	1.67×10^{-1}
20	1.29×10^{-12}	
40	1.98×10^{-12}	
80	5.03×10^{-12}	

Relative errors in the recursion coefficients α_k, β_k for Example 5.3.

stability of algorithm (5.25). In the second case, all accuracy is lost by the time n reaches 12, which confirms the severely ill-conditioned character of the problem of generating orthogonal polynomials from ordinary moments.

The underlying condition numbers, specifically the condition of the map H_n as computed in Gautschi [16], Equation (3.8)f and the estimates of cond $\tilde{G}_n$ in (5.16) and of cond G_n in (5.10), are displayed in Table 5.3. Recall that Table 5.2 illustrates the accuracy of the map $K_n = H_n \circ G_n$; see (5.7). The algorithm (5.25) based on Legendre moments performs somewhat better than the condition numbers of H_n and $\tilde{G}_n$ in Table 5.3 would suggest. For power moments, the rapid loss of accuracy evidenced in Table 5.2 correlates very well (at least for $n \leqslant 12$) with the rapid growth of the condition number of G_n in Table 5.3.

		Legendre moments	power moments
n	cond H_n	cond $\tilde{G}_n \leqslant$	cond $G_n \geqslant$
2	2.37	5.58	5.42
4	5.31	1.89×10^1	4.27×10^3
8	1.52×10^1	6.55×10^1	1.69×10^9
12	2.43×10^1	1.29×10^2	1.06×10^{15}
20	4.56×10^1	2.79×10^2	7.25×10^{26}
40	1.09×10^2	7.03×10^2	7.92×10^{56}
80	2.54×10^2	1.66×10^3	3.55×10^{117}

TABLE 5.3
The condition of the maps $H_n, \tilde{G}_n$ and G_n for Example 5.3

The gradual (but slow) increase of cond $\tilde{G}_n$ can be ascribed to a phenomenon similar to the one observed in Example 5.1, except that this time not all nodes ξ_j are weak, but only about the first two-thirds of them (when ordered increasingly). All remaining nodes are strong, giving rise to the development of spikes and final upward surges as in Example 5.1. The severity of these spikes, though, is much less here than shown in Table 5.1. The maximum peak is of the order of magnitude $1 \times 10, 7 \times 10^2, 2 \times 10^4, 5 \times 10^5$ for $n = 10, 20, 40, 80$, respectively, and the corresponding global maxima $g_n(1)$ have orders of magnitude $2 \times 10^4, 1 \times 10^6, 4 \times 10^7$ and 1×10^9.

EXAMPLE 5.4: The half-range Hermite measure $d\sigma(x) = e^{-x^2} dx$ on $[0, \infty]$.

This example illustrates the potential ill-conditioning of the map $\tilde{G}_n$ in the case of measures supported on an infinite interval. Modified Hermite and Laguerre moments can be readily computed from the explicit power representations of the (monic) Hermite and Laguerre polynomials. One finds:

$$m_k = 2^{-k} \int_0^\infty e^{-x^2} H_k(x) \, dx$$

$$= \begin{cases} \frac{1}{2}\sqrt{\pi}, & k = 0, \\ \frac{1}{2} \sum_{r=0}^{\kappa} (-1)^r \prod_{i=r+1}^{\kappa} i \prod_{i=\kappa-r+1}^{\kappa} (i + \frac{1}{2}), & k = 2\kappa + 1, \\ 0, & k = 2\kappa, \end{cases}$$

(5.28)

and

$$m_k = (-1)^k k! \int_0^\infty e^{-x^2} L_k(x) \, dx = \frac{(-1)^k}{2} \sum_{r=0}^{k} (-1)^r \frac{k! \Gamma\left(\frac{r+1}{2}\right)}{(k-r)! r!^2}.$$

(5.29)

As expected, the algorithm (5.25), in both cases, loses accuracy rather quickly, but perhaps unexpectedly, the loss is about twice as

TABLE 5.4

n	Hermite moments		Laguerre moments	
	cond $\tilde{G}_n$	$\varepsilon_n(\alpha, \beta)$	cond $\tilde{G}_n$	$\varepsilon_n(\alpha, \beta)$
2	1.29×10^1	4.06×10^{-14}	7.27×10^1	1.63×10^{-13}
4	2.11×10^3	3.28×10^{-12}	1.35×10^6	1.05×10^{-8}
6	5.66×10^5	7.75×10^{-10}	1.30×10^{11}	1.19×10^{-3}
8	1.86×10^9	3.39×10^{-6}	3.12×10^{16}	4.42×10^2
10	6.76×10^{10}	4.28×10^{-3}	1.42×10^{22}	—

The condition of $\tilde{G}_n$ in Example 5.4 for modified moments based on Hermite and Laguerre polynomials and mean square errors in the coefficients α_k, β_k, $k = 0, 1, \ldots, n-1$.

large for Laguerre moments than for Hermite moments. An explanation is provided by the respective estimates (5.16) of cond $\tilde{G}_n$ shown in Table 5.4 together with the mean square (relative) error $\varepsilon_n(\alpha, \beta)$ of the α_k, β_k.[†]

The initial nodes ξ_j are again weak, as in the previous examples. All remaining nodes are strong and produce the now familiar peaking and surge phenomenon on $[0, \infty]$. On $[-\infty, 0]$, g_n takes off to ∞. Even though the Hermite measure $ds(x) = e^{-x^2} dx$ has its support on $[-\infty, \infty]$ and, therefore, also contributes to cond $\tilde{G}_n$ through the values of $g_n(x)$ for $x < 0$, the damping power of e^{-x^2} is much stronger for large $|x|$ than the damping power of e^{-x} for large $x > 0$, which is the reason why the condition of $\tilde{G}_n$ turns out to be significantly smaller for Hermite moments than for Laguerre moments.

EXAMPLE 5.5:

$$
d\sigma(x) = \begin{cases} \dfrac{1}{\pi} \dfrac{|x - \frac{1}{2}|}{\left\{ x(1-x)(\frac{1}{3} - x)(\frac{2}{3} - x) \right\}^{1/2}} dx, & x \in (0, \frac{1}{3}) \cup (\frac{2}{3}, 1), \\ 0 & \text{elsewhere.} \end{cases}
$$

$$(5.30)$$

[†]The values of cond $\tilde{G}_n$ in the case of Hermite moments, as given in Gautschi [16], Table 4.8, are in error, being consistently somewhat too large. The error is due to an incorrect computation of $\|\tilde{m}\|_2$ using m_0 instead of $\tilde{m}_0 = \pi^{-1/4} m_0$.

This measure arises in the study of the diatomic linear chain (Wheeler [35]) and corresponds to the mass ratio $m/M = 1/2$, where m and M are the masses of the two kinds of particles alternating along the chain.

Wheeler [35] applies algorithm (5.25) to generate the associated orthogonal polynomials, using two choices of modified moments: Chebyshev moments, with $ds(x) = \pi^{-1}[x(1-x)]^{-1/2} dx$ on $[0,1]$, on the one hand, and modified moments with

$$ds(x) =$$

$$\begin{cases} \dfrac{18}{\pi|x-1/2|} \left\{ x(1-x)\left(x-\dfrac{1}{3}\right)\left(x-\dfrac{2}{3}\right) \right\}^{1/2}, & x \in \left(0, \tfrac{1}{3}\right) \cup \left(\tfrac{2}{3}, 1\right), \\ 0 \quad \text{elsewhere,} \end{cases} \tag{5.31}$$

on the other. He observes exponentially increasing instability (as n increases) in the first case, and perfect stability in the second. An explanation of this can be given on the basis of (5.16).

All zeros ξ_j of the orthogonal polynomial π_n associated with $d\sigma$, except possibly one, are known to congregate on the two support intervals $[0, 1/3]$ and $[2/3, 1]$ (Szegö [32], Theorem 3.41.2 and the sentence following it). In fact, by symmetry, the "hole" $[1/3, 2/3]$ contains exactly one zero at $x = 1/2$, if n is odd, and none, if n is even. It was determined numerically that all zeros on the two support intervals are weak, and the one at $x = 1/2$ (if n is odd) is strong. The function g_n is wiggling on both support intervals, remaining ≤ 1 there, and shoots up to a large single peak on the hole if n is even, and to a twin peak if n is odd. The peak values for $n = 5, 10, 20, 40$ are approximately $1.4, 6.5 \times 10^2, 2.4 \times 10^8, 1.1 \times 10^{20}$, respectively. In the case of Chebyshev moments, the integral $\int_0^1 g_n(x) \, ds(x)$ becomes large with increasing n, since the measure $ds(x)$ is supported on the entire interval $[0,1]$, including the hole $[1/3, 2/3]$ where g_n is large. The condition of $\tilde{G}_n$ therefore gradually deteriorates; the condition numbers for $n = 5, 10, 20, 40$, in fact, are $7.3 \times 10^{-1}, 4.1, 1.5 \times 10^3, 6.2 \times 10^8$, respectively. In contrast, the measure $ds(x)$ in (5.31) is zero on the hole and thus gives rise to an integral $\int_0^1 g_n(x)\,ds(x)$ which is bounded uniformly in n. Accordingly, the condition numbers remain quite small, namely $1.18, 1.07, 1.07, 1.07$, respectively, for $n = 5, 10, 20, 40$.

The orthogonal polynomials relative to the measure $d\sigma$ in (5.30) have since been obtained explicitly (Gautschi [17]). Example 5.5, nevertheless, continues to be of interest, as it shows the importance of matching the supports of the two measures $d\sigma$ and ds in cases where $d\sigma$ is supported on separate intervals.

REFERENCES

1. J. L. Blue, "A Legendre polynomial integral," *Math. Comput.* **33** (1979), 739–741.

2. C. de Boor, "On calculating with B-splines," *J. Approximation Theory* **6** (1972), 50–62.

3. _____, "On local linear functionals which vanish at all B-splines but one," *Theory of Approximation with Applications* (Law, A. G. & Sahney, B. N., eds.), pp. 120–145, Academic Press, New York-San Francisco-London, 1976.

4. _____, personal communication, 1978.

5. _____, *A Practical Guide to Splines*, Springer-Verlag, New York-Heidelberg-Berlin, 1978.

6. M. Branders, "Application of Chebyshev polynomials in numerical integration" (Flemish), Thesis, Catholic University of Leuven, Belgium, 1976.

7. H. Buchholz, *Die konfluente hypergeometrische Funktion*, Springer-Verlag, Berlin-Göttingen-Heidelberg, 1953.

8. P. L. Chebyshev, "Sur l'interpolation par la méthode des moindres carrés," Mém. Acad. Impér. Sci. St. Pétersbourg (7) **1**, no. 15 (1859), 1–24. [Oeuvres I, 471–498]

9. L. Gatteschi, "On some orthogonal polynomial integrals," *Math. Comput.* **35** (1980), 1291–1298.

10. W. Gautschi, "Construction of Gauss-Christoffel quadrature formulas," *Math. Comput.* **22** (1968), 251–270.

11. _____, "On the construction of Gaussian quadrature rules from modified moments," *Math. Comput.* **24** (1970), 245–260.

12. _____, "The condition of orthogonal polynomials," *Math. Comput.*, **26** (1972), 923–924.

13. _____, "On the condition of algebraic equations," *Numer. Math.*, **21** (1973), 405–424.

14. _____, "The condition of polynomials in power form," *Math. Comput.*, **33** (1979), 343–352.

15. _____, "On the preceding paper, 'A Legendre polynomial integral' by J. L. Blue," *Math. Comput.*, **33** (1979), 742–743.

16. _____, "On generating orthogonal polynomials," *SIAM J. Sci. Statist. Comput.* **3** (1982), 289–317.

17. _____, "On some orthogonal polynomials of interest in theoretical chemistry," *BIT* **25** (1985), to appear.

18. _____, "On the sensitivity of orthogonal polynomials to perturbations in the moments," in preparation.

19. A. J. Geurts, "A contribution to the theory of condition," *Numer. Math.* **39** (1982), 85–96.

20. G. H. Golub and J. H. Welsch, "Calculation of Gauss quadrature rules," *Math. Comput.*, **23** (1969), 221–230.

21. S. Lewanowicz, "Construction of a recurrence relation for modified moments," *J. Comput. Appl. Math.*, **5** (1979), 193–206.

22. Y. L. Luke, *Algorithms for the Computation of Mathematical Functions*, Academic Press, New York-San Francisco-London, 1977.

23. T. Lyche, "A note on the condition numbers of the B-spline basis," *J. Approximation Theory*, **22** (1978), 202–205.

24. I. P. Natanson, *Constructive Function Theory*, Vol. III, Frederick Ungar Publ. Co., New York, 1965.

25. J. M. Ortega and W. C. Rheinboldt, *Iterative Solution of Nonlinear Equations in Several Variables*, Academic Press, New York-London, 1970.

26. R. Piessens and M. Branders, "The evaluation and application of some modified moments," *BIT*, **13** (1973), 443–450.

27. J. R. Rice, "A theory of condition," *SIAM J. Numer. Anal.*, **3** (1966), 287–310.

28. T. J. Rivlin, *The Chebyshev Polynomials*, John Wiley & Sons, London-Sydney-Toronto, 1974.

29. R. A. Sack, and A. F. Donovan, "An algorithm for Gaussian quadrature given modified moments," *Numer. Math.*, **18** (1971–72), 465–478.

30. A. Schönhage, *Approximationstheorie*, Walter de Gruyter & Co., Berlin-New York, 1971.

31. G. W. Stewart, *Introduction to Matrix Computations*, Academic Press, New York-London, 1973.

32. G. Szegö, *Orthogonal Polynomials*, AMS Colloquium Publications, Vol. 23, 4th ed., Providence, RI, 1975.

33. E. V. Voronovskaja, *The Functional Method and its Applications*, Translations of Mathematical Monographs, Vol. 28, American Mathematical Society, Providence, R.I., 1970.

34. J. C. Wheeler, "Modified moments and Gaussian quadratures," *Rocky Mountain J. Math.*, **4** (1974), 287–296.

35. _____, "Modified moments and continued fraction coefficients for the diatomic linear chain," *J. Chem. Phys.* **80** (1984), 472–476.

36. J. C. Wheeler and C. Blumstein, "Modified moments for harmonic solids," *Phys. Rev.*, B6 (1972), 4380–4382.

37. H. S. Wilf, *Mathematics for the Physical Sciences*, John Wiley & Sons, New York-London, 1962.

38. J. H. Wilkinson, *Rounding Errors in Algebraic Processes*, Prentice-Hall, Englewood Cliffs, N.J., 1963.

39. _____, *The Algebraic Eigenvalue Problem*, Clarendon Press, Oxford, 1965.

A GENERALIZED CONJUGATE GRADIENT METHOD FOR THE NUMERICAL SOLUTION OF ELLIPTIC PARTIAL DIFFERENTIAL EQUATIONS*

Paul Concus, Gene H. Golub, and Dianne P. O'Leary

Abstract

We consider a generalized conjugate gradient method for solving sparse, symmetric, positive-definite systems of linear equations, principally those arising from the discretization of boundary value problems for elliptic partial differential equations. The method is based on splitting off from the original coefficient matrix a symmetric, positive-definite one that corresponds to a more easily solvable system of equations, and then accelerating the associated iteration using conjugate gradients. Optimality and convergence

---

*Note added in reprinting: The *generalized conjugate gradient method* discussed here has become known more popularly as the *preconditioned conjugate gradient method*.

properties are presented, and the relation to other methods is discussed. Several splittings for which the method seems particularly effective are also discussed and, for some, numerical examples are given.

0. INTRODUCTION

In 1952, Hestenes and Stiefel [0] proposed the conjugate gradient method (CG) for solving the system of linear algebraic equations

$$A\mathbf{x} = \mathbf{b},$$

where A is an $n \times n$, symmetric, positive-definite matrix. This elegant method has as one of its important properties that in the absence of round-off error the solution is obtained in at most n iteration steps. Furthermore, the entire matrix A need not be stored as an array in memory; at each stage of the iteration it is necessary to compute only the product $A\mathbf{z}$ for a given vector $\mathbf{z}$.

Unfortunately the initial interest and excitement in CG was dissipated, because in practice the numerical properties of the algorithm differed from the theoretical ones; viz., even for small systems of equations ($n \leqslant 100$) the algorithm did not necessarily terminate in n iterations. In addition, for large systems of equations arising from the discretization of two-dimensional elliptic partial differential equations, competing methods such as successive over-relaxation (SOR) required only $O(\sqrt{n})$ iterations to achieve a prescribed accuracy [1]. It is interesting to note that in the proceedings of the Conference on Sparse Matrices and Their Applications held in 1971 [2] there is hardly any mention of the CG method.

In 1970, Reid [3] renewed interest in CG by giving evidence that the method could be used in a highly effective manner as an iterative procedure for solving large sparse systems of linear equations. Since then a number of authors have described the use of CG for solving a variety of problems (cf. [4], [5], [6], [7], [8]). Curiously enough, although CG was generally discarded during the sixties as a useful method for solving linear equations, except in conjunction with other methods [9], there was considerable interest in it for solving nonlinear equations (cf. [10]).

The conjugate gradient method has a number of attractive properties when used as an iterative method:

(i)   It does not require an estimation of parameters.
(ii) It takes advantage of the distribution of the eigenvalues of the iteration operator.
(iii) It requires fewer restrictions on the matrix A for optimal behavior than do such methods as SOR.

Our basic view is that CG is most effective when used as an iteration acceleration technique.

In this paper, we derive and show how to apply a generalization of the CG method and illustrate it with numerical examples. Based on our investigations, we feel that the generalized CG method has the potential for widespread application in the numerical solution of boundary value problems for elliptic partial differential equations. Additional experience should further indicate how best to take full advantage of the method's inherent possibilities.

1.   DERIVATION OF THE METHOD

Consider the system of equations

$$A\mathbf{x} = \mathbf{b}, \tag{1.1}$$

where A is an $n \times n$, symmetric, positive-definite matrix and $\mathbf{b}$ is a given vector. It is frequently desirable to rewrite (1.1) as

$$M\mathbf{x} = N\mathbf{x} + \mathbf{c}, \tag{1.2}$$

where M is positive-definite and symmetric and N is symmetric. In section 4 we describe several decompositions of the form (1.2). We are interested in those situations for which it is a much simpler computational task to solve the system

$$M\mathbf{z} = \mathbf{d} \tag{1.3}$$

than it is to solve (1.1).

We consider an iteration of the form

$$\mathbf{x}^{(k+1)} = \mathbf{x}^{(k-1)} + \omega_{k+1}\left(\alpha_k \mathbf{z}^{(k)} + \mathbf{x}^{(k)} - \mathbf{x}^{(k-1)}\right), \qquad (1.4)$$

where

$$M\mathbf{z}^{(k)} = \mathbf{c} - (M - N)\mathbf{x}^{(k)}. \qquad (1.5)$$

Many iterative methods can be described by (1.4); e.g. the Chebyshev semi-iterative method and the Richardson second-order method (cf. [11]). The generalized CG method is also of this form.

For the Richardson or Chebyshev methods, the optimal parameters (ω_{k+1}, α_k) are given as simple, easy-to-compute functions of the smallest and largest eigenvalues of the iteration matrix $M^{-1}N$ [11]; thus good estimates of these eigenvalues are required for the methods to be efficient. The methods do not take into account the values of any of the interior eigenvalues of the iteration matrix.

The CG method, on the other hand, needs no *a priori* information on the extremal eigenvalues and does take into account the interior ones, but at a cost of increased computational requirements for evaluating ω_{k+1} and α_k. In section 3, we describe a technique to provide directly from the CG method good estimates for the extreme eigenvalues of the iteration matrix.

From equations (1.4) and (1.5), we obtain the relation

$$M\mathbf{z}^{(k+1)} = M\mathbf{z}^{(k-1)} - \omega_{k+1}\left(\alpha_k(M - N)\mathbf{z}^{(k)} + M\left(\mathbf{z}^{(k-1)} - \mathbf{z}^{(k)}\right)\right).$$

$$(1.6)$$

For the generalized CG method the parameters $\{\alpha_k, \omega_{k+1}\}$ are computed so that

$$\mathbf{z}^{(p)^T}M\mathbf{z}^{(q)} = 0 \qquad \text{for } p \neq q \text{ and } p, q = 0, 1, \dots, n-1. \quad (1.7)$$

Since M is $n \times n$ positive-definite, (1.7) implies that for some $k \leqslant n$

$$\mathbf{z}^{(k)} = \mathbf{0}$$

and, hence,

$$\mathbf{x}^{(k)} = \mathbf{x}. \tag{1.8}$$

That is, the iteration converges in no more than n steps.

We derive the above result by induction. Assume

$$\mathbf{z}^{(p)^T}M\mathbf{z}^{(q)} = 0 \quad \text{for } p \neq q \quad \text{and} \quad p, q = 0, 1, \ldots, k. \tag{1.9}$$

Then if

$$\alpha_k = \mathbf{z}^{(k)^T}M\mathbf{z}^{(k)} / \mathbf{z}^{(k)^T}(M - N)\mathbf{z}^{(k)}, \tag{1.10}$$

there holds

$$\mathbf{z}^{(k)^T}M\mathbf{z}^{(k+1)} = 0,$$

and if

$$\omega_{k+1} = \left(1 - \alpha_k \frac{\mathbf{z}^{(k-1)^T}N\mathbf{z}^{(k)}}{\mathbf{z}^{(k-1)^T}M\mathbf{z}^{(k-1)}}\right)^{-1} \tag{1.11}$$

then

$$\mathbf{z}^{(k-1)^T}M\mathbf{z}^{(k+1)} = 0.$$

We can simplify the above expression for ω_{k+1} as follows. From (1.6) we obtain

$$M\mathbf{z}^{(k)} = M\mathbf{z}^{(k-2)} - \omega_k\big(\alpha_{k-1}(M - N)\mathbf{z}^{(k-1)} + M(\mathbf{z}^{(k-2)} - \mathbf{z}^{(k-1)})\big),$$

and then from (1.9)

$$\mathbf{z}^{(k)^T}N\mathbf{z}^{(k-1)} = \mathbf{z}^{(k)^T}M\mathbf{z}^{(k)} / (\omega_k\alpha_{k-1}).$$

Since

$$\mathbf{z}^{(k-1)^T}N\mathbf{z}^{(k)} = \mathbf{z}^{(k)^T}N\mathbf{z}^{(k-1)},$$

it follows

$$\omega_{k+1} = \left(1 - \frac{\alpha_k}{\alpha_{k-1}} \frac{\mathbf{z}^{(k)^T} M \mathbf{z}^{(k)}}{\mathbf{z}^{(k-1)^T} M \mathbf{z}^{(k-1)}} \frac{1}{\omega_k}\right)^{-1}.$$

From (1.6), for $j < k - 1$

$$\mathbf{z}^{(j)^T} M \mathbf{z}^{(k+1)} = \alpha_k \omega_{k+1} \mathbf{z}^{(j)^T} N \mathbf{z}^{(k)}.$$

But

$$M \mathbf{z}^{(j+1)} = M \mathbf{z}^{(j-1)} - \omega_{j+1} \left(\alpha_j (M - N) \mathbf{z}^{(j)} + M (\mathbf{z}^{(j-1)} - \mathbf{z}^{(j)})\right),$$

so that

$$\mathbf{z}^{(k)^T} N \mathbf{z}^{(j)} = 0.$$

Thus, since $N = N^T$,

$$\mathbf{z}^{(j)^T} M \mathbf{z}^{(k+1)} = 0 \qquad \text{for} \quad j < k - 1.$$

Hence by induction we obtain (1.7) and (1.8).

The *generalized CG method* is summarized as follows.

ALGORITHM.

Let $\mathbf{x}^{(0)}$ be a given vector and arbitrarily define $\mathbf{x}^{(-1)}$. For $k = 0, 1, \ldots$

(1) Solve $M \mathbf{z}^{(k)} = \mathbf{c} - (M - N) \mathbf{x}^{(k)}$.
(2) Compute

$$\alpha_k = \frac{\mathbf{z}^{(k)^T} M \mathbf{z}^{(k)}}{\mathbf{z}^{(k)^T} (M - N) \mathbf{z}^{(k)}},$$

$$\omega_{k+1} = \left(1 - \frac{\alpha_k}{\alpha_{k-1}} \frac{\mathbf{z}^{(k)^T} M \mathbf{z}^{(k)}}{\mathbf{z}^{(k-1)^T} M \mathbf{z}^{(k-1)}} \cdot \frac{1}{\omega_k}\right)^{-1} \qquad (k \geq 1),$$

$$\omega_1 = 1.$$

(3) Compute

$$\mathbf{x}^{(k+1)} = \mathbf{x}^{(k-1)} + \omega_{k+1}\left(\alpha_k \mathbf{z}^{(k)} + \mathbf{x}^{(k)} - \mathbf{x}^{(k-1)}\right).$$

Note that the algorithm can be viewed as an acceleration of the underlying first-order iteration $(\omega_{k+1} \equiv 1)$, $\mathbf{x}^{(k+1)} = \mathbf{x}^{(k)} + \alpha_k \mathbf{z}^{(k)}$. As with other higher order methods, the storage requirements of the algorithm are greater than those of the underlying first-order iteration being accelerated.

The algorithm presented above is given primarily for expository purposes. For actual computation, the following equivalent form can be more efficient in terms of storage [3].

ALGORITHM (alternative form).

Let $\mathbf{x}^{(0)}$ be a given vector and arbitrarily define $\mathbf{p}^{(-1)}$. For $k = 0, 1, \ldots$

(1) Solve $M\mathbf{z}^{(k)} = \mathbf{c} - (M - N)\mathbf{x}^{(k)}$.
(2) Compute

$$b_k = \frac{\mathbf{z}^{(k)^T} M \mathbf{z}^{(k)}}{\mathbf{z}^{(k-1)^T} M \mathbf{z}^{(k-1)}}, \qquad k \geqslant 1,$$

$$b_0 = 0,$$

$$\mathbf{p}^{(k)} = \mathbf{z}^{(k)} + b_k \mathbf{p}^{(k-1)}.$$

(3) Compute

$$a_k = \frac{\mathbf{z}^{(k)^T} M \mathbf{z}^{(k)}}{\mathbf{p}^{(k)^T}(M - N)\mathbf{p}^{(k)}},$$

$$\mathbf{x}^{(k+1)} = \mathbf{x}^{(k)} + a_k \mathbf{p}^{(k)}.$$

In the computation of the numerators of a_k and b_k one need not recompute $M\mathbf{z}^{(k)}$, since it can be saved from step (1). Also, instead of computing the right-hand side of step (1) explicitly at each iteration, it is often advantageous to compute it recursively from

$$\left[\mathbf{c} - (M - N)\mathbf{x}^{(k+1)}\right] = \left[\mathbf{c} - (M - N)\mathbf{x}^{(k)}\right] - a_k(M - N)\mathbf{p}^{(k)},$$

$$(1.12)$$

which equation is obtained from step (3). The vector $(M - N)\mathbf{p}^{(k)}$ appearing in (1.12) may be saved from the computation of a_k. Similar remarks hold for the algorithm in its first form as well. There is evidence that the use of (1.12) is no less accurate than use of the explicit computation (see [18], [3] for particular examples).

The calculated vectors $\{\mathbf{z}^{(k)}\}_{k=0}^{n-1}$ will not generally be M-orthogonal in practice because of rounding errors. One might consider forcing the newly calculated vectors to be M-orthogonal by a procedure such as Gram-Schmidt. However, this would require the storage of all the previously obtained vectors.

Our basic approach is to permit the gradual loss of orthogonality and with it the finite termination property of CG. We consider primarily the iterative aspects of the algorithm. In fact, for solving large sparse systems arising from the discretization of elliptic partial differential equations, the application of principal interest for us and for which the generalized CG method seems particularly effective, convergence to desired accuracy often occurs within a number of iterations small compared with n.

2. OPTIMALITY PROPERTIES

From (1.6), we obtain

$$\mathbf{z}^{(k+1)} = \mathbf{z}^{(k-1)} - \omega_{k+1}\left(\alpha_k(I - M^{-1}N)\mathbf{z}^{(k)} + \mathbf{z}^{(k-1)} - \mathbf{z}^{(k)}\right).$$

$$(2.1)$$

Define

$$K = I - M^{-1}N. \qquad (2.2)$$

We have $\mathbf{z}^{(1)} = (I - \alpha_0 K)\mathbf{z}^{(0)}$, and there follows by induction that

$$\mathbf{z}^{(l+1)} = \left[I - KP_l(K) \right] \mathbf{z}^{(0)} \tag{2.3}$$

where

$$P_l(K) = \sum_{j=0}^{l} \beta_j^{(l)} K^j. \tag{2.4}$$

We denote

$$p_l(\lambda) = \sum_{j=0}^{l} \beta_j^{(l)} \lambda^j \tag{2.5}$$

and from (2.1) we have for $k = 2, 3, \ldots, l$

$$p_k(\lambda) = \omega_{k+1}(1 - \alpha_k \lambda) p_{k-1}(\lambda) - (\omega_{k+1} - 1) p_{k-2}(\lambda) + \alpha_k \omega_{k+1},$$

and

$$p_0(\lambda) = \alpha_0, \qquad p_1(\lambda) = \omega_2(\alpha_0 + \alpha_1 - \alpha_0 \alpha_1 \lambda).$$

The coefficients $\{ \beta_j^{(l)} \}_{j=0}^{l}$ can be generated directly. From (2.3) and the relation $\mathbf{z}^{(l+1)} = \mathbf{z}^{(0)} + K(\mathbf{x}^{(l+1)} - \mathbf{x}^{(0)})$, there follows

$$\mathbf{x}^{(l+1)} = \mathbf{x}^{(0)} + P_l(K)\mathbf{z}^{(0)}.$$

Then if

$$Z = [\mathbf{z}^{(0)}, K\mathbf{z}^{(0)}, \ldots, K^l \mathbf{z}^{(0)}], \tag{2.6}$$

$$\mathbf{x}^{(l+1)} = \mathbf{x}^{(0)} + Z\boldsymbol{\beta}^{(l)}. \tag{2.7}$$

Consider the *weighted error function*:

$$E(\mathbf{x}^{(l+1)}) = \tfrac{1}{2}(\mathbf{x} - \mathbf{x}^{(l+1)})^T (M - N)(\mathbf{x} - \mathbf{x}^{(l+1)}). \tag{2.8}$$

Assuming that $(M - N)$ is nonsingular, we obtain, using

$$\mathbf{z}^{(0)} = K(\mathbf{x} - \mathbf{x}^{(0)}),$$

the relations

$$E(\mathbf{x}^{(l+1)}) = \tfrac{1}{2}\mathbf{z}^{(0)^T}(I - KP_l(K))^T M(M - N)^{-1}M(I - KP_l(K))\mathbf{z}^{(0)}$$

$$= \tfrac{1}{2}\mathbf{e}^{(0)^T}(I - KP_l(K))^T(M - N)(I - KP_l(K))\mathbf{e}^{(0)},$$

$$(2.9)$$

where

$$\mathbf{e}^{(0)} = \mathbf{x} - \mathbf{x}^{(0)}.$$

Equivalently, we can use (2.7) and re-write (2.8) as

$$E(\mathbf{x}^{(l+1)}) = \tfrac{1}{2}(K^{-1}\mathbf{z}^{(0)} - Z\boldsymbol{\beta}^{(l)})^T(M - N)(K^{-1}\mathbf{z}^{(0)} - Z\boldsymbol{\beta}^{(l)}).$$

$$(2.10)$$

The quantity $E(\mathbf{x}^{(l+1)})$ is minimized when we choose $\boldsymbol{\beta}^{(l)}$ so that

$$G\boldsymbol{\beta}^{(l)} = \mathbf{h},$$

where

$$G = Z^T(M - N)Z, \qquad \mathbf{h} = Z^T M\mathbf{z}^{(0)}.$$

Let

$$\kappa = \lambda_{\max}(K)/\lambda_{\min}(K).$$

Then using arguments similar to those given in [12], the following can be shown:

(A) $$\frac{E(\mathbf{x}^{(l+1)})}{E(\mathbf{x}^{(0)})} \leqslant 4\left(\frac{\sqrt{\kappa} - 1}{\sqrt{\kappa} + 1}\right)^{2(l+1)}. \qquad (2.11)$$

(B) The generalized CG method is optimal in the class of all algorithms for which

$$\mathbf{x}^{(l+1)} = \mathbf{x}^{(0)} + P_l(K)\mathbf{z}^{(0)}.$$

That is, the approximation $\mathbf{x}^{(l+1)}$ generated by the generalized

CG method satisfies

$$E(\mathbf{x}^{((l+1))}) = \min_{P_l} \tfrac{1}{2}\mathbf{e}^{(0)^T}\big(I - KP_l(K)\big)^T$$

$$\times (M - N)\big(I - KP_l(K)\big)\mathbf{e}^{(0)},$$

where the minimum is taken with respect to all polynomials P_l of degree l.

Recall that we have assumed that M and $(M - N)$ are positive definite and symmetric. Thus the eigenvalues of $K = (I - M^{-1}N)$ are all real and K is similar to a diagonal matrix. Hence, if K has only $p < n$ distinct eigenvalues, there exists a matrix polynomial $Q_p(K)$ so that

$$Q_p(K) = 0.$$

In this case, $E(\mathbf{x}^{(p)}) = 0$ and hence

$$\mathbf{x}^{(p)} = \mathbf{x},$$

so that the iteration converges in only p steps. The same result also holds if K has a larger number of distinct eigenvalues but $\mathbf{e}^{(0)}$ lies in a subspace generated by the eigenvectors associated with only p of these eigenvalues.

We remark also that statement (B) implies CG is optimal for the particular eigenvector mix of the initial error $\mathbf{e}^{(0)}$, taking into account interior as well as extremal eigenvalues. As will be discussed in the next section, the extremal eigenvalues are approximated especially well as CG proceeds, the iteration then behaving as if the corresponding vectors are not present. Thus the error estimate (2.11), which is based on the extremal eigenvalues, tends to be pessimistic asymptotically. One often observes, in practice (see section 5), a superlinear rate of convergence for the CG method.

3. EIGENVALUE COMPUTATIONS

The CG method can be used in a very effective manner for computing the extreme eigenvalues of the matrix $K = I - M^{-1}N$.

We write (see (2.1))

$$\mathbf{z}^{(k+1)} = \mathbf{z}^{(k-1)} - \omega_{k+1}\left(\alpha_k K\mathbf{z}^{(k)} + \mathbf{z}^{(k-1)} - \mathbf{z}^{(k)}\right) \quad (3.1)$$

as

$$K\mathbf{z}^{(k)} = c_{k-1}\mathbf{z}^{(k-1)} + a_k\mathbf{z}^{(k)} + b_{k+1}\mathbf{z}^{(k+1)},$$

or

$$K\left[\mathbf{z}^{(0)}, \mathbf{z}^{(1)}, \ldots, \mathbf{z}^{(n-1)}\right]$$

$$= \left[\mathbf{z}^{(0)}, \mathbf{z}^{(1)}, \ldots, \mathbf{z}^{(n-1)}\right] \begin{bmatrix} a_0 & c_0 & & & \bigcirc \\ b_1 & a_1 & c_1 & & \\ & \ddots & \ddots & \ddots & \\ & & & & c_{n-2} \\ \bigcirc & & & b_{n-1} & a_{n-1} \end{bmatrix}$$

thus defining a_k, b_k, and c_k. In matrix notation, the above equation can be written as

$$KZ = ZJ. \quad (3.2)$$

Assuming that the columns of Z are linearly independent, there follows from (3.2) that

$$K = ZJZ^{-1},$$

hence the eigenvalues of K are equal to those of J. As pointed out in section 2, if K has repeated eigenvalues or if the vector $\mathbf{z}^{(0)}$ is deficient in the direction of some eigenvectors of K, iteration (3.1) will terminate in $k < n$ steps.

The process described by (3.1) is essentially the Lanczos algorithm [13]. It has been shown by Kaniel [14] and by Paige [15] that good estimates of the extreme eigenvalues of K often can be obtained from the truncated matrix

$$J_k = \begin{bmatrix} a_0 & c_0 & & & \bigcirc \\ b_1 & a_1 & c_1 & & \\ & \ddots & \ddots & \ddots & \\ & & & & c_{k-2} \\ \bigcirc & & & b_{k-1} & a_{k-1} \end{bmatrix}$$

where k is considerably less than n. This result holds even in the presence of round-off error [16].

It was pointed out in section 1 that the equation describing the CG method is of the same form as that describing the Chebyshev semi-iterative method and Richardson second order method, but that a knowledge of the extreme eigenvalues of K is required for obtaining parameters for the latter two methods. Thus one could construct a polyalgorithm in which the CG method is used initially to obtain good approximations to the solution and to the extreme eigenvalues of K, after which the Chebyshev semi-iterative method, say, is used, thereby avoiding the additional work of repeatedly calculating CG parameters. This technique has been used in an effective manner by O'Leary [17].

4. CHOICE OF M

For the splitting $M = I$, $N = I - A$ one obtains the basic, un-modified CG algorithm, for which

$$\mathbf{z}^{(k)} = \mathbf{r}^{(k)} = \mathbf{b} - A\mathbf{x}^{(k)}$$

is simply the residual at the kth step. Since the rate of convergence of the generalized CG method, as given by the estimate (2.11), decreases with increasing

$$\kappa = \lambda_{\max}(K)/\lambda_{\min}(K),$$

it is desirable to choose a splitting for which κ is as small as possible. If $A = L + D + U$, where D consists of the diagonal elements of A and $L(U)$ is a strictly lower (upper) triangular matrix, then it is reasonable to consider the choice

$$M = D, \qquad N = -(L + U).$$

This M, which is equivalent to a rescaling of the problem, is one for which (1.3) can be solved very simply for $\mathbf{z}$. It has been shown by Forsythe and Straus [19] that if A is two-cyclic then among all diagonal matrices this choice of M will minimize κ.

In many cases, the matrix A can be written in the form

$$A = \left(\begin{array}{c|c} M_1 & F \\ \hline F^T & M_2 \end{array} \right),$$
(4.3)

where the systems

$$M_1 z_1 = d_1 \quad \text{and} \quad M_2 z_2 = d_2$$

are easy to solve, and for such matrices, it is convenient to choose

$$M = \left(\begin{array}{c|c} M_1 & 0 \\ \hline 0 & M_2 \end{array} \right) \quad \text{and} \quad N = \left(\begin{array}{c|c} 0 & -F \\ \hline -F^T & 0 \end{array} \right).$$

Using (4.3), we can write the system (1.1) in the form

$$M_1 x_1 + F x_2 = b_1$$
(4.4a)

$$F^T x_1 + M_2 x_2 = b_2.$$
(4.4b)

Let the initial approximation for x_1 be $x_1^{(0)}$, and obtain $x_2^{(0)}$ as the solution to (4.4b) so that

$$M_2 x_2^{(0)} = b_2 - F^T x_1^{(0)}.$$

This implies that

$$z_2^{(0)} = 0,$$

and, hence, by (1.10)

$$\alpha_0 = 1,$$

and thus

$$x^{(1)} = \left(\begin{array}{c} x_1^{(0)} + z_1^{(0)} \\ x_2^{(0)} \end{array} \right).$$

A short calculation shows that $z_1^{(1)} = 0$ and hence $\alpha_1 = 1$. Using (1.6), a simple inductive argument then yields that for $j = 0, 1, 2, \ldots$

$$\alpha_j \equiv 1, \qquad z_1^{(2j+1)} = 0, \qquad z_2^{(2j)} = 0. \qquad (4.5)$$

This result was first observed by Reid [8] for the case in which M_1 and M_2 are diagonal, i.e., in which the matrix A has "Property A" and is suitably ordered. Other cases for elliptic boundary value problems in which matrices of the form (4.3) arise will be discussed in section 5. For these cases convergence can be rapid because K has only a few distinct eigenvalues, even though κ is not especially small.

Various other splittings of the matrix A can occur quite naturally in the solution of elliptic partial differential equations. For example, if one wishes to solve

$$-\Delta u + \sigma(x, y)u = f \qquad (x, y) \in R$$

$$u = g \qquad (x, y) \in \partial R,$$

where R is a rectangular region and $\Delta u = \partial^2 u / \partial x^2 + \partial^2 u / \partial y^2$, it is convenient to choose M as the finite difference approximation to a separable operator, such as the Helmholtz operator $-\Delta + C$, for which fast direct methods can be used [23]. A numerical example for this case is discussed in section 5. If one wishes to solve a separable equation, but on a nonrectangular region S, then by extending the problem to one on a rectangle R in which S is embedded, M can be chosen as the discrete approximation to the separable operator on R, for which fast direct methods can be used. Such a technique provides an alternative to the related capacitance matrix method [25] for handling such problems. Forms of this method utilizing CG, but in a different manner than here, are described in [26] and [27].

Several authors [4], [20], [21] have used CG in combination with symmetric successive overrelaxation (SSOR). For this method the solution of the equation $Mz^{(k)} = c - (M - N)x^{(k)}$ reduces to the solution of

$$(D + \omega L)D^{-1}(D + \omega U)z^{(k)} = \omega(2 - \omega)r^{(k)}$$

where D, L, and U are as described previously in this section (although D may be block diagonal), $\mathbf{r}^{(k)} = \mathbf{b} - A\mathbf{x}^{(k)}$, and ω is a parameter in the open interval $(0,2)$. SSOR is particularly effective in combination with CG because of the distribution of the eigenvalues of K (cf. [22]).

Meijerink and van der Vorst [7] have proposed that the following factorization of A be used:

$$A = FF^T + E,$$

so that

$$M = FF^T, \qquad N = -E.$$

The matrix F is chosen with a sparsity pattern resembling that of A. This splitting appears to yield a matrix K with eigenvalues that also are favorably distributed for CG. A block form of this technique recently developed by Underwood [24] achieves a more accurate approximate factorization of A with less computer storage and about the same number of arithmetic operations per iteration.

Generally, in addition to the requirement that (1.5) be "easy" to solve, M should have the following features if the generalized CG algorithm is to be computationally efficient. For rapid convergence one seeks a splitting so that

(i) $M^{-1}N$ has small or nearly equal eigenvalues

or (ii) $M^{-1}N$ has small rank.

Often a choice for M satisfying these restrictions comes about naturally from the inherent features of a given problem.

5. NUMERICAL EXAMPLES

For the first example, we consider the test problem discussed in [23]

$$-\operatorname{div}(a(x, y)\nabla u) = f \qquad (x, y) \in R$$

$$u = g \qquad (x, y) \in \partial R,$$

where $a(x, y) = [1 + \tfrac{1}{2}(x^4 + y^4)]^2$ and R is the unit square

$0 < x, \ y < 1$. After a transformation the problem becomes

$$- \Delta w + \sigma(x, y)w = a^{-1/2}f \qquad (x, y) \in R$$

$$w = a^{1/2}g \qquad (x, y) \in \partial R, \qquad (5.1)$$

where $\sigma(x, y) = 6(x^2 + y^2)/a^{1/2}$. As in [23] we discretize (5.1) on a uniform mesh of width h, using for Δ the standard five-point approximation Δ_h (see Henrici's paper in this study for approximations to Δ), and we choose the splitting

$$M = A + N = - \Delta_h + CI$$

with $C = 0 = \sigma_{\min}$ or $C = 3 = \frac{1}{2}(\sigma_{\max} + \sigma_{\min})$.

In [23] Chebyshev acceleration was used, which requires an estimate of the ratio of the extremal eigenvalues of the iteration matrix. Here we use the modified CG algorithm of section 1. For an initial guess $\mathbf{W}^{(0)} \equiv \mathbf{0}$ and choice of f and g corresponding to the solution $w = 2[(x - 1/2)^2 + (y - 1/2)^2]$, the results are given in Table 1 for $h = 1/64$. The results obtained for $h = 1/32$ were essentially identical, as the iteration is basically independent of h for this problem (see [23]).

Note that the Chebyshev method is sensitive both to the value of C and to the accuracy of the eigenvalues from which the parameters are calculated. The parameters used for the middle column were based on Gerschgorin estimates from the Rayleigh quotient,

	Chebyshev (from [23])			CG	
	$C = 0$ exact eigenvalues	$C = 3$ approximate eigenvalues	$C = 3$ exact eigenvalues		
iteration				$C = 0$	$C = 3$
1			1.6(−2)	4.5(−2)	1.6(−2)
2			7.4(−4)	2.6(−3)	6.7(−4)
3			1.1(−5)	3.0(−5)	1.0(−5)
4			2.7(−7)	5.7(−7)	1.1(−7)
5	2.4(−6)	1.1(−6)	4.3(−9)	5.1(−9)	8.2(−10)
6			1.2(−10)	4.4(−11)	5.7(−12)

TABLE 1

Maximum error vs. iteration number for first example.

which gave a ratio of largest to smallest eigenvalue about three times too large. The CG method appears to be less sensitive to the value of C. After several iterations CG begins to converge more rapidly than does the optimal Chebyshev method, which behavior is typical of the CG superlinear convergence property discussed in section 2. This example is one for which rapid convergence results because the eigenvalues of $M^{-1}N$ are small.

We give as the second example

$$-\Delta u = f \qquad (x, y) \in T$$

$$u = g \qquad (x, y) \in \partial T$$

where T is the domain shown in Figure 1. For a uniform square mesh of width h, and $0 < l < (2h)^{-1}$ a whole number, so that all boundary segments are mesh lines, the coefficient matrix A for the standard five-point discretization and natural ordering has the form (4.3). M_1 and M_2 correspond to the mesh points in each of the two squares, T_1 and T_2, and F to the coupling between them. F has non-zero entries in only $p = 2l - 1$ of its rows.

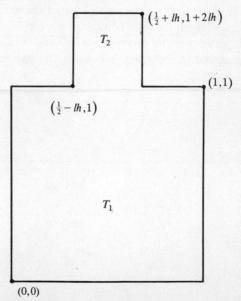

FIG. 1. T-shaped domain.

According to the discussion following (4.3) we choose

$$M = \left(\begin{array}{c|c} M_1 & 0 \\ \hline 0 & M_2 \end{array} \right)$$

and for initial approximation

$$\mathbf{U}^{(0)} = \left(\frac{\mathbf{U}_1^{(0)}}{M_2^{-1}(\mathbf{b}_2 - F^T \mathbf{U}_1^{(0)})} \right).$$

Then for the generalized CG algorithm, there holds $\alpha_k \equiv 1$ and that $\mathbf{z}_1$ and $\mathbf{z}_2$ are alternately zero, thereby reducing computational and storage requirements. We use a fast direct Poisson solver for the systems involving M_1 and M_2.

The results for $\mathbf{U}_1^{(0)}$ uniformly distributed random numbers in $(0,2)$ and $f(x, y)$ and $g(x, y)$ such that $u = x^2 + y^2$ is the solution are given in Table 2. Here the average error per point, the two norm of the error divided by the square root of the number of interior mesh points, is given for each of the test problems.

For this example, the eigenvalues of $M^{-1}N$ are not especially small in magnitude; however, since $M^{-1}N$ has rank of only $2p$,

	Case I	Case II	Case III
h	1/32	1/64	1/64
l	4	4	8
p	7	7	15
iteration	ave. error/pt	ave. error/pt	ave. error/pt
1	8.58(−2)	3.70(−2)	1.08(−1)
2	7.05(−2)	3.13(−2)	9.82(−2)
3	1.30(−2)	6.66(−3)	4.94(−2)
4	3.35(−3)	2.53(−3)	1.80(−2)
5	2.71(−4)	6.03(−4)	4.28(−3)
10	2.65(−7)	5.13(−8)	7.35(−5)
15	1.14(−13)	5.60(−13)	4.71(−8)

TABLE 2
Average error per point vs. iteration number.

convergence is obtained in only a moderate number of iterations. For Case I and Case II the last row represents full convergence to machine accuracy subject to rounding errors, as would be expected since $2p = 14$ for these cases.

We wish to thank Myron Stein of the Los Alamos Scientific Laboratory for his careful computer programming of the second test problem. This work was supported in part by the Energy Research and Development Administration, by the Hertz Foundation, and by the National Science Foundation.

REFERENCES

0. M. Hestenes and E. Stiefel, "Methods of conjugate gradients for solving linear systems," *J. Research NBS* **49** (1952), 409–436.*

1. D. Young, *Iterative Solution of Large Linear Systems*, Academic Press, New York, 1971.

2. D. J. Rose and R. A. Willoughby (ed.), *Sparse Matrices and Their Applications*, Plenum Press, New York-London, 1972.

3. J. K. Reid, "On the method of conjugate gradients for the solution of large sparse systems of linear equations," Proc. Conference on *Large Sparse Sets of Linear Equations*, Academic Press, New York, 1971.

4. O. Axelsson, "On preconditioning and convergence acceleration in sparse matrix problems," Report CERN 74-10 of the CERN European Organization for Nuclear Research, Data Handling Division, Laboratory I, 8 May 1974.

5. R. Bartels and J. W. Daniel, "A conjugate gradient approach to nonlinear elliptic boundary value problems in irregular regions," Proc. Conf. on *Numerical Solution of Differential Equations*, Springer-Verlag, Berlin, 1974.

6. R. Chandra, S. C. Eisenstat, and M. H. Schultz, "Conjugate gradient methods for partial differential equations," *Advances in Computer Methods for Partial Differential Equations*, R. Vichnevtsky (ed), Publ. A.I. C.A.-1975.

7. J. A. Meijerink and H. A. van der Vorst, "Iterative solution of linear systems arising from discrete approximations to partial differential equations," Academisch Computer Centrum, Utrecht, The Netherlands, 1974.

8. J. Reid, "The use of conjugate gradients for systems of linear equations possessing 'Property A'," *SIAM J. Numer. Anal.* **9** (1972), 325–332.

9. E. L. Wachspress, "Extended applications of alternating direction implicit iteration model problem theory," *SIAM J.* **11** (1963), 994–1016.

---

*Note added in reprinting: See also M. Hestenes, "The conjugate-gradient method for solving linear systems," Proc. Symposia in Applied Math VI, McGraw-Hill, New York, 1956, 83–102.

198   *Paul Concus, Gene H. Golub, and Dianne P. O'Leary*

10. R. Fletcher and C. M. Reeves, "Function minimization by conjugate gradients," *Comput. J.*, **7** (1964), 145–154.
11. G. H. Golub and R. S. Varga, "Chebyshev semi-iterative methods, successive over-relaxation iterative methods, and second order Richardson iterative methods," *Numer. Math.* **3** (1961), 147–168.
12. D. K. Faddeev and V. N. Faddeeva, *Computational Methods of Linear Algebra*, W. H. Freeman and Co., San Francisco, and London, 1963.
13. C. Lanczos, "An iteration method for the solution of the eigenvalue problem of linear differential and integral operators," *J. Research NBS* **45** (1950), 255–282.
14. S. Kaniel, "Estimates for some computational techniques in linear algebra," *Math. Comp.* **20** (1966), 369–378.
15. C. C. Paige, "The computation of eigenvalues and eigenvectors of very large sparse matrices," Ph.D. Thesis, London Univ., Institute of Computer Science, 1971.
16. C. C. Paige, "Computational variants of the Lanczos method for the eigenproblem," *J. Inst. Math. Appl.* **10** (1972), 373–381.
17. D. O'Leary, "Hybrid conjugate gradient algorithms," Ph.D. Thesis, Computer Science Dept., Stanford Univ., 1975.
18. M. Engeli, T. Ginsburg, H. Rutishauser and E. Stiefel, *Refined Iterative Methods for Computation of the Solution and the Eigenvalues of Self-Adjoint Boundary Value Problems*, Birkhäuser Verlag, Basel/Stuttgart, 1959.
19. G. Forsythe and E. G. Straus, "On best conditioned matrices," *Proc. Amer. Math. Soc.* **6** (1955), 340–345.
20. D. M. Young, L. Hayes, and E. Schleicher, "The use of the accelerated SSOR method to solve large linear systems," Abstract, 1975 SIAM Fall Meeting, San Francisco.
21. L. W. Ehrlich, "On some experience using matrix splitting and conjugate gradient," Abstract, 1975 SIAM Fall Meeting, San Francisco.
22. L. W. Ehrlich, "The block symmetric successive overrelaxation method," *J. SIAM* **12** (1964), 807–826.
23. P. Concus and G. H. Golub, "Use of fast direct methods for the efficient numerical solution of nonseparable elliptic equations," *SIAM J. Numer. Anal.* **10** (1973), 1103–1120.
24. R. R. Underwood, "An approximate factorization procedure based on the block Cholesky factorization and its use with the conjugate gradient method," Tech. Rept., General Electric Nuclear Energy Division, San Jose, CA (to appear).
25. B. L. Buzbee, F. W. Dorr, J. A. George, and G. H. Golub, "The direct solution of the discrete Poisson equation in irregular regions," *SIAM J. Num. Anal.* **8** (1971), 722–736.
26. J. A. George, "The use of direct methods for the solution of the discrete Poisson equation on non-rectangular regions," Report CS-70-159, Computer Science Dept., Stanford Univ., Stanford, CA, (1970).
27. W. Proskurowski and O. Widlund, "On the numerical solution of Laplace's and Helmholtz's equations by the capacitance matrix method," Tech. Rept., Courant Institute, NYU (to appear).

SOLVING DIFFERENTIAL EQUATIONS ON A HAND HELD PROGRAMMABLE CALCULATOR*

J. Barkley Rosser

Dedicated to Prof. Dr. Johannes Weissinger on his 65th birthday.

1. PRELIMINARIES

The present discussion is limited to initial value ordinary differential equations. One wishes to solve a system of equations

$$y_1' = f_1(x, y_1, y_2, \ldots, y_p)$$
$$y_2' = f_2(x, y_1, y_2, \ldots, y_p) \tag{1.1}$$
$$\ldots\ldots\ldots\ldots\ldots\ldots$$
$$y_p' = f_p(x, y_1, y_2, \ldots, y_p),$$

being given the values of $y_1, y_2, \ldots, y_p$ at $x = x_0$; here y_i' denotes dy_i/dx. With the better hand held calculators, one can handle a system of up to three equations. The best available hand held calculator can possibly handle up to six equations, depending on the complexity of the functions f_i. We may look forward to having still better calculators in the future.

We will mostly limit our discussion to the special case of a single equation, namely,

$$y' = f(x, y), \tag{1.2}$$

being given the value of y at $x = x_0$; here y' denotes dy/dx. The discussion of (1.2) can easily be generalized to a system of equations. See Conte and de Boor [4], pp. 365–366. Incidentally, sets of higher-order equations can be reduced to a system of first-order equations, such as (1.1). See Conte and de Boor [4], p. 365.

---

*Sponsored by the United States Army under Contract No. DAAG29-75-C-0024.

The person who only occasionally has to solve numerically a differential equation may never have had a course in numerical analysis, or may have forgotten much of it. So it seems necessary to make the present paper reasonably self-contained, presupposing little previous experience. Even if the reader has had experience, it was presumably on a large, fast computer. For the hand held calculator, considerations are sufficiently different that one cannot rely too much on previous experience with a large, fast computer. The discussion to follow is comprehensive enough to cover the important differences.

The overall procedure for solution is the classic one, namely, to try step by step to calculate approximations $y_1, y_2, y_3, \ldots$ for y corresponding to the values $x_1, x_2, x_3, \ldots$, where $x_0 < x_1 < x_2 < x_3 < \cdots$. We take y_0 as the value given for y at $x = x_0$.

We denote $x_{n+1} - x_n$ by h_n, called the step length. Usually h_n will be taken the same for a considerable succession of steps, and we write just h. A change in h is occasionally called for, but not uncommonly this involves some travail, and in the main one tries to avoid it. Indeed, if one wishes a table of y at equal intervals of x, a constant h is most convenient. On the other hand, all authorities agree that tests should be made frequently to see if a change of step length is needed, shorter to keep errors under control, or longer to avoid an unduly extended calculation that would result from taking h smaller than necessary.

2. THE EULER METHOD

If one has proceeded to a good approximation y_n for the value of y at $x = x_n$, a Taylor's expansion will give

$$y_{n+1} = y_n + hy_n' + \frac{h^2}{2} y''(\xi), \qquad (2.1)$$

where $x_n < \xi < x_{n+1}$. We evaluate y_n' by (1.2), and get

$$y_{n+1} = y_n + hf(x_n, y_n) + \frac{h^2}{2} y''(\xi). \qquad (2.2)$$

Unless one already knows the solution of (1.2), one has no way to calculate precisely the final term on the right of (2.2). Certainly, if h

is taken to be small enough, and one is not at a singularity of y'', the final term will be quite small, and the approximation

$$y_{n+1} \cong y_n + hf(x_n, y_n) \tag{2.3}$$

will give a sufficiently accurate value for y_{n+1}, after which one gets y_{n+2} by a similar formula, then y_{n+3}, etc. This is called the Euler method. A problem with which we must cope is being sure that we have taken h small enough that repeated use of (2.3) instead of (2.2) does not lead to serious error in the end.

Indeed one has to give serious attention to error propagation. Naturally, this depends on the behavior of solution curves of (1.2) near to the one of interest. Thus (2.2) is exact (if y_n is exact) but not very helpful, since no way comes easily to mind to determine $y''(\xi)$, short of knowing the solution of (1.2). As soon as we go to (2.3), we have probably wandered off to another solution curve of (1.2). We hope we have not wandered far, and we hope that this other solution curve does not subsequently diverge greatly from the one of interest. These pose difficult questions.

To get a feeling for this, let us attempt to solve

$$y' = -y, \tag{2.4}$$

given that $y = 1$ when $x = 0$. Let us try to approximate the value of y when $x = 6$. As the answer for (2.4) is obviously $y = e^{-x}$, we can say that for $0 < \xi \leqslant 6$ we have $|y''(\xi)| < 1$. Hence the error in (2.3) will be less than $h^2/2$. If we take steps of constant length h, we will require $6/h$ steps to get from $x = 0$ to $x = 6$. With an error less than $h^2/2$ at each step, and $6/h$ steps, the final error might be expected to add up to something less than

$$\left(\frac{h^2}{2} \right)\left(\frac{6}{h} \right) = 3h.$$

From this, it is tempting to conclude that the method is first order; that is, the overall error is roughly proportional to h. For example, if we decrease h by a factor of 2, we would expect to decrease the error at $x = 6$ by approximately a factor of 2.

Within bounds, the conclusion is correct. However, the argument given above to support it is fallacious. To see this, let us look at a specific example. Take $h = 0.1$. By (2.3), we will get

$$y_1 = 0.9.$$

This is too small by about

$$0.0048374.$$

(In accordance with (2.2), this error is close to $h^2/2$). However, the value of y at $x = 6$ is

$$e^{-6} \cong 0.0024888. \tag{2.5}$$

Thus our final answer is less than the error on the first step. To suggest that we can approximate the overall error at $x = 6$ by adding up such errors as shown above for each of the 60 steps required to get from 0 to 6 is simply not sound.

In fact, for the equation (2.4), use of (2.3) with $h = 0.1$ gives

$$y_{n+1} = (0.9)\, y_n.$$

Using this 60 times would give an estimate for y at $x = 6$ of

$$(0.9)^{60} \cong 0.0017970.$$

This is in error by less than 28%. For a procedure in which the error on the first step was almost twice the total final answer, this is not bad.

By (2.3), we get

$$y_{n+1} \cong (1 - h)\, y_n \tag{2.6}$$

for the equation (2.4). This gives

$$
\begin{aligned}
y_{n+1} &\cong \left\{ e^{-h} - h^2/2! + h^3/3! - \cdots \right\} y_n \\
&= \left\{ e^{-h} - \frac{h^2}{2} \left(1 - h/3 + h^2/12 - \cdots \right) \right\} y_n \\
&= e^{-h} \left\{ 1 - \frac{h^2}{2} e^h \left(1 - h/3 + h^2/12 - \cdots \right) \right\} y_n.
\end{aligned}
$$

The factor

$$e^h(1 - h/3 + h^2/12 - \cdots) \tag{2.7}$$

tends to remain constant close to unity for $|h|$ small, since as h increases the first factor increases and the second decreases. So we replace (2.7) by unity, getting

$$y_{n+1} \cong e^{-h}\{1 - h^2/2\}\, y_n. \tag{2.8}$$

So at $x = 6$, we get

$$y = \left[e^{-h}\{1 - h^2/2\} \right]^{6/h}$$

$$= e^{-6}\left[\{1 - h^2/2\}^{-\frac{2}{h^2}} \right]^{-3h}$$

$$\cong e^{-6}e^{-3h}.$$

For $h = 0.1$, we get

$$y \cong e^{-6}\{0.74082\}.$$

Hence, by the above analysis, we expect the calculated value to be too low by about 26%: it is actually too low by about 28%.

The final formula,

$$y \cong e^{-6}e^{-3h} \tag{2.9}$$

shows that (for a reasonable range of h) the relative error is indeed of order h. Actually for $h = 0.1$ and the equation (2.4), we see by (2.2) that we have about 0.5% relative error at each step. Accumulating such relative errors for 60 steps can give an overall relative error of 30%, which is about what we got.

What happens here is that when we jump to another solution curve of (2.4) at the first step (and again and again at subsequent steps), these various solution curves approach each other strongly as x increases, so that what seemed a large absolute error at the first step fades to very little. It is the relative error that one must take account of here.

It is easy to fall into the fallacy that the total (or global) absolute error at the end is the sum of the absolute errors at each step. In fact, we can cite a text on numerical analysis which strongly suggests that this is the case. This points up two morals:

1.   Do not believe everything you read in a text on numerical analysis.

2.   Sometimes it can be the global RELATIVE error that is the sum of the relative errors at each step.

Of course, the second moral is seldom strictly true. Even for the present exposition, one must heed the first moral. However, as we showed above, the second moral points the way to proceed for equation (2.4) if one wishes to come anywhere near a good estimate of the global error.

For programmers who try to write general purpose differential equation solvers for large computers with error control built in, the question of how to handle error accumulation poses formidable difficulties. See Hull, et al. [15], pp. 607–608. When one is solving on a hand held calculator, one sees the progress of the solution, step by step. In regions where it is best to reckon by accumulating relative errors, one can do so. However, when it would be better to accumulate absolute errors (for example, if one is going to pass through a zero value of y), the change to accumulating absolute errors is easily made. Such flexibility is hard to arrange in a preset program for a large computer.

By (2.9), if we should wish to get to $x = 6$ with 0.1% accuracy, we should have to take h about $1/3000$. Thus we would require 18,000 steps. This would not be too bad on a large, fast computer, but on a hand held calculator it could require hours, especially if our equation to be solved were more complicated than (2.4). So we need something better than the Euler method.

We included the Euler method in the present discussion because, in spite of its simplicity, it introduces the reader to some of the important considerations in solving differential equations. One such is the importance of maintaining flexibility about whether one is accumulating relative errors or absolute errors. With a hand held calculator, such flexibility is easily maintained. With a preset program for a large, fast computer, such flexibility is hard to attain. Thus, in Hull, et al. [15], the decision was made to proceed only by accumulating absolute errors (see their p. 607). To get to $x = 6$ with a 0.1% error by this scheme would require far more than 18,000 steps. On the other hand, the large computers are so very fast that it is still a practical scheme.

It will occur to the reader that it would be helpful to have some general idea of how the solution is going to go in future steps, for instance to decide whether to accumulate absolute errors or relative errors. We will later discuss ways to "look ahead." One possible way is to take a few steps with the Euler method with large h. This would be very quick and would give a general idea of what one might expect to encounter.

3. A RUNGE-KUTTA METHOD

The text by Henrici [14] explains how to do many calculations on the HP-25 programmable calculator. On p. 182, he suggests that for solving the differential equation (1.2) one might use the second order Runge-Kutta method

$$y_{n+1} = y_n + \tfrac{1}{2}(k_1 + k_2), \tag{3.1}$$

where

$$k_1 = hf(x_n, y_n),$$

$$k_2 = hf(x_n + h, y_n + k_1).$$

Applied to

$$y' = ky, \tag{3.2}$$

this gives

$$y_{n+1} \cong \left(1 + hk + (hk)^2/2\right) y_n.$$

An analysis like that in the previous section shows that if we set $y = 1$ at $x = 0$, then at $x = X$ we will get approximately

$$y \cong e^{kX} e^{-h^2 k^3 X/6}. \tag{3.3}$$

The first factor on the right is what y should be, and the second factor shows about how far off we are from the true value. The error is proportional to h^2 for h in a reasonable range, so that we do have here a method of second order. To get down to specifics, let us take $k = -1$ (as in (2.4)) and $X = 6$, and ask for 0.1% accuracy. We need to take h about 1/32. Thus it will take 192 steps to get to $x = 6$. However (note (3.1)), each step requires TWO evaluations of the function $f(x, y)$. So we require 384 function evaluations. Actually, we require more evaluations than that if we are going to try to exercise some control over error.

We will now give explanations of two popular ways to try to exercise error control. This is done by properly adjusting the value of h; indeed it is sometimes advantageous to vary the step size from one step to the next (this can happen if the final y is all that is needed, rather than a table of y). In any case, tests should be made frequently to see if one is using the best value of h.

Such tests are made by working with the "local" or "step by step" error. This is defined as follows. There is a solution curve of (1.2) which passes through the point (x_n, y_n). This will not be the solution curve we are seeking unless we have miraculous luck in estimating y_n. If we set $x_{n+1} = x_n + h_n$, then the said solution curve will have a value y_{n+1} corresponding to x_{n+1}. However, our integration procedure will calculate an approximation $\hat{y}_{n+1}$ to go with x_{n+1}. The absolute local error will be

$$y_{n+1} - \hat{y}_{n+1}. \tag{3.4}$$

The relative local error will be

$$\frac{y_{n+1} - \hat{y}_{n+1}}{y_{n+1}}. \tag{3.5}$$

Suppose our path of integration is of length L, and we are going to cover this in a succession of steps of lengths h_i. Thus

$$L = \sum h_i. \tag{3.6}$$

If we are in a situation where the global absolute error is the sum of the local absolute errors, and we wish to insure that the global absolute error is bounded by a preassigned quantity, *tol*, it suffices to ensure that at the nth step the local absolute error is bounded by

$$\frac{h_n tol}{L}. \tag{3.7}$$

Clearly, the previous sentence holds just as well if we replace "absolute" throughout by "relative."

So, except for the sticky question whether we should be accumulating absolute errors or relative errors, we can say how much local error we can allow in a step of length h_n if we are to keep the global error bounded by *tol*.

By comparing (2.2) with (2.3), we see that the local absolute error in the Euler method is

$$\frac{h_n^2}{2} y''(\xi). \tag{3.8}$$

So, unless we are encountering a singularity of y'', we can certainly make the local absolute error as small as need be at any given step. Unless we are encountering a zero of y, we can equally make the local relative error as small as need be.

But usually we have little direct clue as to the size of $y''(\xi)$, so that (3.8) alone does not seem to be much help. Analogous to (2.2), there is a formula (see pp. 192–200 of Ralston [17]) that gives the absolute error of (3.1) as

$$h^3 K + h^4 L + \cdots, \tag{3.9}$$

where K, L, etc. are complicated functions of derivatives of y and partial derivatives of $f(x, y)$. For small h, we may say that the absolute error of (3.1) is about $h^3 K$. As we customarily take h rather small, it is customary to affirm in texts on numerical analysis that the absolute error, about $h^3 K$, of (3.1) is smaller than the absolute error, $h^2 y''(\xi)/2$, of (2.2). Obviously this fails in case we should have $|hK|$ greater than $|y''(\xi)/2|$, as can happen if we are near a zero of y''. We remind the reader of our first moral.

Nevertheless, it is standard doctrine to affirm that the absolute error of (3.1) is less than that of (2.2). Obviously the same is true for relative error, indeed, not merely less, but much less; so much so that by comparison with (2.2) we may consider (3.1) as giving the correct answer. It will then follow that the difference between (3.1) and (2.2) is the absolute error of (2.2). We can calculate the difference between (3.1) and (2.2) for a given h_n. Recall that the absolute error of (2.2) is (3.8). So, since we are dealing with a given h_n, we can calculate an approximation for $y''(\xi)$. If the error (absolute or relative) of (2.2) is not what we wish it to be for the given h_n, it is now trivial to choose a value of h_n that gives a local error of the desired size.

So there emerges a method for error control for the Euler method. After the estimation of y_n, we proceed to an estimation of y_{n+1} at $x = x_n + h_n$. We are called upon to do this with a local error bounded by (3.7). (Whether this local error is to be absolute or relative is a sticky question, always.) So we take an h_n and calculate y_{n+1} by each of (2.2) and (3.1). The difference is taken to be the absolute error in (2.2), namely (3.8). So we can calculate an approximation for $y''(\xi)$. Then it is easy to see what we should have taken h_n to be to accord with (3.7). We take it so, and calculate an approximation for y_{n+1}. Now we apply a similar procedure to get to y_{n+2}, then $y_{n+3}, \ldots$.

Actually, one would not usually have to make a blind guess for the first choice of h_n. If one has been monitoring the values of $y''(\xi)$ from step to step, one can make a fair guess what it will be for y_n. It is usual to choose h_n deliberately to make the anticipated local error for y_{n+1} a bit less than allowable. Then, when one checks the error with the h_n of one's choice, it seldom turns out too big, so that one seldom has to do the step over again.

Though the above program rests on the shaky assumption that (3.1) is far more accurate than (2.2), the program actually is likely to succeed fairly well. While (3.1) can be more in error than (2.2), this very rarely happens. Usually (3.1) will be appreciably better than (2.2). Of course, it is seldom so much better that one is really entitled to equate the error in (2.2) to the difference between (3.1) and (2.2), which we did to estimate $y''(\xi)$. However, it can often be that the error of (3.1) is, say, less than one fourth the error of (2.2).

Then one will only miscalculate the size of $y''(\xi)$ by 25% or less. If, as suggested above, we choose h_n deliberately to make the local error for y_{n+1} a bit less than allowable, we probably compensate for a moderate miscalculation of $y''(\xi)$. At least, the global error should not come out too much larger than *tol*. And, if we have taken *tol* a bit small, to be on the safe side, we are probably O.K. Error control is more of an art than a science.

For higher-order Runge-Kutta methods there is a variation, known as the Runge-Kutta-Fehlberg method, which uses an error control procedure based on essentially the principles outlined above. For details, see Burden, et al. [3], Section 6.5, pp. 249–258. This RKF method, with its error control procedure, has been widely adopted, and seems to work out well enough.

Actually, though the above procedure should work out well enough as an error control for the Euler method, we must remember that the Euler method, being only of first order, requires far too many steps to be practical, except possibly for look ahead. So, as we said above, something better is desired.

Although (3.1) could occasionally give a poorer result than (2.2), it usually gives a much better result. So why not follow the suggestion of Henrici [14] and use it? Well, what about error control? By (3.9), the local absolute error of (3.1) is about h^3K. But how do we estimate K? One way would be to try to copy the idea we explained above for the Euler method. Find a third order method, and hope it gives a result enough better than (3.1) that we may consider it essentially correct. Then we can calculate the absolute error of (3.1) for the given h_n, and so get an estimate for K. We will suggest an alternate procedure.

As we said, if we use (3.1) to go from y_n to y_{n+1}, the local absolute error is about h^3K. If we take two successive steps, from y_n to y_{n+2}, and if K does not change much from one step to the next, then we can accumulate a local absolute error of $2h^3K$ (over and above whatever error we had at y_n) in getting to y_{n+2}. Now apply (3.1), with $2h$ in place of h, to get from y_n to $\hat{y}_{n+2}$ in a single step. Assuming that K is not fluctuating badly, we will make an absolute error of about $(2h)^3K$ in getting to $\hat{y}_{n+2}$, or four times the error we made in getting to y_{n+2} in two steps. So now we have two approximations, of which $\hat{y}_{n+2}$ has an error about four times the two step y_{n+2}. From this, we can estimate that the two step y_{n+2} is in absolute error by about

$$(y_{n+2} - \hat{y}_{n+2})/3. \tag{3.10}$$

But, for a given h_n, the absolute error in the two step y_{n+2} is about $2h_n^3 K$. So we get an estimate for K. If the absolute error, (3.10), for the two step y_{n+2} is not small enough, we go back to y_n and use (3.1) two more times, using our estimate for K to choose a value for h_n that will make the error come out right. (In determining the correct value for h_n, do not forget that there are two steps from y_n to y_{n+2}, so that we must get a local error appropriate to the sum of errors for two steps.) See Hull, et al. [15], bottom of p. 616, for remarks about this method.

If (3.10) is really the absolute error of the two step y_{n+2}, why not add (3.10) to the two step y_{n+2} and get the true y_{n+2}? That is, set

$$y_{n+2}^{(\text{true})} = \frac{4y_{n+2} - \hat{y}_{n+2}}{3}. \tag{3.11}$$

Actually, this gives a third-order method for integrating (1.2), but based on two steps instead of one. And of course, it does not give the exact value of y_{n+2}, because (3.10) is only an approximation. We shall give some numerical examples later, and have included the result of using (3.11) for interest. It is not too bad. However, we have no reasonable way to exercise error control if we use (3.11) to determine y_{n+2} instead of the two step y_{n+2} (for which (3.10) is an estimate of the local absolute error). Also, we would have no improved value for y_{n+1}. So we recommend against use of (3.11), and suggest using (3.10) only to estimate the error of (3.1).

Suppose we have a system of two equations

$$y' = f(x, y, z) \tag{3.12}$$

$$z' = g(x, y, z), \tag{3.13}$$

being given the values of y and z for $x = x_0$. We easily generalize

(3.1) to handle this case. Set

$$y_{n+1} = y_n + \tfrac{1}{2}(k_1 + k_2), \tag{3.14}$$

$$z_{n+1} = z_n + \tfrac{1}{2}(l_1 + l_2), \tag{3.15}$$

where

$$k_1 = hf(x_n, y_n, z_n),$$

$$l_1 = hg(x_n, y_n, z_n),$$

$$k_2 = hf(x_n + h, y_n + k_1, z_n + l_1),$$

$$l_2 = hg(x_n + h, y_n + k_1, z_n + l_1).$$

Notice that if z is missing from $f(x, y, z)$, then (3.14) is the same as (3.1).

Error control is easy. For both (3.14) and (3.15) the absolute error is given in the form (3.9), except different K's, L's, etc. for y and z. We go from y_n to y_{n+2} in two steps of length h each, and from y_n to $\hat{y}_{n+2}$ in one step of length $2h$. Then the two step y_{n+2} is in absolute error by about (3.10), which gives an estimate for the K that goes with y. We proceed similarly for z_n.

Extension to a set of three or more equations is immediate.

Going back to (3.1), for the pair of steps of length h, we require two function evaluations per step. For the step of length $2h$, we also require two evaluations, except that the evaluation of $f(x, y)$ at $x = x_n$ has already been made. So, for the evaluation of y_{n+1} and y_{n+2}, with an estimate of error at x_{n+2}, we require altogether five function evaluations. (If the memory locations are too limited to store $f(x_n, y_n)$ while the other function evaluations are being made, we will require six function evaluations.) So for the 192 steps to get from $x = 0$ to $x = 6$, that is, 96 pairs of steps, we require 480 function evaluations. While this is a great improvement over the 18,000 function evaluations required by the Euler method, it could be rather time consuming if $f(x, y)$ is at all complex.

So we wish for something better. We cannot manage anything better on the HP-25. It has only 50 program steps, and implementation of (3.1) uses 39 of these, leaving only 11 program steps for the calculation of $f(x, y)$. In many cases, 11 program steps will not be adequate. So, if we are to do much with differential equations, we will need a more capable calculator than the HP-25.

4.   RUNGE-KUTTA METHODS OF HIGHER ORDER

Let us suppose that we have a calculator at least as capable as the HP-65. Many programmable calculators now on the market are appreciably more capable than the HP-65, but it sufficed for the calculations in this paper. With it, one can carry out higher-order Runge-Kutta methods than (3.1).

A convenient third-order method is given by

$$y_{n+1} = y_n + \tfrac{1}{6}(k_1 + 4k_2 + k_3), \tag{4.1}$$

where

$$k_1 = hf(x_n, y_n),$$

$$k_2 = hf\left(x_n + \frac{h}{2}, y_n + \frac{k_1}{2}\right),$$

$$k_3 = hf(x_n + h, y_n - k_1 + 2k_2);$$

see (5.6-46) on p. 199 of Ralston [17] or 25.5.8 on p. 896 of Abramowitz and Stegun [1].

The absolute error of (4.1) is given by

$$h^4 K + h^5 L + \cdots. \tag{4.2}$$

For small h, we can say that the absolute error is approximately $h^4 K$. We can take two steps of length h to get from y_n to y_{n+2} or a single step of length $2h$ to get from y_n to $\hat{y}_{n+2}$. Then the absolute error of the two step y_{n+2} should be about

$$(y_{n+2} - \hat{y}_{n+2})/7. \tag{4.3}$$

This enables us to get an estimate for K, so that we can control the local error. Although taking y_{n+2} to be

$$\frac{8y_{n+2} - \hat{y}_{n+2}}{7} \tag{4.4}$$

instead of the two step y_{n+2} will give a fourth-order method, we recommend against its use because there seems no convenient way to exercise error control. However, we shall tabulate some results of using (4.4), just for interest.

We trust it is obvious how to generalize (4.1) to a set of two or more equations. See (3.14) and (3.15).

If we wish to solve $y' = -y$ from 0 to 6 with an accuracy of 0.1%, we should take $h \leqslant 0.181712$. It will do well enough to take $h = 2/11$, so that we require 33 steps to get to $x = 6$. For (4.1), we require three function evaluations per step. To estimate the error after two steps, we need a single step of length $2h$. This also takes three evaluations, but one has been done before. So we require eight function evaluations for two steps. For the first 32 steps, namely, 16 pairs, that takes 128 evaluations. Then we have three evaluations for the 33rd step, for a total of 131 evaluations. This is quite an improvement over the 480 evaluations that we needed for the Runge-Kutta method of order two.

Can we do still better with a fourth-order Runge-Kutta method? The most popular fourth-order Runge-Kutta method is given by:

$$y_{n+1} = y_n + \tfrac{1}{6}(k_1 + 2k_2 + 2k_3 + k_4), \tag{4.5}$$

where

$$k_1 = hf(x_n, y_n),$$

$$k_2 = hf\left(x_n + \frac{h}{2}, y_n + \frac{k_1}{2}\right),$$

$$k_3 = hf\left(x_n + \frac{h}{2}, y_n + \frac{k_2}{2}\right),$$

$$k_4 = hf(x_n + h, y_n + k_3);$$

see (5.6-48) on p. 200 of Ralston [17] or 25.5.10 on p. 896 of Abramowitz and Stegun [1].

The absolute error of (4.5) is given by

$$h^5 K + h^6 L + \cdots . \qquad (4.6)$$

So, as with the lower-order Runge-Kutta methods, we can exercise error control by taking two steps of length h to get from y_n to y_{n+2} or a single step of length $2h$ to get from y_n to $\hat{y}_{n+2}$. The absolute error of the two step y_{n+2} should be about

$$\frac{y_{n+2} - \hat{y}_{n+2}}{15} . \qquad (4.7)$$

So we can estimate K. Taking y_{n+2} to be

$$\frac{16 y_{n+2} - \hat{y}_{n+2}}{15} \qquad (4.8)$$

would give a fifth-order method, but we recommend against its use. Besides lack of a reasonable method for error control when using (4.8), a tabulation which we shall give will disclose other shortcomings.

If it is not obvious how to generalize (4.5) to a set of two or more equations, see Conte and de Boor [4], pp. 365–366, or Abramowitz and Stegun [1], formula 25.5.18 on p. 897.

If we wish to solve $y' = -y$ from 0 to 6 with 0.1% accuracy, we must take h slightly less than $3/8$. Taking $h = 3/8$ is probably close enough, which would require 16 steps to get to $x = 6$. There are four function evaluations per step. To estimate the error of y_{n+2} after a pair of steps, we must also make a single step of length $2h$ from y_n to $\hat{y}_{n+2}$. This also takes four evaluations, but one has been done before. So we require 11 function evaluations for each pair of steps. So we need 88 function evaluations.

In Burden, et al., [3], on p. 254 is given a Runge-Kutta method of order four, set up for error control by the Fehlberg technique. It requires six function evaluations per step, including the error control. This is not quite as good as we do with (4.5).

The usual Runge-Kutta methods of order five require at least six evaluations per step. (See Zurmuhl [23], who gives the coefficients for Runge-Kuttas up to order ten.) This would make 17 evaluations per pair of steps if error control is exercised after each pair of steps. The Runge-Kutta-Fehlberg method of order five is slightly better, requiring eight evaluations per step, including error control. Actually, for the problem we have been considering, that of solving $y' = -y$ from 0 to 6 with 0.1% accuracy, if we are to improve on what we can do with (4.5) we must take h so large that the methods of error control become of very dubious validity. The higher-order Runge-Kutta-type methods are generally of use only if a considerably higher degree of accuracy is desired.

So it appears we cannot do much better with Runge-Kutta-type methods for the problem we have been considering than to use (4.5). As we saw, this requires 88 function evaluations.

Can we do better? In Enright and Hull [6] testing is reported of four sorts of methods:
1. Runge-Kutta-Fehlberg methods.
2. Rational extrapolation methods.
3. Adams-type predictor-corrector methods.
4. Milne-type predictor-corrector methods.

The Runge-Kutta-Fehlberg methods are Runge-Kutta-type methods improved according to ideas in Fehlberg, [7], [8], and [9], to give easier ways of estimating step by step errors. (See Burden, et al. [3], pp. 252–256.) This is purported to reduce somewhat the number of function evaluations. In view of certain advantages of Runge-Kutta methods, the Fehlberg improvements make the Runge-Kutta methods fairly competitive when the function evaluations can be done quickly. However, on p. 949 of Enright and Hull [6] some disadvantages with the Fehlberg methods are reported. By phone (March, 1978) T. E. Hull of Toronto informed the present writer that new improvements had eliminated certain of these disadvantages, and that IMSL had recently embodied these new improvements in its Runge-Kutta software package, DVERK. Actually, these disadvantages arise only when one is using a Runge-Kutta-Fehlberg method of order five or higher. It is unlikely that one would be using an order this high on a hand held calculator. More to the point, these disadvantages arise only when the $f(x, y)$ of (1.2) does not depend on y. While such a circumstance could fail to be taken account of in a prepackaged ODE solver for a large computer, it would be noted immediately by an individual using a hand held calculator. He would then use an approximate quadrature rule (see Abramowitz and Stegun [1], p. 885 ff.) rather than one of the types of methods listed above.

However, at best the Runge-Kutta-Fehlberg methods are competitive only when the function evaluations can be done quite quickly.

The rational extrapolation methods derive from ideas of Gragg [10] and were developed carefully in Bulirsch and Stoer [2]. In Enright and Hull [6] these methods are given a fairly good rating when the function evaluations can be done quickly. They seem complex to program for a hand held calculator, and so we will say no more about them.

Enright and Hull [6] give their highest ratings to certain Adams-type predictor-corrector methods. However, their conclusions do not necessarily hold for hand held calculators because of the very limited number of memory locations of hand held calculators. So we will make a study of Adams-type and Milne-type predictor-corrector methods for hand held calculators.

Incidentally, one cannot merely take one of the programs tested in Enright and Hull [6] and put it on his hand held calculator. In order to accommodate the wide diversity of vagaries that can arise in solutions of differential equations, these programs have a large amount of "overhead", and are beyond the capabilities of present hand held calculators, besides being so long as to be very time consuming at the slow speed of hand held calculators. However, unless one has a very large system of equations to solve, one can proceed step by step on a hand held calculator. One can "look ahead" as an aid to planning the calculation, one can monitor the accumulation of errors as one goes, and one should be alert for idiosyncracies that might arise. If the latter do arise, one can try shifting methods; if worse comes to worst, one can try a change of variables, or more complex stratagems.

5. PREDICTOR-CORRECTOR METHODS

The classical predictor-corrector method is that of Milne (see (28.1) and (28.2) on p. 65 of Milne [16], or (5.5-12) on p. 182 of

Ralston [17]):

$$y_{n+1}^{(p)} = y_{n-3} + \frac{4h}{3}\left(2y_n' - y_{n-1}' + 2y_{n-2}'\right), \tag{5.1}$$

$$y_{n+1} = y_{n-1} + \frac{h}{3}\left(f\left(x_{n+1}, y_{n+1}^{(p)}\right) + 4y_n' + y_{n-1}'\right). \tag{5.2}$$

This uses two function evaluations per step. There is the obvious one in (5.2), and when one comes to use (5.1) for the next step, one needs y_{n+1}', which calls for a second function evaluation. Incidentally, it is required that

$$x_{n+1-i} - x_{n-i} = h \quad \text{for } i = 0, 1, 2, 3.$$

To use this method, we must have y_{n-3}, y_{n-1}, y_{n-2}', y_{n-1}', and y_n' stored, and for the next step we will have to have y_{n-2} and y_n. So our storage requirements amount to four values of y and three of y'. In addition, one has to carry the current values of x_n and h. This uses up nine memory locations, which are all there on the HP-65. This leaves no memory locations to use in evaluating $f(x, y)$. So use of the classical Milne predictor-corrector would often not be possible on the HP-65. For calculators with more memory locations, the large requirement for memory locations could preclude use of the Milne predictor-corrector if one wishes to solve a system of even as few as three equations. A more compelling reason not to use the Milne method is that it is now known to be unstable in certain circumstances which arise not too infrequently. We shall produce predictor-corrector methods that have much more modest memory requirements and are stable besides. Meanwhile the Milne-predictor-corrector given by (5.1) and (5.2) will be used to illustrate typical features of all predictor-corrector methods.

As the name would indicate, a predictor-corrector method embodies a predictor and a corrector. The predictor is (5.1) and the corrector is

$$y_{n+1}^{(c)} = y_{n-1} + \frac{h}{3}\left(y_{n+1}' + 4y_n' + y_{n-1}'\right), \tag{5.3}$$

which is nothing more than Simpson's rule for integrating y' approximately. The obvious disadvantage of (5.3) is that one needs the value of y'_{n+1} to get the value of $y^{(c)}_{n+1}$, whereas by (1.2) one needs the value of $y^{(c)}_{n+1}$ to get the corresponding value of y'_{n+1}. Actually, if h is reasonably small and $f(x, y)$ is reasonably well behaved, one can get out of this impasse by an iteration scheme; after making a guess for y'_{n+1}, successively substitute the current y'_{n+1} into (5.3) to get a $y^{(c)}_{n+1}$ and substitute $y^{(c)}_{n+1}$ into (1.2) to get a better value for y'_{n+1}. In the usual case that will arise, this will converge to a y'_{n+1} and $y^{(c)}_{n+1}$ that satisfy both (5.3) and (1.2). Unfortunately, it is prodigal with function evaluations.

So we adopt a compromise. A predictor is given, in this case (5.1), which produces a reasonably close approximation for y_{n+1}. This is substituted into (1.2) to get a guess for y'_{n+1}. This is then substituted into the corrector; the net result is embodied in (5.2). There the process is stopped. We do not have as much accuracy as we can get with a few more iterations, but we have held the total number of function evaluations to two per step.

Since we need four values of y to use the predictor (and at equally spaced values of x), these four values must somehow be obtained before we can start to use this process. On pp. 61–64 of Milne [16] a scheme is given to get four starting values. It is now generally agreed that the best way to get started is to calculate y_1, y_2, and y_3 by Runge-Kutta.

One attractive feature of predictor-corrector methods is the ease with which one gets an estimate of the step by step error. The error for (5.1) is

$$\frac{14}{45} h^5 y^{(5)}(\xi) \tag{5.4}$$

and that for (5.3) is

$$-\frac{1}{90} h^5 y^{(5)}(\xi), \tag{5.5}$$

provided that one has values of y'_{n+1} and $y^{(c)}_{n+1}$ that satisfy both (5.3) and (1.2). While the values of ξ in (5.4) and (5.5) will scarcely ever be the same, one should find by taking h small enough that $y^{(5)}(\xi)$ will seldom change radically for ξ between x_n and x_{n+1}. So it would appear likely that $y_{n+1}^{(p)}$ is about 28 times as far from the true y_{n+1} as $y_{n+1}^{(c)}$, and on the opposite side. So the error in (5.3) will be about

$$-\frac{1}{29}\left\{ y_{n+1}^{(c)} - y_{n+1}^{(p)} \right\}. \tag{5.6}$$

Actually, the value of y_{n+1} given by (5.2) is close enough to the $y_{n+1}^{(c)}$ that one would get by iterating with (5.3) that the error in y_{n+1} from (5.2) is approximately what one would get by using y_{n+1} from (5.2) in place of $y_{n+1}^{(c)}$ in (5.6). That is, one is seldom very far off in saying that the error of y_{n+1}, as given by (5.2), is about

$$-\frac{1}{29}\left\{ y_{n+1} - y_{n+1}^{(p)} \right\}, \tag{5.7}$$

using the value of y_{n+1} from (5.2).

Since one needs four consecutive values of y at equal step sizes for the Milne method, a change of step size (should it be required) is not easy. If one has as many as seven preceding consecutive steps of equal length, one can double the step size by picking every other value from the present and preceding y's. However this would require recovering the values of y_{n-6}, y_{n-4}, and y'_{n-4} besides the values that one usually stores. Or one could carry on for three more steps before doubling, being careful to save the key values.

If one wishes to halve the step size, one can use

$$y_{n-1/2} = \frac{1}{128}\left\{ 45y_n + 72y_{n-1} + 11y_{n-2} \right.$$

$$\left. + h\left(-9y'_n + 36y'_{n-1} + 3y'_{n-2} \right) \right\} \tag{5.8}$$

$$y_{n-3/2} = \frac{1}{128}\left\{ 11y_n + 72y_{n-1} + 45y_{n-2} \right.$$

$$\left. - h\left(3y'_n + 36y'_{n-1} - 9y'_{n-2} \right) \right\} \tag{5.9}$$

(see p. 208 of Hamming [12] or (A57) and (A58) on p. 451 of Rosser [18]).

Unfortunately, doubling or halving the step size is often not the most efficient change to make. For a change of a different size, one can use interpolation formulas, of which (5.8) and (5.9) are examples, but it might be simpler just to make a fresh start, generating the next three values of y by Runge-Kutta.

6. ADAMS PREDICTOR-CORRECTORS

For the Adams method of order r, the predictor (which is commonly called an Adams-Bashforth formula) is

$$y_{n+1}^{(p)} = y_n + h \sum_{i=0}^{r-1} \alpha_i y_{n-i}' + K_p h^{r+1} y^{(r+1)}(\xi), \qquad (6.1)$$

and the corrector (which is commonly called an Adams-Moulton formula) is

$$y_{n+1}^{(c)} = y_n + h \sum_{i=0}^{r-1} \beta_i y_{n+1-i}' - K_c h^{r+1} y^{(r+1)}(\xi). \qquad (6.2)$$

The h that appears is

$$h = x_{n+1-i} - x_{n-i},$$

which is required to be the same for $i = 0, \ldots, r-1$.

The error term on the right of (6.2) is based on the assumption that

$$y_{n+1}' = f(x_{n+1}, y_{n+1}^{(c)}). \qquad (6.3)$$

As indicated with (5.3), values of $y_{n+1}^{(c)}$ and y_{n+1}' that satisfy both (6.2) and (6.3) can usually be found by an iterative process. That is, if one chooses a $\bar{y}_{n+1}$ that is near the limiting $y_{n+1}^{(c)}$, and then forms

$$\bar{y}_{n+1}' = f(x_{n+1}, \bar{y}_{n+1}),$$

and substitutes this $\bar{y}_{n+1}'$ for y_{n+1}' on the right of (6.2), one will get a value nearer to $y_{n+1}^{(c)}$ than $\bar{y}_{n+1}$. Repetition of this converges to $y_{n+1}^{(c)}$. However, this is costly in function evaluations, and what is mostly done is to define

$$y_{n+1}^{(t)} = y_n + h\beta_0 f\left(x_{n+1}, y_{n+1}^{(p)}\right) + h \sum_{i=1}^{r-1} \beta_i y_{n+1-i}', \qquad (6.4)$$

after which one takes $y_{n+1} = y_{n+1}^{(t)}$. This holds the number of function evaluations to two per step. (One may think of the superscript t as standing for "traditional.")

The requirements for memory locations are one for y_n, r for y_{n-i}' $(i = 0, \ldots, r-1)$, one for x_n, and one for h. Actually, one can reduce the memory requirements to one fewer for most calculators. Suppose we are at x_n, and have y_n, but have not yet calculated y_n'. Let us have y_{n-i}' stored at memory location i, for $i = 1, \ldots, r-1$, with y_n at location r, x_n at location $r+1$, and h at location $r+2$. We accumulate

$$\sum_{i=1}^{r-1} \alpha_i y_{n-i}', \qquad (6.5)$$

simultaneously with moving y_{n-i}' to location $i+1$ for $i = r-2$, $r-3, \ldots, 1$. This is done as follows. Put y_{n-r+1}' into the display, and multiply by α_{r-1}. Then $\alpha_{r-1} y_{n-r+1}'$ is in the display.

Consider a reverse polish calculator. When

$$\sum_{i=s}^{r-1} \alpha_i y_{n-i}' \qquad (6.6)$$

is in the display (with $s > 1$), bring y_{n-s+1}' into the display. This pushes (6.6) up into the y-register. Store y_{n-s+1}' into location s. Multiply by α_{s-1}, and add. Then

$$\sum_{i=s-1}^{r-1} \alpha_i y_{n-i}' \qquad (6.7)$$

is in the display. Proceed this way until (6.5) is in the display. Store it in location 1. Now calculate

$$y_n' = f(x_n, y_n),$$

finishing the calculation with y_n' in the display. Bring (6.5) into the display, interchange the x-register and y-register, and store y_n' in location 1. We now have y_n' in the x-register and (6.5) in the y-register. Multiply by α_0 and add. Then bring h into the display. Add it to x_n in location $r + 1$, by register arithmetic. Now multiply and add y_n. By (6.1), this gives $y_{n+1}^{(p)}$. Calculate $f(x_{n+1}, y_{n+1}^{(p)})$, finishing with this quantity in the display. One can then easily calculate $y_{n+1}^{(t)}$ by (6.4), and store it in location r. We are now ready to proceed to the next step of the integration.

Or consider an algebraic entry calculator. When (6.6) is in the display (with $s > 1$), press the $+$ key and the left parenthesis key. (If the calculator is hierarchical, one need not press the left parenthesis key.) Bring y_{n-s+1}' into the display. Store y_{n-s+1}' into location s. Press the x key, and bring α_{s-1} into the display. Press the right parenthesis key. (Omit this if the calculator is hierarchical.) Then press the $=$ key. Then (6.7) will be in the display. Proceed this way until (6.5) is in the display. Store it in location 1. Now calculate

$$y_n' = f(x_n, y_n),$$

finishing the calculation with y_n' in the display. We now wish to interchange y_n' and (6.5). For this, add y_n' into location 1 by register arithmetic. Then subtract the contents of location 1 from what is in the display. Then add the contents of the display into location 1 by register arithmetic. Finally, change the sign of the quantity in the display. (This will inflict round off errors on both y_n' and (6.5). It might be desirable to avoid this. "Borrow" a couple of extra locations, p and q. Store y_n' in p. Call (6.5) into the display, and then store it in q. Call y_n' into the display from p and then store it into location 1. Finally, call (6.5) into the display from q. Alternatively, suppose the calculator has the capability to interchange the contents of the display and location m. There will not be anything of importance in location m, since this must be left free in case such interchanges are necessary in calculating $f(x_n, y_n)$. Store y_n' into location m. Now bring (6.5) into the display from location 1.

Interchange the display and location m. Store y_n' into location 1 from the display. Then interchange the display and location m again.) Now, with (6.5) in the display and y_n' in location 1, we can and do add $\alpha_0 y_n'$ to (6.5). Press the x key. Bring h into the display. Add it to x_n in location $r + 1$, by register arithmetic. Press the = key. Then add y_n, which gives $y_{n+1}^{(p)}$, by (6.1). Calculate $f(x_{n+1}, y_{n+1}^{(p)})$, finishing with this quantity in the display. One can then easily calculate $y_{n+1}^{(t)}$ by (6.4), and store it in location r. We are now ready to proceed to the next step of the integration.

In the above, it is assumed that the α's and β's are part of the program. The α's and β's are fairly simple numbers, so that this does not seem to overload the program, unless one has a very large value of r, in which case one hopes that additional memory locations will be available to store the α's and β's.

The Adams formulas of any order, with an error term, can be derived by use of the Newton backward difference formula. Details are given on pp. 340–342 and 350–351 of Conte and de Boor [4] where the Adams method of order 4 is derived. A rather diffuse explanation is scattered through a number of pages of Milne [16], but on p. 50 is a table from which the coefficients can be calculated up to order 9. Henrici [13] gives coefficients of the predictors on p. 194 and of the correctors on p. 199, both up to order 6. Neither the tables of Milne nor of Henrici give the error terms, but these can be derived easily by taking $y = x^{r+1}$ in (6.1) and (6.2), which will give the values of K_p and K_c. Correctors up to order 9, with error terms, can be got by a trivial modification of formulas (A2), (A4), (A7), (A12), (A20), (A26), (A34), and (A42) on pp. 446–449 of Rosser [18].

In order to get started with the integration, we need the r values $y_0, y_1, \ldots, y_{r-1}$.

7. ADAMS OF ORDER 2

We have the predictor and corrector

$$y_{n+1}^{(p)} = y_n + \frac{h}{2}\left(3y_n' - y_{n-1}'\right) + \frac{5h^3}{12} y'''(\xi) \qquad (7.1)$$

$$y_{n+1}^{(c)} = y_n + \frac{h}{2}\left(y_{n+1}' + y_n'\right) - \frac{h^3}{12} y'''(\xi). \qquad (7.2)$$

J. Barkley Rosser

We recognize (7.2) as the trapezoid rule. As noted for the general case, we define

$$y_{n+1}^{(t)} = y_n + \frac{h}{2}\left(f\left(x_{n+1}, y_{n+1}^{(p)}\right) + y_n'\right) \tag{7.3}$$

and take $y_{n+1} = y_{n+1}^{(t)}$.

As y_0 is given, we need to obtain a value only for y_1 to get started. If one wishes to avoid use of Runge-Kutta, the value of y_1 can be obtained by iterating with (7.2). The same applies if one needs to make a restart after changing the length of the step.

While the values of ξ in the error terms of (7.1) and (7.2) will scarcely ever be the same, if y''' is slowly varying then the absolute error of $y_{n+1}^{(c)}$ will be about one fifth that of $y_{n+1}^{(p)}$ if we have iterated on (7.2) until both (7.2) and (6.3) are satisfied. In practice, the value of $y_{n+1}^{(t)}$ is close enough to this limiting value of $y_{n+1}^{(c)}$ that one can say that the error of $y_{n+1}^{(t)}$ is about one fifth that of $y_{n+1}^{(p)}$. In other words the step by step absolute error of the new y_{n+1} is about

$$-\tfrac{1}{6}\left(y_{n+1}^{(t)} - y_{n+1}^{(p)}\right). \tag{7.4}$$

If it were not so, then one is likely using too large a value of h.

Let us stress again that it is advisable to have an estimate for the error, step by step. There seems no way to use (7.4) for this purpose except by allocating an additional memory location to hold the value of $y_{n+1}^{(p)}$ until the value of $y_{n+1}^{(t)}$ is calculated. Even if one is very short on memory locations, this should be done. To run through an integration without some estimates of the local errors one is incurring is courting disaster. If there absolutely is not an extra memory location available, arrange the program so that it stops part way through each step with $y_{n+1}^{(p)}$ in the display. Then copy down $y_{n+1}^{(p)}$. Of course, if the calculator has a print out attachment, one can have both $y_{n+1}^{(p)}$ and $y_{n+1}^{(t)}$ printed out.

If we take (7.4) as the absolute error in $y_{n+1}^{(t)}$, then since $y_{n+1}^{(t)}$ and $y_{n+1}^{(c)}$ are nearly equal, we can use (7.2) to estimate $y'''(\xi)$. Then, if the error (7.4) is not sufficiently small, we can use (7.2) to determine how much smaller h should be. Then we have to start over with the smaller h. It is probably good strategy to keep h a bit smaller than necessary. But do not take h excessively small, since this will greatly increase the number of steps needed.

One is tempted to try for more accuracy by defining

$$y_{n+1}^{(e)} = y_{n+1}^{(t)} - \tfrac{1}{6}\left(y_{n+1}^{(t)} - y_{n+1}^{(p)}\right), \qquad (7.5)$$

and then taking y_{n+1} to be $y_{n+1}^{(e)}$. (We use the superscript e to denote "extrapolated".) This does indeed seem to give more accuracy, mostly. If one would use

$$y_{n+1}^{(c)} - \tfrac{1}{6}\left(y_{n+1}^{(c)} - y_{n+1}^{(p)}\right),$$

with the limiting value of $y_{n+1}^{(c)}$, one would have a third-order method. See Ralston [17] p. 186. As $y_{n+1}^{(t)}$ is close to $y_{n+1}^{(c)}$, one has close to a third-order method. So one should have more accuracy, but one has no way to tell how much more; there is no way to estimate the step by step error. As we shall shortly show, numerical examples disclose a somewhat erratic behavior of $y_{n+1}^{(e)}$. Indeed, use of $y_{n+1}^{(t)}$ throughout gives better results sometimes than use of $y_{n+1}^{(e)}$ throughout.

This is disconcerting. It seems to indicate that, at least for some steps, use of (7.5) is not as good as just using $y_{n+1}^{(t)}$. If this is so, it would seem to indicate that (7.4) must be a poor estimate of the absolute error. This is not necessarily serious. If the absolute error is less than half of (7.4) (in absolute value), then use of (7.5) would put one further from the true answer than use of $y_{n+1}^{(t)}$.

In short, while one should not put too much confidence in (7.4) as an estimate of the absolute error of $y_{n+1}^{(t)}$, as an indication of the order of magnitude of the absolute error, it seems to do fairly well. It has been used in this way for a generation, with what seems to have been reasonable success.

The doctrine on use of $y_{n+1}^{(e)}$ is not clear. In Ralston [17] on page 186, such use is not favored. Indeed, it is there stated that it affects the stability properties, and this is also affirmed on p. 210 in Hamming [12]. However, in Hamming [12], there is advocacy of more complicated schemes which make use of an equivalent of $y_{n+1}^{(e)}$. For the Adams methods, which are strongly stable, the present writer has found no evidence that use of $y_{n+1}^{(e)}$ causes stability to deteriorate, though it does indeed for some other types of predictor-corrector methods. However, we strongly recommend against use of $y_{n+1}^{(e)}$ because there is no reasonable way to estimate the step by step error. Granted that (7.4) is only an indication as to the size of the error for $y_{n+1}^{(t)}$, it has a long record of reasonable success for that purpose.

Since we have brought up the question of stability, we should note that usage of the term is not uniform. Many people, following Dahlquist [5], say that stability requires that all roots of a certain difference equation should be less than unity in absolute value. In Hamming [12] on p. 191, it is pointed out that according to that definition no method for solving $y' = ky$ could be stable for $k > 0$. So Hamming [12] introduces (on p. 191) the notion of relative stability, in which roots can be larger than unity in absolute value provided they are less in absolute value than a certain selected root. In Definition 5.2 on p. 176 of Ralston [17], we find Ralston taking this as the definition of stability; what Hamming calls "relative stability" is called "stability" by Ralston. We concur with Ralston in our use of the word "stable." If a method is stable in this sense, one can safely carry out a numerical integration of indefinite extent by means of it.

It turns out that to achieve stability for predictor-corrector methods, one must maintain certain bounds for

$$h \frac{\partial f(x, y)}{\partial y}. \tag{7.6}$$

The same is claimed to be true for some Runge-Kutta methods. See Shampine and Watts [20] p. 270. Writers of texts on numerical analysis have been very lax about calling this to the attention of their readers, and most people probably harbor the illusion that Runge-Kutta methods are stable under all circumstances. Fortunately, the Runge-Kutta's of orders two, three, and four given in this paper happen to be relatively stable (in the sense of Hamming) in all circumstances, and hence give no trouble about stability. Far and away most people who use a Runge-Kutta use one of these three, which is probably why very few cases of instability have been encountered when using Runge-Kutta.

A discussion of why one needs to set bounds for (7.6) to avoid instability is given in Hamming [12] on pp. 189–190 and in Ralston [17] on pp. 169–178.

Certainly, when one is using predictor-corrector methods, one should make routine estimates of $\partial f(x, y)/\partial y$ to check whether (7.6) is remaining in bounds. A numerical example of what can happen if one fails to do this will be given in Section 10. If one has two values y and $y + \varepsilon$, and ε is fairly small, then we get a good estimate by

$$\frac{\partial f(x, y)}{\partial y} \cong \frac{f(x, y + \varepsilon) - f(x, y)}{\varepsilon} \tag{7.7}$$

For the calculations proposed, we can take $x = x_{n+1}$, $y + \varepsilon = y_{n+1}^{(t)}$ and $y = y_{n+1}^{(p)}$. Thus we can get our estimate by (7.7) without having to perform any additional function evaluations, since $f(x_{n+1}, y_{n+1}^{(t)})$ must be calculated to provide the value of y'_{n+1} required in the predictor for the next step.

It does appear that we need still another memory location to calculate (7.7), in addition to the one called for at (7.4). With our suggested values for $y + \varepsilon$ and y, we have ε equal to -6 times (7.4). To get (7.4) (which we certainly should calculate), when we get to $y^{(t)}$ in the calculation, we first subtract $y^{(t)}$ from $y^{(p)}$, which we have stored in memory location s (if such is available; if not, we have written it on a piece of paper), which we do by register arithmetic. Then we divide by 6, giving us (7.4), located at memory location s. When we get to the calculation of $f(x_{n+1}, y_{n+1}^{(p)})$ (which comes before the calculation of $y^{(t)}$), we store this value in memory location t (if such is available; otherwise we write it on a piece of paper). When we pass the calculation of $y^{(t)}$, we are starting on the next step, which could involve a new value of h_n (usually not, since we try to keep h_n constant, as far as possible). So, before we change h_n, we multiply (7.4), which is located in memory location s (or written on a piece of paper) by 6 and divide by h_n; we put this in memory location s (or write it on a piece of paper). During the next step, when we finally get to the calculation of y'_{n+1}, which is $f(x_{n+1}, y^{(t)})$, we subtract this y'_{n+1} from $f(x_{n+1}, y_{n+1}^{(p)})$ (which we have stored in memory location t, or written on a piece of paper). This gives the negative of the numerator of (7.7). Dividing by what we have stored in location s (or written on a piece of paper) will give us an approximation to (7.6).

In an effort to get a bit more accuracy without requiring more function evaluations, Hamming devised various modifications of predictor-corrector methods. These won some favor. Selected ones are recommended in Hamming [11], Hamming, [12], and Ralston [17]. They are considered with care in Rosser [19] where various examples are given of the results of using them. They complicate the procedures considerably. They require allocation of an additional memory location. With them it is harder to get started or to change the step size. Some of them provide no estimate of the step by step error. Worst of all, for perhaps a third of the examples worked in Rosser [19] they give poorer results than simply using $y_{n+1}^{(t)}$, as is proposed above. Even when they give better results the improvement is rarely very marked. So it does not seem worthwhile to present them.

In Table 1, we present the results of some calculations using the procedures given above. The entries in the table are the relative errors at $x = 6$. This has been done for the three equations $y' = y$, $y' = -y$, and $y' = -2xy^2$, all with $y(0) = 1$, and for three cases $h = 0.1$, 0.2, and 0.3. Needless to say, for $y' = -2xy^2$, the solution is $y = (1 + x^2)^{-1}$. The rows labelled R-K in Table 1 give the results

TABLE 1
Relative error at $x = 6$ for some second-order methods.

		$y' = y$	$y' = -y$	$y' = -2xy^2$
$h = 0.1$	R-K	9.24×10^{-3}	-1.08×10^{-2}	-1.52×10^{-3}
	R-Ke	8.86×10^{-4}	1.13×10^{-3}	1.10×10^{-4}
	t	-3.65×10^{-3}	6.65×10^{-3}	6.12×10^{-4}
	e	7.10×10^{-4}	8.84×10^{-4}	8.91×10^{-5}
$h = 0.2$	R-K	3.39×10^{-2}	-4.76×10^{-2}	-6.56×10^{-3}
	R-Ke	6.26×10^{-3}	1.01×10^{-2}	7.05×10^{-4}
	t	-1.02×10^{-2}	3.46×10^{-2}	3.21×10^{-3}
	e	5.09×10^{-3}	7.90×10^{-3}	7.63×10^{-4}
$h = 0.3$	R-K	6.96×10^{-2}	-1.19×10^{-1}	-1.60×10^{-2}
	R-Ke	1.89×10^{-2}	3.78×10^{-2}	1.22×10^{-3}
	t	-1.47×10^{-2}	9.89×10^{-2}	9.47×10^{-3}
	e	1.54×10^{-2}	2.98×10^{-2}	2.81×10^{-3}

of the Runge-Kutta of order two that we discussed earlier, namely (3.1). The rows labelled R-Ke give the results of using (3.11). The rows labelled t or e refer to the cases where $y_{n+1}^{(t)}$ or $y_{n+1}^{(e)}$ are used for y_{n+1} (see (7.3) and (7.5)).

Reference to Table 1 will verify some comments which were made before. The behavior of $y_{n+1}^{(e)}$ is distinctly erratic, the most extreme case being for $y' = y$ and $h = 0.3$, where $y_{n+1}^{(e)}$ gives a poorer result than $y_{n+1}^{(t)}$.

To get started, we calculated y_1 and y_2 from the known solutions of the equations. Table 1 is condensed from a table in Rosser [19]. This contained rows for some of the Hamming modifications which we are not considering here. To get started with them, we needed three starting values, y_0, y_1, and y_2.

If we use these methods to solve $y' = ky$, we get stability in the ranges shown in Table 2. This means that for stability the bounds shown in Table 2 must be satisfied by (7.6). The fact that one has stability for all positive values of hk is surprising. This fact is not particularly useful, since for large hk the step by step errors would be so large as to render the method of little value.

Observe that in Table 2, use of extrapolated values does not diminish the region of stability. If anything, the reverse is true.

To show how the Adams method would work for a system of equations, we give the details when there are two equations:

$$y' = f(x, y, z), \tag{7.8}$$

$$z' = g(x, y, z). \tag{7.9}$$

We set

$$y_{n+1}^{(p)} = y_n + \frac{h}{2}\left(3y_n' - y_{n-1}'\right), \tag{7.10}$$

$$z_{n+1}^{(p)} = z_n + \frac{h}{2}\left(3z_n' - z_{n-1}'\right), \tag{7.11}$$

$$y_{n+1}^{(t)} = y_n + \frac{h}{2}\left(f\left(x_{n+1}, y_{n+1}^{(p)}, z_{n+1}^{(p)}\right) + y_n'\right), \tag{7.12}$$

$$z_{n+1}^{(t)} = z_n + \frac{h}{2}\left(g\left(x_{n+1}, y_{n+1}^{(p)}, z_{n+1}^{(p)}\right) + z_n'\right). \tag{7.13}$$

We then take y_{n+1} and z_{n+1} to be $y_{n+1}^{(t)}$ and $z_{n+1}^{(t)}$, respectively.

TABLE 2

Stability ranges for $y' = ky$ for Adams second order.

t	$-0.6 \leqslant hk$	e	$-0.8 \leqslant hk$

To achieve error control, we take (7.4) as giving the local absolute error of $y_{n+1}^{(t)}$. Similarly for $z_{n+1}^{(t)}$. Both (7.6) and a similar formula for z must stay within the ranges indicated in Table 2.

8. ADAMS OF ORDER 3

The third-order Adams is given by

$$y_{n+1}^{(p)} = y_n + \frac{h}{12}\left(23y_n' - 16y_{n-1}' + 5y_{n-2}'\right) + \frac{9h^4}{24}y^{(4)}(\xi), \quad (8.1)$$

$$y_{n+1}^{(c)} = y_n + \frac{h}{12}\left(5y_{n+1}' + 8y_n' - y_{n-1}'\right) - \frac{h^4}{24}y^{(4)}(\xi). \quad (8.2)$$

Analogously to the second-order Adams, we set

$$y_{n+1}^{(t)} = y_n + \frac{h}{12}\left(5f\left(x_{n+1}, y_{n+1}^{(p)}\right) + 8y_n' - y_{n-1}'\right), \quad (8.3)$$

$$y_{n+1}^{(e)} = y_{n+1}^{(t)} - \tfrac{1}{10}\left(y_{n+1}^{(t)} - y_{n+1}^{(p)}\right). \quad (8.4)$$

From the error terms given in (8.1) and (8.2), we judge that

$$-\tfrac{1}{10}\left(y_{n+1}^{(t)} - y_{n+1}^{(p)}\right) \quad (8.5)$$

is a reasonable estimate for the absolute error of $y_{n+1}^{(t)}$. Hence one is tempted to think that $y_{n+1}^{(e)}$ would be a better candidate for y_{n+1} than $y_{n+1}^{(t)}$. This is not so, for reasons analogous to those for the Adams of order two. By all means, take y_{n+1} to be $y_{n+1}^{(t)}$.

Some numerical results are given in Table 3. The rows labelled R-Ke give the results of using (4.4). This theoretically produces a method of order 4. What this means is that the leading term in its error (analogous to (4.2)) is a multiple of h^5. However, there are other terms, which can predominate if h is not small enough. In the

TABLE 3
Relative error at $x = 6$ for some third-order methods.

		$y' = y$	$y' = -y$	$y' = -2xy^2$
$h = 0.1$	R-K	2.31×10^{-4}	2.71×10^{-4}	3.31×10^{-5}
	R-Ke	2.01×10^{-5}	-2.59×10^{-5}	-5.51×10^{-6}
	t	-1.50×10^{-4}	-3.80×10^{-4}	-2.06×10^{-5}
	e	5.90×10^{-5}	-7.98×10^{-5}	-9.36×10^{-7}
$h = 0.2$	R-K	1.70×10^{-3}	2.35×10^{-3}	3.03×10^{-4}
	R-Ke	2.85×10^{-4}	-4.71×10^{-4}	-1.34×10^{-4}
	t	-5.96×10^{-4}	-4.38×10^{-3}	-1.27×10^{-4}
	e	8.14×10^{-4}	-1.49×10^{-3}	9.04×10^{-6}
$h = 0.3$	R-K	5.30×10^{-3}	8.56×10^{-3}	1.19×10^{-3}
	R-Ke	1.28×10^{-3}	-2.71×10^{-3}	-1.10×10^{-3}
	t	-4.58×10^{-4}	-2.06×10^{-2}	-3.48×10^{-5}
	e	3.56×10^{-3}	-8.84×10^{-3}	2.52×10^{-4}

rows labelled R-Ke, the errors for $h = 0.2$ should theoretically be about 16 times those for $h = 0.1$, and the errors for $h = 0.3$ should theoretically be about 81 times those for $h = 0.1$. This is far from being the case. Furthermore, for the equation $y' = -2xy^2$ and $h = 0.3$, the purported fourth-order method is scarcely better than the third-order method. For the Adams methods, one should note how erratic the e values are; also, in three cases they are poorer than the t values.

The column headed $y' = -2xy^2$ is quite erratic. This bears out the remark at the top of p. 210 in Hamming, [12], that the corresponding solution is often troublesome to approximate by polynomials.

For the equation $y' = ky$, we get stability in the ranges shown in Table 4. These bounds should be satisfied by (7.6). For the case $h = 0.3$ for $y' = -2xy^2$, the bounds are not satisfied by (7.6) for a region near $x = 1$. However, we got through the region of instability in two or three steps, which were not enough for the instability to

TABLE 4
Stability ranges for $y' = ky$ for Adams third order.

t	$-0.8 \leqslant hk$	e	$-0.9 \leqslant hk$

build up appreciably. For most of the range of integration, (7.6) was well within the bounds of stability.

As with the second-order Adams method, use of extrapolated values does not diminish the region of stability.

We trust it is obvious how to generalize (8.1) and (8.3) to a set of two or more equations. See (7.8) through (7.13).

Error control is handled analogously to the treatment for second-order Adams.

9. ADAMS OF ORDER 4

The fourth-order Adams method (see Conte and de Boor, [4], p. 342 and p. 351), is given by

$$y_{n+1}^{(p)} = y_n + \frac{h}{24}\left(55y_n' - 59y_{n-1}' + 37y_{n-2}' - 9y_{n-3}'\right)$$

$$+ \frac{251h^5}{720} y^{(5)}(\xi), \tag{9.1}$$

$$y_{n+1}^{(c)} = y_n + \frac{h}{24}\left(9y_{n+1}' + 19y_n' - 5y_{n-1}' + y_{n-2}'\right)$$

$$- \frac{19h^5}{720} y^{(5)}(\xi). \tag{9.2}$$

Analogously to the previous Adams's, we set

$$y_{n+1}^{(t)} = y_n + \frac{h}{24}\left(9f\left(x_{n+1}, y_{n+1}^{(p)}\right) + 19y_n' - 5y_{n-1}' + y_{n-2}'\right), \tag{9.3}$$

$$y_{n+1}^{(e)} = y_{n+1}^{(t)} - \tfrac{19}{270}\left(y_{n+1}^{(t)} - y_{n+1}^{(p)}\right). \tag{9.4}$$

From the error terms given in (9.1) and (9.2) we judge that

$$-\tfrac{19}{270}\left(y_{n+1}^{(t)} - y_{n+1}^{(p)}\right) \tag{9.5}$$

is a reasonable estimate for the absolute error of $y_{n+1}^{(t)}$. Hence one is tempted to think that $y^{(e)}$ would be a better candidate for y_{n+1} than $y_{n+1}^{(t)}$. This is not so, for reasons analogous to those for the Adams of order two. By all means, take y_{n+1} to be $y_{n+1}^{(t)}$.

Some numerical results are given in Table 5. The rows labelled R-Ke give the results of using (4.8). This theoretically produces a

		$y' = y$	$y' = -y$	$y' = -2xy^2$
$h = 0.1$	R-K	4.62×10^{-6}	-5.44×10^{-6}	-8.10×10^{-7}
	R-Ke	3.72×10^{-7}	5.04×10^{-7}	4.72×10^{-8}
	t	-7.65×10^{-6}	2.73×10^{-5}	9.87×10^{-7}
	e	5.11×10^{-6}	7.50×10^{-6}	-5.63×10^{-7}
$h = 0.2$	R-K	6.77×10^{-5}	-9.46×10^{-5}	-1.37×10^{-5}
	R-Ke	1.10×10^{-5}	1.84×10^{-5}	8.14×10^{-7}
	t	-2.98×10^{-5}	6.93×10^{-4}	-5.46×10^{-5}
	e	1.34×10^{-4}	2.94×10^{-4}	-4.97×10^{-5}
$h = 0.3$	R-K	3.16×10^{-4}	-5.21×10^{-4}	-7.19×10^{-5}
	R-Ke	7.38×10^{-5}	1.60×10^{-4}	-5.60×10^{-6}
	t	1.71×10^{-4}	5.30×10^{-3}	-9.46×10^{-4}
	e	8.43×10^{-4}	2.74×10^{-3}	-6.92×10^{-4}

TABLE 5

Relative error at $x = 6$ for some fourth-order methods.

method of order 5. The errors for $h = (0.1)N$ should theoretically be about N^5 times those for $h = 0.1$. This fails dismally. See the discussion in connection with Table 3. Not only are the e values extremely erratic, but the t values are erratic enough to be quite disquieting. Evidently the error estimate (9.5) will not be very trustworthy unless h is near 0.1, or less. By contrast, the Runge-Kutta values behave quite well even up to $h = 0.3$, despite the poor results for R-Ke.

For the equation $y' = ky$, we get stability in the ranges shown in Table 6. These bounds should be satisfied by (7.6). For $h = 0.3$ for $y' = -2xy^2$, this was not the case for a short interval, not long enough to cause any trouble.

Use of extrapolated values does not diminish the region of stability.

If it is not obvious how to generalize (9.1) and (9.3) to a set of two or more equations, see Conte and de Boor, [4], p. 366. The $y_{n+1}^{(0)}$, $z_{n+1}^{(0)}$, $y_{n+1}^{(1)}$, $z_{n+1}^{(1)}$ appearing there would be $y_{n+1}^{(p)}$, $z_{n+1}^{(p)}$, $y_{n+1}^{(t)}$, and $z_{n+1}^{(t)}$ in our notation.

t	$-0.6 \leqslant hk$	e	$-0.6 \leqslant hk$

TABLE 6

Stability ranges for $y' = ky$ for Adams fourth order.

TABLE 7
Relative error by double precision.

		$y' = y$	$y' = -y$	$y' = -2xy^2$
$h = 0.1$	R-K	4.62×10^{-6}	-5.46×10^{-6}	-8.18×10^{-7}
	R-Ke	4.13×10^{-7}	4.83×10^{-7}	3.98×10^{-8}
$h = 0.2$	R-K	6.77×10^{-5}	-9.46×10^{-5}	-1.37×10^{-5}
	R-Ke	1.10×10^{-5}	1.84×10^{-5}	8.06×10^{-7}

Error control is handled analogously to the treatment for second-order Adams.

The values appearing in Table 5 were what we got on the HP-65. This is only a 10-decimal digit calculator. For the smaller values of h, the calculations were quite extended (more than a few minutes for $h = 0.1$ for the equation $y' = -2xy^2$). So there were considerable accumulations of round off error, enough to cast doubt on the values listed in Table 5. So a double precision calculation was performed on a major computer. The results are listed in Table 7. For $h = 0.3$ the only discrepancy was trivial, -5.61×10^{-6} for $h = 0.3$ with $y' = -2xy^2$. Comparison of Table 5 and Table 7 shows one place where the accumulation of round off errors amounted to about 4 in the eighth decimal place (for a quantity whose mantissa begins with 9, a relative error of about 4×10^{-8}), but otherwise considerably less. For a 10-decimal-place calculator, this is quite reassuring. On the other hand, it points up one limitation of a hand held calculator, where a double-precision calculation of any extent is fairly impractical.

10. OTHER PREDICTOR-CORRECTORS

Although the Adams predictor-correctors require only two function evaluations per step, the higher-order ones begin to put heavy demands on memory locations. They are also difficult to get started, or to change step size with. Can one improve on these aspects? We succeeded in doing this, only to find out that our results had been anticipated long ago.

One method that we devised was first propounded in Southard and Yowell [22]. This was written when even the largest computers were very limited as to memory capacity, and considerable efforts were made to hold down requirements for memory capacity.

The Southard and Yowell method is given by

$$y_{n+1}^{(p)} = -4y_n + 5y_{n-1} + h\left(4y_n' + 2y_{n-1}'\right) + \tfrac{h^4}{6}y^{(4)}(\xi),$$

$$(10.1)$$

$$y_{n+1}^{(c)} = y_n + \tfrac{h}{12}\left(5y_{n+1}' + 8y_n' - y_{n-1}'\right) - \tfrac{h^4}{24}y^{(4)}(\xi).$$

$$(10.2)$$

One will recognize that the corrector is the same as for the third-order Adams, to wit (8.2). So the formula for $y_{n+1}^{(t)}$ would be the same as (8.3). We would have

$$y_{n+1}^{(e)} = y_{n+1}^{(t)} - \tfrac{1}{5}\left(y_{n+1}^{(t)} - y_{n+1}^{(p)}\right). \qquad (10.3)$$

From the error terms given in (10.1) and (10.2), we judge that

$$-\tfrac{1}{5}\left(y_{n+1}^{(t)} - y_{n+1}^{(p)}\right) \qquad (10.4)$$

is a reasonable estimate for the absolute error of $y_{n+1}^{(t)}$. By all means, take y_{n+1} to be $y_{n+1}^{(t)}$.

This predictor-corrector has some good features. It is third order, but has modest memory requirements, and is easy to get started, or

TABLE 8

Relative error at $x = 6$ for Southard and Yowell.

		$y' = y$	$y' = -y$	$y' = -2xy^2$
$h = 0.1$	R-K	2.31×10^{-4}	2.71×10^{-4}	3.31×10^{-5}
	t	-1.64×10^{-4}	-4.18×10^{-4}	-1.50×10^{-5}
	e	1.18×10^{-5}	-1.52×10^{-5}	-1.82×10^{-7}
$h = 0.2$	R-K	1.70×10^{-3}	2.35×10^{-3}	3.03×10^{-4}
	t	-8.99×10^{-4}	-7.16×10^{-3}	4.54×10^{-3}
	e	1.70×10^{-4}	-2.84×10^{-4}	4.00×10^{-6}
$h = 0.3$	R-K	5.30×10^{-3}	8.56×10^{-3}	1.19×10^{-3}
	t	-2.09×10^{-3}	-2.79×10^{-1}	unstable
	e	7.79×10^{-4}	-1.72×10^{-3}	1.59×10^{-4}

get restarted after changing the length of the step. However, it behaves poorly as to stability, as can be seen from Table 8 and Table 9.

Because the range of stability for $y_{n+1}^{(t)}$ is so limited, one finds (7.6) outside that range in the case $h = 0.3$ and $y' = -2xy^2$ for an extended period, and the solution really blows up. Already at $x = 3$, one gets a negative value for y, after which the errors compound catastrophically. Before reaching $x = 6$, the calculator stops on an overflow, because the numbers are too large for its capacity.

If one had been monitoring (7.6), this instability could have been avoided by using a smaller h through the critical region.

The analysis of the stability for $y_{n+1}^{(t)}$ and $y_{n+1}^{(e)}$ would be worthy of special attention by anyone interested in stability. Neither the usual Dahlquist criterion of stability nor the Hamming-Ralston criterion for relative stability is applicable, and a special definition had to be contrived. Never mind the details. The reader can trust that if (7.6) stays too long outside the bounds given in Table 9, there will be trouble; see the case $h = 0.3$ for $y' = -2xy^2$, for example.

In Hamming [11] on p. 47, concern is expressed about the stability of Southard and Yowell, and it is proposed to remedy the situation by using a different predictor

$$y_{n+1}^{(p)} = y_n + y_{n-1} - y_{n-2} + 2h\left(y_n' - y_{n-1}'\right) + \frac{h^4}{3}y^{(4)}(\xi). \quad (10.5)$$

This appreciably increases the number of memory locations needed and requires more starting values to get started or restarted, thus cancelling out the two good features of Southard and Yowell. It does appear to improve the stability, for $y_{n+1}^{(t)}$ it is increased to $-0.5 \le hk$. However, this is considerably less than for the Adams method of order three. If we compare the Adams of order three with what results from (10.5) we find that both are of order three,

TABLE 9

Stability ranges for $y' = ky$ for Southard and Yowell.

t	$-0.31 \le hk$	e	$-0.7 \le hk$

TABLE 10
Relative error at $x = 6$ for Hamming.

		$y' = y$	$y' = -y$
$h = 0.1$	R-K	4.62×10^{-6}	-5.44×10^{-6}
	t	-8.63×10^{-6}	4.55×10^{-5}
	e	4.24×10^{-6}	7.52×10^{-6}
$h = 0.2$	R-K	6.77×10^{-5}	-9.46×10^{-5}
	t	-3.87×10^{-5}	2.37×10^{-3}
	e	1.07×10^{-4}	3.73×10^{-4}
$h = 0.3$	R-K	3.16×10^{-4}	-5.21×10^{-4}
	t	1.05×10^{-4}	unstable
	e	6.63×10^{-4}	5.90×10^{-3}

both need two extra starting values to get started or restarted, but the Adams is considerably more stable and requires fewer memory locations.

In Hamming [11] it was proposed to render the classical Milne predictor-corrector stable by using a different corrector, namely,

$$y_{n+1}^{(c)} = \frac{1}{8}\left(9y_n - y_{n-2} + 3h\left(y_{n+1}' + 2y_n' - y_{n-1}'\right)\right) - \frac{h^5}{40}y^{(5)}(\xi);$$

$$(10.6)$$

the old predictor, (5.1), was retained. The results of this are shown in Table 10 and Table 11. In Table 10 there is no column $y' = -2xy^2$ because the memory requirements exceeded the capacity of the HP-65 with which the calculations of this paper were performed. Although there is a region of stability, it is less than for the Adams method of comparable order, namely order four. All in all, the Hamming method is not a contender for use with hand held calculators.

From Ralston [17] one would get the impression that a further Hamming modification of the above is very superior; see p. 189 of

TABLE 11
Stability ranges for $y' = ky$ for Hamming.

t	$-0.26 \leqslant hk$	e	$-0.38 \leqslant hk$

Ralston [17]. Despite Ralston's endorsement of this particular variation by Hamming, Enright and Hull [6] give it a very low rating on pp. 954–955. Interestingly enough, though Hamming invented the method, and devotes a lot of space to discussing it in Hamming [12], he finally (in pp. 206–210) seems to favor something related to an Adams method.

Amongst the disadvantages of the Hamming method (or one of the further Hamming modifications thereof) for hand held calculators is its excessive requirement for memory locations. To improve this, one could change the predictor to

$$y_{n+1}^{(p)} = -9y_n + 9y_{n-1} + y_{n-2} + h\left(6y_n' + 6y_{n-1}'\right) + \frac{h^5}{10} y^{(5)}(\xi).$$

(10.7)

When used with the Hamming corrector, (10.6), this still gives a fourth-order method. However, it requires two fewer memory locations. Also, it requires fewer starting values to get a start or restart, being in this respect also superior to the Adams method of order four.

This predictor is one of a set which, in a footnote on p. 171 of Hamming [12] is dismissed as not being worth consideration. The combination of (10.7) and (10.6) does turn out to have very poor stability characteristics. For $y' = ky$ one must have $-0.13 \leqslant hk$. This is pretty stringent, and Hamming's dismissal seems valid.

All in all, for a hand held calculator the Adams methods seem superior to any other of the predictor-corrector methods.

11. CONCLUSIONS

So you have a differential equation (or set of same), and wish to get a solution. We have described a number of ways to proceed. All you have to do is to pick the one that is best for your particular problem. Naturally, the choice is going to depend on how long an interval of integration you have, what accuracy you require, etc.

If the interval of integration is short, but you need high accuracy, the high-order Adams methods have the advantage that they require fewer function evaluations than the high-order Runge-Kutta methods. You can use a value of h with the Adams methods half as large as with the Runge-Kutta methods and still do as well or better as to the number of function evaluations. With the smaller value of h, you will have considerably less error, the error control will be more trustworthy, and you should not have much trouble keeping (7.6) within bounds. However, there are still other considerations, as we shall see below.

On the other hand, if the interval of integration is long, but very modest accuracy is required, you would wish to use large h. With an Adams method, you must keep (7.6) within bounds. This may preclude use of a large h. With a Runge-Kutta method, there is not this hindrance to use of a large h. Of course, you cannot take h excessively large, or you will not meet even a modest accuracy requirement. By all means, exercise error control. It may turn out that you can meet the accuracy requirement with a lower order Runge-Kutta, which would reduce the number of function evaluations.

And of course, you have to decide whether to control error by accumulating absolute errors, or relative errors, or a mixture. For this it helps to know in a general way how the solution behaves overall. This information can also be useful in deciding between an Adams method or a Runge-Kutta. If you are going to have to change the step size two or three times because of variabilities in the performance of the solution, this will weigh against the use of Adams, perhaps enough so to compensate for the larger number of function evaluations of the Runge-Kutta. How difficult the function evaluation is becomes relevant at a point like this.

So, unless the problem is so easy that you can scarcely go wrong, it seems almost necessary to have some idea of the general overall behaviour of the solution. So a quick, approximate run through the integration seems a very useful preliminary before you settle down to a serious integration. To get through as quickly as possible, you take h as large as possible without completely compromising accuracy. As we indicated above, such use of a large h indicates that the quick run through had best be done with a Runge-Kutta method. The second-order one is most economical of function evaluations. You should use error control, so as to learn if appreciable changes in the size of h will be required. This requires setting a value of *tol*. This can be only a blind guess at this point, but taking *tol* to be $|y_0|/10$ might do all right. Exercise error control by accumulating absolute errors, but watch the size of $|y|$. If it gets very large, error control will be cutting down the size of h, so that you will not be getting a quick run through. You can remedy this by switching to accumulating relative errors, using *tol* $=10\%$.

For a blind guess for the starting h, you could try $L/20$, where L is the length of the path of integration. After the first two steps, the size of h will be fixed by error control; this might require that the first two steps be done over again with a much smaller h. If error control says you could have used a larger h, just do that for the next two steps. After all, for the quick run through, we are not trying to exercise very strict error control.

We will be using (3.1). The k_2 for a given step will be

$$hf(x_n + h, y_n + k_1)$$

while the k_1 for the next step will be

$$hf(x_n + h, y_n + (y_{n+1} - y_n)).$$

These can be used in (7.7) to get an estimate for

$$\frac{\partial f(x, y)}{\partial y}.$$

A record of these will be useful in determining ahead of time what values of h you can use so that (7.6) will remain within bounds, in case you are contemplating use of an Adams method.

All in all, you can get quite a bit of useful information out of a very quick run through, after which you should be able to make the decisions how to do a serious run through that will give the sort of solution that you desire.

12. SUPPLEMENT

You should be warned that some differential equation problems are of such difficulty that they should be attempted only on a large, fast computer. For example, at the present time this is the case if you have a set of ten simultaneous equations. Or suppose you have only one equation, but wish a solution of considerable accuracy from $x = 0$ to $x = 200$. Because good accuracy is required, you will likely have to take h fairly small. So the integration could extend to 2000 steps, or even much more. This is not impossible on a hand held calculator, but it should definitely be avoided if practical. Or you can have the misfortune to encounter an equation (or system) of the kind called "stiff." If so, unless you use some special techniques that we have not mentioned, the error control will force you to have a very small h, so small that the number of steps needed to complete the solution is more than is reasonable to undertake with a hand held calculator as slow as the present models. We hope you discovered this fairly early, perhaps when your attempt to make a quick approximate run through threatened to become interminably long.

Frankly, solving differential equations on a large, fast computer is quite a complicated affair, in spite of the fact that some programs are advertised as being intended for users who know nothing about differential equations. If you find you must turn to a large, fast computer, two possibilities are open. One is to spend quite a bit of time learning how to use such a computer for ODE's, for instance, by reading Shampine, Watts, and Davenport [21] (which is already a bit obsolete, since there is no reference to the recent work of Shintani), or books referred to therein. The other possibility (which we strongly recommend) is to seek expert assistance.

REFERENCES

1. M. Abramowitz and I. A. Stegun, *Handbook of Mathematical Functions*, Applied Mathematics Series 55, Natl. Bureau of Standards, Washington, D.C., 1964.
2. R. Bulirsch and R. Stoer, "Numerical treatment of ordinary differential equations by extrapolation methods," *Numer. Math.*, **8** (1966), pp. 1–13.
3. Richard L. Burden, J. Douglas Faires, and Albert C. Reynolds, *Numerical Analysis*, Prindle, Weber & Schmidt, Boston, Mass., 1978.
4. S. D. Conte and Carl de Boor, *Elementary Numerical Analysis*, second edition, McGraw-Hill, New York, 1972.
5. G. Dahlquist, "Convergence and stability in the numerical integration of ordinary differential equations," *Math. Scand.*, **4** (1956), pp. 33–53.

6. W. H. Enright and T. E. Hull, "Test results on initial value methods for non-stiff ordinary differential equations," *SIAM Jour. Num. Anal.*, **13** (1976), pp. 944–961.

7. E. Fehlberg, "Classical fifth-, sixth-, seventh-, and eighth-order Runge-Kutta formulas with stepsize control," NASA Tech. Report No. 287, 1968, Huntsville.

8. _____, "Klassische Runge-Kutta-Formeln fünfter und siebenter Ordnung mit Schrittweiten-Kontrolle," *Computing*, **4** (1969), pp. 93–106.

9. _____, "Klassische Runge-Kutta-Formeln vierter und niedriger Ordnung mit Schrittweiten-Kontrolle und ihre Anwendung auf Wärmeleitungsprobleme," *Computing*, **6** (1970), pp. 61–71.

10. W. B. Gragg, "On extrapolation algorithms for ordinary initial value problems," *SIAM Jour. Num. Anal.*, **2** (1965), pp. 384–403.

11. R. W. Hamming, "Stable predictor-corrector methods for ordinary differential equations," *Jour. of ACM*, **6** (1959), pp. 37–47.

12. _____, *Numerical Methods for Scientists and Engineers*, McGraw-Hill, New York, 1962.

13. Peter Henrici, *Discrete Variable Methods in Ordinary Differential Equations*, John Wiley and Sons, New York, 1962.

14. _____, *Computational Analysis with the HP-25 Pocket Calculator*, John Wiley and Sons, New York, 1977.

15. T. E. Hull, W. H. Enright, B. M. Fellen, and A. E. Sedgwick, "Comparing numerical methods for ordinary differential equations," *SIAM Jour. Num. Anal.*, **9** (1972), pp. 603–637.

16. W. E. Milne, *Numerical Solution of Differential Equations*, John Wiley and Sons, New York, 1953.

17. Anthony Ralston, *A First Course in Numerical Analysis*, McGraw-Hill, New York, 1965.

18. J. Barkley Rosser, "A Runge-Kutta for all seasons," *SIAM Review*, **9** (1967), pp. 417–452.

19. _____, "Solving differential equations on a hand held programmable calculator," MRC Technical Summary Report #1848, 1978; copies are available by citing the code number AD A060 661 from: National Technical Information Service, 5285 Port Royal Road, Springfield, Virginia 22151.

20. L. F. Shampine and H. A. Watts, "The art of writing a Runge-Kutta code," Part I, *Mathematical Software III*, ed. by John R. Rice, Academic Press, New York, 1977.

21. L. F. Shampine, H. A. Watts and S. M. Davenport, "Solving nonstiff ordinary differential equations—the state of the art," *SIAM Review*, **18** (1976), pp. 376–411.

22. T. H. Southard and E. C. Yowell, "An alternative 'predictor-corrector' process," *Math. Tables and Other Aids to Comp.*, **6** (1952), pp. 253–254.

23. R. Zurmühl, "Runge-Kutta-Verfahren zur numerischen Integration von Differentialgleichungen n-ter Ordnung," *Zeitschrift Angew. Math. Mech.*, **28** (1948), pp. 173–182.

FINITE DIFFERENCE SOLUTION OF BOUNDARY VALUE PROBLEMS IN ORDINARY DIFFERENTIAL EQUATIONS

V. Pereyra

1. INTRODUCTION

The field of ordinary boundary value problems has experienced a great deal of activity in the last fifteen years, in both theoretical and numerical aspects. We will try to give in this article a partial view of the state of the art in the Numerical Analysis of Two Point Boundary Value Problems (2PBVP).

This is not a survey, and therefore we make no claims to completeness, but we have rather chosen some representative material in order to give a feeling of what is going on today, without overwhelming the reader with technical details.

For additional information we recommend the survey [60]. A very important class of methods not included here is collocation. For a detailed treatment of its theory and associated methods we suggest [59]. The book [57] contains descriptions of most of the methods and software in the public domain and a wealth of references.

In this paper we start considering a simple linear system of the first order with only two equations, and then we indicate how, with an appropriate notation, we can pass to general systems with little difficulty. Nonlinear problems are then considered, although not in as much detail, as one of the most active areas of research today warrants.

A short section on theoretical foundations leads us to the main types of numerical methods based on finite difference approximations. The fundamental building blocks of most modern theories of discretization methods are given in Section 4, while Sections 5 and 6 touch upon the more practical aspects of increasing the rate of convergence of simple, low-order methods, and the very important problem of mesh selection. These two are the main ingredients in current computer software for solving these problems.

Finally in Section 7 we refer briefly to some applications. The reader can rest assured that there are many more, and that she (or he) will not waste her time by learning about this fascinating subject.

2. THEORY OF ORDINARY BOUNDARY VALUE PROBLEMS

In this section we will develop the basic elements of the theory of existence, uniqueness and representation of solutions of ordinary differential equations subject to two point boundary conditions.

This theory can be developed for first-order nonlinear systems of the form

$$\mathbf{y}' = \mathbf{f}(t, \mathbf{y}), \qquad t \in [a, b], \tag{1}$$

subject to the nonlinear conditions

$$\mathbf{g}(\mathbf{y}(a), \mathbf{y}(b)) = \mathbf{O}, \tag{2}$$

where $\mathbf{y} = (y_1, y_2, \ldots, y_m)^T$, $\mathbf{y}^1 = (dy_1/dt, \ldots, dy_m/dt)^T$, $\mathbf{f} = (f_1, \ldots, f_m)^T$, $\mathbf{g} = (g_1, \ldots, g_m)^T$, are m-vector functions of their arguments, and the exponent T means vector transposition (i.e., all vectors are column vectors).

However, for simplicity of exposition and because it has most of the important features, we will limit ourselves to consider mainly the case of two linear equations:

$$x_1' = a_{11}(t)x_1 + a_{12}(t)x_2 + r_1(t),$$
$$x_2' = a_{21}(t)x_1 + a_{22}(t)x_2 + r_2(t), \tag{3}$$

subject to separated linear conditions

$$\alpha_1 x_1(a) + \alpha_2 x_2(a) = \Theta_1$$
$$\gamma_1 x_1(b) + \gamma_2 x_2(b) = \Theta_2. \tag{4}$$

We will assume in what follows that all the coefficient functions appearing in (3) are continuous in $[a, b]$. In vector notation we have

$$\mathbf{x}' = A(t)\mathbf{x} + \mathbf{r},$$
$$\langle \boldsymbol{\alpha}, \mathbf{x}(a) \rangle = \Theta_1, \tag{5}$$
$$\langle \boldsymbol{\gamma}, \mathbf{x}(b) \rangle = \Theta_2,$$

where

$$\mathbf{x} = \begin{pmatrix} x_1(t) \\ x_2(t) \end{pmatrix}, \qquad \mathbf{r} = \begin{pmatrix} r_1(t) \\ r_2(t) \end{pmatrix}, \qquad A(t) = \begin{pmatrix} a_{11}(t) & a_{12}(t) \\ a_{21}(t) & a_{22}(t) \end{pmatrix},$$

$$\boldsymbol{\alpha} = \begin{pmatrix} \alpha_1 \\ \alpha_2 \end{pmatrix}, \qquad \boldsymbol{\gamma} = \begin{pmatrix} \gamma_1 \\ \gamma_2 \end{pmatrix}$$

and $\langle , \rangle$ denotes the inner product of vectors.

Once we have adopted vector notation, there is really no great difficulty in considering systems of arbitrary order m. The only difference will be that we shall have to have m independent boundary conditions, say, $0 < p < m$ at the left boundary, and remaining $q = m - p$ conditions at the right boundary. More general linear coupled conditions of the form

$$B_a \mathbf{x}(a) + B_b \mathbf{x}(b) = \Theta \tag{5'}$$

can also be considered.

It is worth pointing out here that higher-order equations are easily reduced to first-order systems. For instance,

$$y'' = c(t)y + d(t)y' + s(t)$$

becomes, by introducing $x_1 \equiv y$, $x_2 \equiv y'$,

$$x_1' = x_2,$$

$$x_2' = c(t)x_1 + d(t)x_2 + s(t).$$

In general, an mth order equation can be rewritten as a first-order system with m equations, by introducing the successive derivatives $y', y'', \ldots, y^{(m-1)}$ as new variables. This is the reason why first-order systems have received so much attention, both theoretically and in practise: they are simple but still sufficiently general for most purposes.

Fundamental Solutions.   In the study of linear differential equations there are some special solutions which are so important that they are called fundamental. A matrix function

$$X(t) = \begin{pmatrix} x_{11}(t) & x_{12}(t) \\ x_{21}(t) & x_{22}(t) \end{pmatrix}$$

is called a *fundamental solution* of equation (5) if it satisfies

$$X'(t) = A(t)X(t)$$

$$X(a) = I, \tag{6}$$

where $X' = \begin{pmatrix} x_{11}' & x_{12}' \\ x_{21}' & x_{22}' \end{pmatrix}$ and $I = \begin{pmatrix} 1 & 0 \\ 0 & 1 \end{pmatrix}$ is the identity matrix. It is not difficult to prove that $X(t)$ is a nonsingular matrix for $t \in [a,b]$ (i.e., the columns $\mathbf{X}_1 = (x_{11}, x_{21})^T$, $\mathbf{X}_2 = (x_{12}, x_{22})^T$ are two independent solutions of the given homogeneous system).

For initial value problems of the form

$$\mathbf{x}' = A(t)\mathbf{x} + \mathbf{r},$$

$$\mathbf{x}(a) = \mathbf{x}_0, \tag{7}$$

one obtains a simple representation for their solution

$$\mathbf{x}(t) = X(t)\left[\mathbf{x}_0 + \int_a^t X^{-1}(\sigma)\mathbf{r}(\sigma)d\sigma\right], \tag{8}$$

where X^{-1} is the inverse of the matrix X:

$$X^{-1} = \frac{1}{\det(X)}\begin{pmatrix} x_{22} & -x_{12} \\ -x_{21} & x_{11} \end{pmatrix}$$

with $\det(X) = x_{11}x_{22} - x_{12}x_{21}$. In order to verify that (8) is in fact a solution, it is enough to replace it in the equation and the initial condition.

Existence, Uniqueness, and Representation of the Solution to the Boundary Value Problem. We will see now that the question of existence and uniqueness of solutions of the BVP (5) can be reduced to an algebraic condition. This stems from the fact that the initial value problem (7) has a unique solution for arbitrary $\mathbf{x}_0$, and, therefore, every solution of the differential equation is in unique correspondence with its initial value vector. In particular, the solution to the BVP, if it exists, will be associated with some (unknown) initial vector. In the next theorem we show when this initial vector exists and how it can be computed in terms of the fundamental solution and the data of the problem.

THEOREM 9. *Let us consider the matrix*

$$Q = \begin{pmatrix} \alpha_1 & \alpha_2 \\ \langle \gamma, \mathbf{X}_1(b) \rangle & \langle \gamma, \mathbf{X}_2(b) \rangle \end{pmatrix} \tag{10}$$

that arises from applying the boundary conditions to the columns of the fundamental solution. Problem (5) has a unique solution if and only if Q is nonsingular. Moreover, this solution can be represented as

$$\mathbf{x}(t) = X(t)Q^{-1}\left(\mathbf{\Theta} - \begin{pmatrix} 0 \\ \langle \gamma, \hat{\mathbf{x}}(b) \rangle \end{pmatrix}\right) + \hat{\mathbf{x}}(t), \tag{11}$$

with

$$\hat{\mathbf{x}}(t) = X(t)\int_a^t X^{-1}(\sigma)\mathbf{r}(\sigma)\,d\sigma.$$

Proof. As we mentioned above, the solution to the BVP, if it exists, can be represented as

$$\mathbf{x}(t) = X(t)\mathbf{x}_0 + \hat{\mathbf{x}}(t), \tag{12}$$

for some unknown initial vector $\mathbf{x}_0$. According to (8), $\mathbf{x}(t)$ certainly satisfies the inhomogeneous equation for any $\mathbf{x}_0$. If we apply the BC (5) to it, we obtain

$$Q\mathbf{x}_0 + \begin{pmatrix} \langle \alpha, \hat{\mathbf{x}}(a) \rangle \\ \langle \gamma, \hat{\mathbf{x}}(b) \rangle \end{pmatrix} = \mathbf{\Theta}.$$

Or reordering and observing that $\hat{\mathbf{x}}(a) = \mathbf{0}$, we get

$$Q\mathbf{x}_0 = \begin{pmatrix} \Theta_1 \\ \Theta_2 - \langle \gamma, \hat{\mathbf{x}}(b) \rangle \end{pmatrix}, \tag{13}$$

which is a 2×2 system of linear equations for the unknowns $\begin{pmatrix} x_{10} \\ x_{20} \end{pmatrix}$. This system will have a unique solution if and only if Q is nonsingular, which proves the first part of Theorem (9). The representation (11) follows from replacing in (12)

$$\mathbf{x}_0 = Q^{-1}\left(\mathbf{\Theta} - \begin{pmatrix} 0 \\ \langle \gamma, \hat{\mathbf{x}}(b) \rangle \end{pmatrix}\right).$$

The representation (11) reduces the BVP to that of solving three initial value problems, namely, the two systems (6) that determine the fundamental solution X (one for each column), and that corresponding to $\hat{\mathbf{x}}$, which can be written as

$$\hat{\mathbf{x}}' = A(t)\hat{\mathbf{x}} + \mathbf{r}(t),$$

$$\hat{\mathbf{x}}(a) = \mathbf{0} \tag{14}$$

(i.e., $\hat{\mathbf{x}}$ is a particular solution of the nonhomogeneous equation with $\mathbf{0}$ initial condition).

In fact, this observation has motivated a number of methods of solution, based on the notion that "it is simpler to solve initial value problems." This belief was strengthened by the early development of very powerful numerical methods and computer implementations for initial value problems (IVP).

Nonlinear Problems. The theory for the general nonlinear case (1)–(2), can be dealt with by the method of successive linearizations associated with the names of Newton and Kantorovich.

Let us introduce the Jacobian matrices

$$\mathbf{f}_{\mathbf{x}} = \left(\frac{\partial f_i}{\partial x_j}\right), \mathbf{g}_{\mathbf{x}(a)} = \left(\frac{\partial g_i}{\partial x_j(a)}\right), \mathbf{g}_{\mathbf{x}(b)} = \left(\frac{\partial g_i}{\partial x_j(b)}\right)$$

$i, j = 1, \ldots, m$. Then, given a trial guess $\mathbf{x}^0(t)$, a new (hopefully improved) iterate can be generated by solving for $\delta\mathbf{x}(t)$, the linearized boundary value problem.

$$\delta\mathbf{x}' = \mathbf{f}_{\mathbf{x}}(t, \mathbf{x}^0(t))\,\delta\mathbf{x} + \left(-\mathbf{x}^{0\prime}(t) + \mathbf{f}(t, \mathbf{x}^0(t))\right)$$

$$\mathbf{g}_{\mathbf{x}(a)}(\mathbf{x}^0(a), \mathbf{x}^0(b))\delta\mathbf{x}(a) + \mathbf{g}_{\mathbf{x}(b)}(\mathbf{x}^0(a), \mathbf{x}^0(b))\,\delta\mathbf{x}(b)$$

$$= -\mathbf{g}(\mathbf{x}^0(a), \mathbf{x}^0(b)), \tag{15}$$

where the Jacobian matrices have taken up the roles of $A(t)$, B_a and B_b in (5), (5′), and then $\mathbf{x}^1(t) = \mathbf{x}^0(t) + \delta\mathbf{x}(t)$.

Under suitable conditions [16], [61] this process can be repeated, replacing x^0 by x^1, and continuing in this fashion, a sequence of functions x^i can be generated which will converge to an isolated solution $x^*(t)$ of the BVP. This is, in fact, a constructive existence theorem, and shows why linear systems are an important component of the problem.

We shall see in the next section how similar ideas allow us to devise a numerical method for the approximate solution of nonlinear BVP's.

3. NUMERICAL METHODS

Shooting and its Variants. Let us consider problem (3), and for simplicity let us assume that in the BC (4) $\alpha_1 = 1$, $\alpha_2 = 0$, i.e., the first condition reduces to $x_1(a) = \Theta_1$. Let us consider the initial value problem consisting of equations (3) and the conditions $x_1(a) = \Theta_1$, $x_2(a) = \lambda$, where λ is an unknown parameter to be determined.

The method of determining λ is similar to the one we used before; simply consider the solution to this problem parametrized by λ: $x(t; \lambda)$, and apply to it the end condition (4):

$$\gamma_1 x_1(b; \lambda) + \gamma_2 x_2(b; \lambda) - \Theta_2 = 0. \tag{16}$$

Actually, for the linear differential equation (3), and through the representation (8) for the solution $x(t; \lambda)$ (with $x_0 = \begin{pmatrix} \Theta_1 \\ \lambda \end{pmatrix}$), (16) is clearly a linear equation in one unknown, and, therefore, we can easily solve for λ under some mild condition ($\gamma_1 X_{12}(b) + \gamma_2 X_{22}(b) \neq 0$).

There are several variants of this method for linear problems that attempt to minimize the number of initial value problems necessary to calculate the solution. It is possible to show that for a problem with p separated conditions at a, and $q = \min(p, m - p)$, an appropriate formulation reduces the number of IVP from $(m + 1)$ to $(q + 2)$ (see [18]).

Shooting is an attractive and simple procedure that extends to nonlinear problems easily by using an auxiliary root solver. In fact, for problem (1)–(2) one states the auxiliary IVP

$$\mathbf{y}' = \mathbf{f}(t, \mathbf{y}),$$

$$\mathbf{y}(a) = \boldsymbol{\lambda}, \tag{17}$$

where now $\boldsymbol{\lambda}$ is a vector of unknowns. If there are some pure initial conditions in (2), then one can use them and thus reduce the number of unknown parameters. The solution to this problem for a fixed $\boldsymbol{\lambda}$ is denoted by $\mathbf{y}(t; \boldsymbol{\lambda})$. Problem (1)–(2) will be solved if we can integrate (17) from a to b, and if we can solve the nonlinear system of equations

$$\mathbf{g}(\boldsymbol{\lambda}, \mathbf{y}(b; \boldsymbol{\lambda})) = \mathbf{0}$$

for $\boldsymbol{\lambda}$; i.e., if we can find $\boldsymbol{\lambda}$ such that the BC (2) are satisfied. This is the essence of most shooting procedures. The main difficulty lies in the fact that it may not be possible numerically to integrate problems (1)–(2) from a to b.

For instance, the linear BVP

$$\begin{aligned}
y_1' &= y_2 & y_1(0) &= 1 \\
y_2' &= 99y_2 + 100y_1 & y_1(1) &= e^{-1}
\end{aligned}$$

has first component of the solution $y_1(t) = e^{-t}$, which is quite nicely behaved in $[0,1]$. However, for the same differential equations with the initial conditions $y_1(0) = 1$, $y_2(0) = \mu$, we have:

$$y_1(t; \mu) = \frac{\mu + 1}{101} e^{100t} + \frac{100 - \mu}{101} e^{-t}.$$

Clearly, the μ that will produce the solution to the BVP is $\mu^* = -1$ (why?), but the idea is that in a real problem we will not know this. Even if we start with a very accurate initial guess, say, $\mu^0 = -0.9$, we see that in most machines $y(t; \mu^0)$ will overflow before we reach $b = 1$ (i.e., a number too large to be represented will be produced). This seems to be an extreme case but it is sufficiently common in the applications of interest to have motivated a lot of activity to circumvent the difficulty.

One of the most fruitful ideas in this direction is that of multiple or parallel shooting [16], [18], [25]. In order to control the growth of exponentially increasing solutions, a number of internal shooting points are introduced, and initial value solutions are started *simultaneously* from each of these points. This process produces a number of pieces of the tentative numerical solution. For each shooting point, a vector of unknown initial conditions has to be calculated. The requirement that the solution be continuous and satisfy the given boundary conditions provides exactly the adequate number of equations necessary to determine those unknown initial vectors. This is the basic principle of multiple shooting. For a detailed formulation see the references above.

In a modern program implementing this method it is important that the choice of the number *and* position of the shooting points be determined automatically, in a fashion that adapts itself optimally to the problem at hand. Although there has been a great deal of activity in this area in recent times, and although there exists considerable understanding of the problem, still there is no entirely satisfactory solution to it.

In the next section we discuss an extreme case of multiple shooting, in which there are possibly many shooting points, but only one integration step is performed from each of them.

A Finite Difference Method. Coming back to the linear second-order system (5), let us consider a mesh Π of points in the interval $[a, b]$:

$$\Pi : a = t_0 < t_1 < \cdots < t_n = b. \tag{18}$$

In order to be able to solve the differential problem on a computer we need to replace all infinite processes by finite ones. Actually, in problem (5) the only infinite process present is the one implied by the derivative $x'(t)(\equiv \lim_{h \to 0}(x(t + h) - x(t))/h)$.

This is discretized very simply by considering the function values x_i associated with the mesh points t_i and replacing x' by $(x_{i+1} - x_i)/h_i$, where $h_i = t_{i+1} - t_i$. If this expression is considered to represent x' at the midpoint $t_i + h_i/2$ (a "centered difference"), then one has, by expanding $x(t_i + h_i/2)$ in Taylor series about $t_i + h_i/2$,

$$(\mathbf{x}_{i+1} - \mathbf{x}_i)/h_i - \mathbf{x}'(t_i + h_i/2) = h_i^2/6\mathbf{x}''(\xi_i),$$

where $\xi_i \in (t_i, t_{i+1})$. One says that the approximation is accurate to order two (with respect to the mesh size).

Using this expression in each mesh interval we obtain a system of algebraic equations for the unknown values $\mathbf{x}_i$, $i = 0, 1, \ldots, n$:

$$(\mathbf{x}_{i+1} - \mathbf{x}_i)/h_i = \frac{1}{2}\big[A(t_{i+1})\mathbf{x}_{i+1} + A(t_i)\mathbf{x}_i + \mathbf{r}(t_{i+1}) + \mathbf{r}(t_i)\big],$$

$$i = 0, \ldots, n-1, \quad (19)$$

where we have averaged the values of the term in the right-hand side of (5) in order to preserve the order of accuracy. The discretization is called the *trapezoidal rule*. For the boundary conditions we simply have

$$\langle \alpha, \mathbf{x}_0 \rangle = \Theta_1, \qquad \langle \gamma, \mathbf{x}_n \rangle = \Theta_2. \quad (20)$$

Expressions (19) and (20) form a system of $2x(n+1)$ equations for the same number of unknowns. Writing everything that contains $\mathbf{x}_i, \mathbf{x}_{i+1}$ on the left-hand side, we see that this linear system has a matrix of coefficients which takes the block form

$$\begin{bmatrix} \alpha^T & 0 & 0 & \cdot & \cdot & \cdot & 0 \\ W_{11} & W_{12} & 0 & \cdot & \cdot & \cdot & 0 \\ 0 & W_{21} & W_{22} & \cdot & \cdot & \cdot & 0 \\ \cdot & \cdot & \cdot & \cdot & \cdot & \cdot & \cdot \\ 0 & \cdot & \cdot & \cdot & \cdot & W_{n1} & W_{n2} \\ 0 & \cdot & \cdot & \cdot & \cdot & \cdot & \gamma^T \end{bmatrix} \begin{bmatrix} \mathbf{x}_0 \\ \mathbf{x}_1 \\ \cdot \\ \cdot \\ \cdot \\ \mathbf{x}_n \end{bmatrix} = \begin{bmatrix} \Theta_1 \\ \mathbf{b}_1 \\ \cdot \\ \cdot \\ \mathbf{b}_n \\ \Theta_2 \end{bmatrix}, \quad (21)$$

where the 2×2 matrices W_{i1}, W_{i2} and the 2-vectors $\mathbf{b}_i$ are obtained from (19)–(20)

$$W_{i1} = -I/h_{i-1} - A(t_{i-1})/2,$$

$$W_{i2} = I/h_{i-1} - A(t_i)/2, \qquad i = 1, \dots, n,$$

$$\mathbf{b}_i = (\mathbf{r}(t_i) + \mathbf{r}(t_{i-1}))/2, \tag{22}$$

and I is the 2×2 identity matrix.

The changes in passing to general mth-order systems are simply that the blocks W_{ij} will be m-dimensional, and that α and γ will be replaced by pxm and qxm matrices, respectively. For the nonlinear systems (1)–(2) the discretization is the same:

$$(\mathbf{y}_{i+1} - \mathbf{y}_i)/h_i = (\mathbf{f}(t_{i+1}, \mathbf{y}_{i+1}) + \mathbf{f}(t_i, \mathbf{y}_i))/2,$$

$$\mathbf{g}(\mathbf{y}_0, \mathbf{y}_n) = \mathbf{0}, \tag{23}$$

and if Newton's method is used in the solution of the algebraic equations, then, at each step of that process, one obtains a system of linear equations with a matrix of coefficients of the form (21). The W_{ij} are formed now with the Jacobian matrices of $\mathbf{f}$ with respect to $\mathbf{y}_i$, etc. This, by the way, will coincide with the discretization of the linearized equations (15).

System (21) is very sparse, especially for moderate to large n. Very efficient and numerically stable, special Gaussian elimination methods have been devised for solving such systems [17], [18], and most successful implementations use some variant of them [20], [32].

We will not explain these methods in detail here, but will say that alternate row and column pivoting is used to preserve both stability and sparseness. It is possible in this fashion to obtain an LU (triangular) decomposition of the matrix in (21) such that *no* new nonzero elements are generated.

4. ERROR ANALYSIS

In order for a method of the complexity we have been describing to be accepted in today's Numerical Analysis world, it needs to have a good theoretical foundation.

Usually, because of the generality of the problem, although one may have a very complete theory, it may be very hard or even impossible to apply it in a particular instance (i.e., to verify the hypotheses). That is one reason why there is the need, when developing modern numerical software, for extensive numerical experimentation. This experimentation serves to complement the theoretical results, to delimit the true areas of application and, last but not least, to validate the program and certify that the designer's claims are fulfilled in actual practise.

The field of Theoretical Numerical Analysis of Differential Equations is fairly new. In our current area of interest, Ordinary Differential Equations, one of the main landmarks is Dahlquist's theory of the 1950s [5], and its popularization in Henrici's classical book of 1962 [13]. From there on the process accelerated, and currently there is a fairly complete theoretical understanding of numerical procedures for both initial and boundary value problems [16]–[18].

Concurrently there was a development of the abstract theory of discretization of equations in function spaces, which was more or less a natural extension of Dahlquist's theory in ODE's and that of Lax [37] in partial differential equations. See for instance [29]–[31].

The key words in this theory and its generalizations are "consistency," "stability," and "convergence."

A discretization is said to be *consistent of order p* (in $h = \max h_i$) if, when replacing $\mathbf{x}_i$ by the exact solution of the continuous problem ((1) or (5)), $\mathbf{x}(t_i)$, the residual is $\mathbf{0}(h^p)$. For instance, the discretization (19) is consistent of order 2 with problem (5). This is easily verified by writing (19) as

$$\big(\mathbf{x}(t_{i+1}) - \mathbf{x}(t_i)\big)/h_i$$

$$-\tfrac{1}{2}\big[A(t_{i+1})\mathbf{x}(t_{i+1}) + A(t_i)\mathbf{x}(t_i) + \mathbf{r}(t_{i+1}) + \mathbf{r}(t_i)\big] = \tau_i$$

expanding in Taylor's series around $t_i + (h_i/2)$, and using the fact that $\mathbf{x}(t)$ is the solution of the differential equation (5). The residual τ_i is called the *local truncation error* of the discretization.

One of the main worries when one performs large-scale computations with (necessarily) finite precision (numbers have to be truncated to fit the finite word length of the computer, infinite processes have to be interrupted sooner or later, and so on) is the propagation and amplification of errors.

This problem has been recognized earlier in the mathematical modeling of natural phenomena, where the observations will necessarily be tainted with measurement errors. Thus, "good" mathematical models are supposed to yield problems which are "well posed" (in the sense of Hadamard). Essentially what this means, without entering into deep technicalities, is that the model behaves continuously with respect to perturbations in the data. In other words, small changes in the input should only produce small changes in the output.

Our numerical modeling is affected in a similar way by the errors mentioned above, and the notion corresponding to well posedness is that of "stability." Depending upon the problem, there are several technically different notions of stability that are in common use. For our current purpose, the idea of the solution to the discrete equations changing little for small changes on the data of the problem is sufficient.

There is, however, an added complication. We must consider not one discretization on a fixed mesh Π, but rather an infinite family of them, $\{\Pi_\nu\}_{\nu=1,...}$ with maximum stepsize h_ν satisfying $h_\nu \to 0$ for $\nu \to \infty$. More precisely, if $F_{\Pi_\nu}(X) = 0$, $\nu = 1, \ldots$, represents a set of difference equations on the mesh Π_ν, we shall say that the method is *stable* if for any family of pairs of mesh functions

$$\{X^\nu\} = \{\mathbf{x}_i^\nu\}, \{Y^\nu\} = \{\mathbf{y}_i^\nu\}$$

we have

$$\|X^\nu - Y^\nu\|_v \leqslant c\|F_{\Pi_\nu}(X^\nu) - F_{\Pi_\nu}(Y^\nu)\|_v \tag{24}$$

where $\|\cdot\|_v$ are vector norms, and c is a constant *independent* of ν.

We finally say that a method is *convergent of order p* if and only if

$$\|X^\nu - \{\mathbf{x}(t_i)\}\|_v = \mathbf{0}(h_\nu^p), \qquad \nu \to \infty. \tag{25}$$

The fundamental result in this theory is:

THEOREM 26. *Let us consider a method F_Π consistent of order p with problem* (1) *(or* (5)). *Let us assume that $F_{\Pi_\nu}(X) = 0$ has solutions X^ν for a family of meshes Π_ν, $\nu = 1, \ldots,$. If F_Π is stable, then it is convergent of order p.*

Proof. Let us write inequality (24) for X^ν and the corresponding discretizations of the exact solution of (1) (or (5)), $\{\mathbf{x}(t_i)\}_{t_i \in \Pi_\nu}$. Then $\|X^\nu - \{\mathbf{x}(t_i)\}\|_v \leqslant c\|F_{\Pi_\nu}\{\mathbf{x}_i\}\|_v = \mathbf{0}(h_\nu^p)$, by the consistency of F_Π. ∎

This is an excellent example of the power of synthesis of abstract formulations and proofs. It is a very important result that has been proved independently for a number of problems and their discretizations. A formulation very similar to the one we have given here permits one to obtain all those particular results in an unified, clean way [29].

The assumption that solutions of the discrete problem exist can be removed, since it is possible to prove their existence, for sufficiently small h_ν, from the existence of the solution to the continuous problem [18].

Asymptotic Expansions. The $\mathbf{0}(h^p)$ relationships give some information about the behavior of a discretization but, with added regularity in the problem, much more can be said.

Using longer Taylor expansions one can obtain more detailed expressions for the local truncation error, and in turn, a general theorem of Stetter [42] allows us to derive detailed asymptotic expansion in powers of h for the global error.

For instance, for the trapezoidal rule applied to problem (1) the local error τ_i can be expressed as

$$\tau_i = \sum_{\nu=1}^{L} \frac{\nu}{2^{2\nu-1}(2\nu+1)} \mathbf{f}_{i+1/2}^{(2\nu)} h_i^{2\nu}/(2\nu)! + \mathbf{0}(h^{2L+2}), \quad (27)$$

where we have assumed sufficient differentiability for $\mathbf{f}(t, \mathbf{y})$.

Stetter's Theorem says in a nutshell that, under appropriate hypotheses, the global error

$$\mathbf{e}_i = X_i - \mathbf{x}(t_i) \tag{28}$$

has a similar expansion in even powers of h:

$$\mathbf{e}_i = \sum_{\nu=1}^{L} \mathbf{E}_\nu(t_i) h^{2\nu} + \mathbf{0}(h^{2L+2}), \tag{29}$$

where the coefficients $\mathbf{E}_\nu(t)$ do not depend upon h, and satisfy the linearized equations

$$\mathbf{E}'_\nu = \mathbf{f}_\mathbf{y}(t, \mathbf{y}(t)) \mathbf{E}_\nu + \Phi_\nu(t), \tag{30}$$

$$\mathbf{g}_\mathbf{y}(a)\mathbf{E}_\nu(a) + \mathbf{g}_\mathbf{y}(b)\mathbf{E}_\nu(b) = \mathbf{0},$$

with $\Phi_\nu(t)$ appropriate forcing functions that are constructed during the proof (and contain in particular the successive terms of the expansion of the local error). This expansion is understood to be valid for h sufficiently small. For more details we refer to [29], [42], [43].

5. VARIABLE ORDER METHODS

Stetter's Theorem seems to have only theoretical interest, since the functions Φ_ν in (30), and consequently the $\mathbf{E}_\nu$ in (29) are very difficult to calculate in any practical situation. However, knowing the *structure* of the error, even if one does not know the precise coefficient functions, has given impulse to a substantial development in the area of high order methods obtained from combinations or improvement of low order ones.

Successive Richardson Extrapolations. This now classical method is a very natural consequence of expansion (29). In fact, assume that we have computed a discrete solution X_Π; then subdivide each interval in Π in two equal parts, call the resulting mesh

$\Pi/2$, and the corresponding discrete solution $X_{\Pi/2}$. Let us also assume that expansion (29) is valid for Π (and therefore for $\Pi/2$). Let h be the maximum stepsize in Π, and $h/2$ that of $\Pi/2$.

Then, the Richardson Extrapolation consists of combining the values of X_{Π} and $X_{\Pi/2}$ at the common mesh points (those of Π), according to the following formula:

$$X_{\Pi}^{(1)} = \frac{4X_{\Pi/2}^{(0)} - X_{\Pi}^{(0)}}{3} \tag{31}$$

where we have called $X_{\Pi}^{(0)} \equiv X_{\Pi}$, and $X_{\Pi/2}^{(0)}$ to the restriction of $X_{\Pi/2}$ to the coarsest mesh.

Observing that the error associated with $X_{\Pi/2}^{(0)}$ satisfies

$$\mathbf{e}_{\Pi/2}(t) = \sum_{\nu=1}^{L} \mathbf{E}_\nu(t)(h/2)^{2\nu} + \mathbf{0}(h^{2L+2}),$$

it is easy to verify that

$$X_{\Pi i}^{(1)} - \mathbf{x}(t_i) = \sum_{\nu=2}^{L} \mathbf{E}_\nu^{(1)}(t_i) h^{2\nu} + \mathbf{0}(h^{2L+2}).$$

That is, the new discrete solution $X_{\Pi}^{(1)}$ is accurate to order h^4. This process can be continued by computing further solutions with the trapezoidal rule for finer and finer meshes $\Pi_j = \Pi/m_j$, with m_j a sequence of increasing integers. For instance, for $m_j = 2^j$, the successive extrapolates are defined by:

$$X_{\Pi j}^{(i)} = \left(4^i X_{\Pi_{j+1}}^{(i-1)} - X_{\Pi_j}^{(i-1)}\right)/(4^i - 1),$$

$$i = 1,\ldots,L \qquad j = 0,\ldots,i-1. \tag{32}$$

$X_{\Pi_j}^{(i)}$ satisfies

$$X_{\Pi_j l}^{(i)} - \mathbf{x}(t_l) = \sum_{\nu=i+1}^{L} \mathbf{E}_\nu^{(i)}(t_l) h^{2\nu} + \mathbf{0}(h^{2L+2}), \tag{33}$$

i.e., any entry in the ith column of this so-called generalized Romberg triangle has order of approximation $0(h^{2i+2})$.

The only fact that has been used to achieve this result is the existence of the expansion (29) in even powers of h. The method can be extended to many other problems, types of expansions and sequences $\{m_i\}$ (see [14], [31]).

Iterated Deferred Corrections. A successive Richardson Extrapolation is simple to implement but has two drawbacks: (a) it requires the computation of discrete solutions in finer and finer meshes to obtain more accurate results only on the coarsest one; (b) considerable storage is required, since it is necessary to keep at least a whole row of solutions in order to progress in the Romberg triangle. These drawbacks are quite noticeable for our problem when $m \times n_0$ is large (we recall that m is the dimensionality of the differential system, while n_0 is the number of mesh points in the coarsest mesh). This is so because the work involved in obtaining the discrete solution is proportional to a power of this quantity, and it gets worse on the finer meshes where $n_i = n_0 \times m_i$.

· The difference correction method of Fox [8], [9] avoids this difficulty. This method permits us to increase the order of accuracy of our discretization by solving problems with the basic operator F_Π on one fixed mesh. The improvement is achieved by computing successive corrections, which are used as forcing terms in the discrete equations. Thus we can employ repeatedly all the software developed for the basic discretization (say the trapezoidal rule), and the dimensionality of the auxiliary problems will never increase.

Fox developed this method, and he and his collaborators have applied it to a number of problems since 1947. We have given an abstract formulation with a complete theory in a number of papers [29], [30], and have applied it to a number of new problems, including the development of usable software [20], [31], [32], [34]. For some more recent results see Keller and Pereyra [19].

It is not difficult to write approximations G_Π of higher order of consistency with (1) than the trapezoidal rule. One way of doing this is to approximate the terms of the local truncation error expansion (27) by finite differences. In this case G_Π will be of the form

$$G_\Pi(X) = F_\Pi(X) - S_\Pi(X), \tag{34}$$

where S_Π is a correction term, and we have used a minus sign for later convenience. Generally, S_Π will involve more mesh points per equation than F_Π (for instance, a second order approximation to $\mathbf{f}_{yy}(t_i + h_i/2)$ will require four mesh points), and it will give rise to systems with a larger band width, which will therefore be more complicated and expensive to solve.

The deferred correction method takes advantage of the particular form of G_Π in order to preserve the simple structure of the basic method, while still obtaining a heightened precision. How to do it is pretty obvious from (34). After solving $F_\Pi(X) = 0$ and obtaining the $\mathbf{0}(h^2)$ approximation $X^{(0)}$, one solves

$$F_\Pi(X) = S_\Pi^{(1)}(X^{(0)}), \qquad (35)$$

where $S_\Pi^{(1)}(\mathbf{x}) = \mathbf{f}_{i+1/2}^{(2)} h_i^2/48 + \mathbf{0}(h^4)$ (see (27)). It is not too difficult to prove that $S_\Pi^{(1)}(X^{(0)}) - S_\Pi^{(1)}(\mathbf{x}) = \mathbf{0}(h^4)$. In fact, since only linear approximations will be considered, it is enough to apply $S_\Pi^{(1)}$ to expansion (29) in order to obtain the result.

The beauty of the process is that it can be repeated, by bringing in successive terms of the local truncation error and thus gaining two orders of accuracy in each correction. Thus, if $S_\Pi^{(\nu)}(\mathbf{x})$ approximates the first ν terms of the local error to order $h^{2\nu+2}$, then the solution $X^{(\nu)}$ to

$$F_\Pi(X) = S_\Pi^{(\nu)}(X^{(\nu-1)}) \qquad (36)$$

will satisfy

$$X_i^{(\nu)} - \mathbf{x}(t_i) = \mathbf{0}(h^{2\nu+2}).$$

We have skipped a number of technical and practical difficulties, which can be found in the references above. Recently, interest in this technique has revived, and a number of new theoretical results have been published [3], [21], [41], [62].

Iterated Defect Corrections. With a different approach, Stetter [44] recently observed that really it suffices to approximate the local truncation error to high order, and that the explicit expansion (27) is not required. Extending ideas of Zadunaisky [46], [47], Stetter and his collaborators [10], [11] have developed several variants of a method that they call "iterated defect corrections." Lindberg [21] established the connection between one of these variants and our procedure. He also applied it to a number of difficult problems for which expansions for the local error were not readily available.

Variable Order Methods and Error Estimation. The methods we have mentioned above have the special feature that they can be employed recursively, and give an effective way of achieving variable order algorithms, capable of adapting themselves in an automatic fashion to the problem at hand.

In order to implement such an adaptive strategy, one must have good error estimators, and it comes as no small blessing that this type of method has natural ones built in.

Actually, it is quite obvious that when one has two approximations X, Z to the solution of a problem on the same mesh Π, if Z is more precise than X, then the difference $(X - Z)$ will give an estimate of the error $(X - x)$ (i.e., Z acts as x, on the mesh Π). A number of error estimators associated with the procedures mentioned above rest on this fact. It is also true that in this area, global methods for BVP fare much better than those based on IV techniques. The fact that we have stored a complete solution makes the a posteriori error estimation and control particularly easy, while with Shooting techniques it is practically impossible.

6. ADAPTATIVE MESH SELECTION

We indicated earlier that the finite difference methods can be applied on general meshes. This gives considerable freedom to choose the location of the mesh points $\{t_i\}$. However, it also imposes upon us the burden of making that choice to minimize their number. This is the problem of mesh selection.

Unfortunately, the optimal location of the mesh points usually depends upon the solution to the problem, and it is therefore not decidable a priori; thus the need for adaptive procedures for obtaining this optimal mesh.

Initial value methods incorporated this feature of dynamically chosen meshes into their implementations long ago, and current software reflects the mature state of that art. However, marching procedures can only adapt locally and they cannot foresee the future. One can only correct a poor choice by backtracking and recomputing. When we consider that IV codes are used to solve BVP's in a complicated fashion, by combining the solution to several problems computed independently, it is not at all clear that the end result is optimal in any respect. This is a matter that requires further study since it has received little attention from the experts.

Ten years ago, the situation with respect to mesh selection was no clearer when using global methods for BVP's. Some practical procedures had been developed for special cases [15], [28], [52], based on the intuitive notion that one wants a finer mesh where the solution varies more rapidly. On a different front, a number of authors (see [35] for detailed references) were producing interesting results for the optimal location of knots in the spline approximation of functions. Some of these ideas were being employed for the optimal mesh location problem both in scalar 2PBVP [6] and in elliptic PDE's [48].

In [35], Pereyra and Sewell connected these results with the general discretization problem by global methods and a number of current developments [39], [45] have followed from it. The main idea in [35] was to equidistribute (approximately) an integral norm of the local truncation error, i.e., to choose the optimal mesh so that on each subinterval $[t_i, t_{i+1}]$ some integral involving the local error has a constant value. Of course, since the local error depends upon the exact solution, we can only implement this idea in an iterative fashion, using estimates based on the discrete solutions. This iterative procedure has the advantage that we can interrupt it whenever an adequate mesh has been found, which is usually much cheaper than trying to find the optimal one.

After this, one of the most interesting contributions is due to White [45], who managed to formulate the equidistribution problem as an enlarged differential system. By means of a change of independent variable that depends upon the exact solution, and a clever manipulation, White showed that a new system could be obtained in which the variable mesh function was one of the unknowns. Thus, by solving this system on a uniform mesh one can

obtain simultaneously the optimal mesh and the solution to the
original problem on it. White showed some impressive evidence of
the power of his procedure in solving problems with very thin
boundary layers with just a few mesh points. This type of problem
is very common in fluid mechanics applications, in which the
solution varies rapidly in a thin layer and then is very tame in the
bulk of the integration region. A uniform mesh procedure would
need thousands of mesh points in order to resolve the thin-layer
behavior, and use an unnecessarily fine mesh in the smooth part of
the integration interval.

White's procedure has two drawbacks, however. First of all, even
linear problems get transformed into nonlinear ones. What is worse,
the resulting nonlinear equations are very hard to solve numerically. This is just another manifestation of the well-known principle
of "conservation of the difficulties" (or "you don't get anything for
nothing"), and it arises because one is in fact solving for the
optimal mesh, a tough problem.

Nonetheless it is appealing to think of transforming the independent variable to smooth out the boundary layer, and then working
with a uniform mesh in the new variable. In [27] we have modified
White's procedure in the following manner. An analytical change of
variable is postulated. The discrete problem is solved in the new
variable with a uniform mesh. Using this discrete solution and the
principle of equidistribution, one can construct an updated change
of variable, and repeat the procedure. The computation of the
discrete solutions can be done on a mesh with a fixed number of
points: the change of variables will distribute them appropriately.

In this way we have been able to remove the drawbacks mentioned above, since linear problems will remain linear under a
known change of variables, and nonlinear ones will not change their
difficulty in any essential way. Also, this two-pass procedure permits us to stop at any suboptimal mesh that seems adequate for our
purposes, without having to go all the way to an optimal one, which
may be expensive and unnecessary.

In the course of formulating our algorithm we had to face the
problem of interpolating monotone data by smooth monotone
splines. This is a problem that has only recently received attention
from the specialists [12], [22], [36], and as a result of this investi-

gation a new algorithm is proposed in [26], which is more suited to our needs than any of the formerly available ones.

7. APPLICATIONS

One question some people ask themselves before starting in a long course of study is "Does this subject have any use?" In our case: "Are there any 2PBVP's that appear when modeling real world situations?" The answer is YES, many times over.

If, a few years ago, one had questioned (as I did) the manager of a scientific computing center, one might have received a fairly negative answer. Customers interested in ODE's used to request only IV solvers. Also, elementary, and even not so elementary, books on Numerical Analysis devoted only token space (or no space at all) to BVP's. This is changing now, as we can see, for instance, in [54].

The lack of interest at the computing center level was really a natural consequence of the lack of knowledge about the problem. Really, many of the users were interested in solving BVP's, but they knew that there was no software available, and so they just went and devised their own, requesting IVP codes, and modifying them in the most obvious and straightforward manner. Then they would usually rediscover over and over the standard difficulties with shooting procedures.

In the past few years some fairly powerful, general codes (PASVA2 [20], PASVA3 [32], COLSYS [55], SUPORT [56], BOUNDS [58]) have found their way into many computing libraries, and a growing number of important problems are being solved by them. We will mention just a few that represent the wide spectrum that the subject covers.

The most classical one is the ballistic problem. The equations of movement of a missile in a gravitational field are ordinary differential equations. If we want it to go from one specified location to another, the resulting problem is a 2PBVP. In current applications, the problem is more complicated because of the need to control the trajectory in order to optimize some cost function. It turns out that through Pontrjagin's Maximum Principle many of these optimal control problems can also be reduced to 2PBVP (see [54] for an elementary description). A very important source of problems is the semidiscretization of partial differential equations. Also, many problems that have symmetries, although originally formulated as PDE's, can be reduced to ODE's, and usually the resulting problem has boundary conditions associated with it. Sometimes however, other not so obvious reductions are possible. For instance, the propagation of elastic waves in a general isotropic medium is modeled by the wave equation, a partial differential equation in three variables. However, a high-frequency approximation allows us to obtain a good deal of information by solving a system of ODE's, the so-called ray equations of geometrical optics. This problem has special interest in seismology, underwater exploration, energy resource prospecting, etc. For details on its formulation and numerical solution see [33].

Mejía and collaborators [23], [24], [40], [53] have worked on problems of modeling the chemical behavior of kidneys. Some of these problems can be expressed as 2PBVP's, and working with a modification of our program, PASVA3, by R. Le Veque and a number of other techniques, they have been able to solve some substantial nonlinear problems.

Also, at the National Institutes of Health, Rinzel and Miller [38] have used a finite difference code for finding periodic solutions to the Hodgkin-Huxley equations that model the space-clamped squid's giant axon. This is a classical problem in mathematical biology.

There are many applications for the modeling of chemical reactors at various places [1], [49]. In [4] we have used PASVA2 for calculating the shape of liquid meniscus under strange gravitational conditions. This work is used at NASA to study the position of fuel in rockets, and thus to aid in the proper design of fuel intakes.

A traditional source of problems is fluid mechanics. For some recent applications in this area see [49]–[51].

REFERENCES

1. T. S. Ahn, School of Chem. Eng., Univ. of New South Wales, Australia. Personal communication (1980).
2. J. P. Aubin, "Approximation des spaces de distributions et des opérateurs différentiels," *Bull. Soc. Math.* **12**, France (1967).

3. J. Christiansen and R. D. Russell, "Deferred corrections using nonsymmetric end formulas," *Numer. Math* **35** (1980), 21–33.

4. P. Concus and V. Pereyra, "A package for calculating axisymmetric menisci," Lawrence-Berkeley Lab. Rep. 8700 (1979). *Acta Cient Venezolana* **34** (1983).

5. G. Dahlquist, "Convergence and stability in the numerical integration of ODE's," *Math, Scand.* **4** (1956), 33–53.

6. C. De Boor, "Good approximation by splines with variable knots II," *Lect. Notes Math* **363**. Springer-Verlag, Berlin (1973), 12–20.

7. L. Fox, "Some improvements in the use of relaxation methods for the solution of ordinary and partial differential equations," Proc. Roy Soc. London **A190** (1947), 31–59.

8. _____, *The Numerical Solution of Two Point Boundary Value Problems in Ordinary Differential Equations*, Oxford U. Press (1957).

9. _____, (editor) *Numerical Solution of Ordinary and Partial Differential Equations*, Pergamon Press, Oxford (1962).

10. R. Frank, "The method of iterated defect correction and its application to 2PBVP's. Part I." *Numer. Math.* **25** (1976), 409–419.

11. R. Frank, J. Hertling, and C. W. Ueberhuber, "Iterated defect correction based on estimates of the local discretization error," Rep. 18/76, Inst. Numer. Math., Techn. Univ. Wien (1976).

12. F. N. Fritsch and R. E. Carlson, "Monotone piecewise cubic interpolation," *SIAM J. Numer. Anal.* **17** (1980), 238–246.

13. P. Henrici, *Discrete Variable Methods in ODE's*, Wiley, New York (1962).

14. D. C. Joyce, "Survey of extrapolation processes in numerical analysis," *SIAM Rev.* **13** (1971), 435–490.

15. E. Kalnay de Rivas, "On the use of nonuniform grids in finite-difference equations." *J. Comp. Phys.* **10** (1972), 202–210.

16. H. B. Keller, *Numerical Methods for Two-Point Boundary Value Problems*, Ginn-Blaisdell, Waltham, Mass. (1968).

17. _____, "Accurate difference methods for nonlinear two-point boundary value problems." *SIAM J. Numer. Anal.* **11** (1974), 305–320.

18. _____, *Numerical Solution of Two Point Boundary Value Problems*, SIAM, Philadelphia (1976).

19. H. B. Keller and V. Pereyra, "Difference methods and deferred corrections for ordinary boundary value problems," *SIAM J. Numer. Anal.* **16** (1979), 241–259.

20. M. Lentini and V. Pereyra, "An adaptive finite difference solver for nonlinear two-point boundary problems with mild boundary layers," *SIAM J. Numer. Anal.* **14** (1977), 91–111.

21. B. Lindberg, "Error estimation and iterative improvement for the numerical solution of operator equations," *BIT* **20** (1980), 486–500.

22. D. Mc Allister, E. Passow, and J. Roulier, "Algorithms for computing shape preserving spline interpolation to data," *Math. Comp.* **31** (1977), 717–725.

23. R. Mejía, "Six-tube vasa recta model," personal communication (1979).

24. R. Mejía and J. L. Stephenson, "Numerical solution of multinephron kidney equations," *J. Comp. Phys.* **32** (1979), 235–246.

25. M. Osborne, "On shooting methods for boundary value problems," *J. Math. Anal. Appl.* **27** (1969), 417–433.

26. G. Pagallo and V. Pereyra, "Smooth monotone spline interpolation," *Lect. Notes Math* **909**, Springer-Verlag, Berlin (1982), 142–146.

27. _____, "Mesh selection by adaptive changes of independent variable," manuscript (1980).

28. C. Pearson, "On a differential equation of the boundary layer type," *J. Math. & Phys.* **47** (1968), 135–154.

29. V. Pereyra, "Iterated deferred corrections for nonlinear operator equations," *Numer. Math.* **10** (1967), 316–323.

30. _____, "Highly accurate discrete methods for nonlinear problems," MRC Tech. Rep. 749, Math. Res. Center, Univ. Wisconsin, Madison (1967).

31. _____, "Accelerating the convergence of discretization algorithms," *SIAM J. Numer. Anal.* **4** (1967), 508–533.

32. _____, "PASVA3: an adaptive finite difference FORTRAN program for first order nonlinear, ordinary boundary problems." In [57] 67–88.

33. V. Pereyra, W. H. K. Lee, and H. B. Keller, "Solving two-point seismic ray tracing problems in a heterogeneous medium," *Bull. Seismological Soc. of America* **70** (1980), 79–99.

34. V. Pereyra, W. Proskurowski, and O. Widlund, "High order fast Laplace solvers for the Dirichlet problem on general regions," *Math. Comp.* **31** (1977), 1–16.

35. V. Pereyra and E. G. Sewell, "Mesh selection for discrete solution of boundary problems in ODE's," *Numer. Math.* **23** (1975), 261–268.

36. S. Pruess, "Alternatives to the exponential spline in tension," *Math. Comp.* **33** (1979), 1273–1281.

37. R. D. Richtmyer and K. W. Morton, *Difference Methods for IVP's*, 2nd ed. Interscience, Wiley, New York, 1967.

38. J. Rinzel and R. N. Miller, "Numerical calculation of stable and unstable periodic solutions to the Hodgkin-Huxley equations," *Math. Biosc.* (1980).

39. R. D. Russell and J. Christiansen, "Adaptive mesh selection strategies for solving BVP's," *SIAM J. Numer. Anal.* **15** (1978), 59–80.

40. J. L. Stephenson, R. P. Tewarson, and R. Mejia, "Quantitative analysis of mass and energy balance in non-ideal models of the renal counterflow system," *Proc. Nat. Acad. Sc.* **71** (1974), 1618–1622.

41. R. D. Skeel, "A theoretical framework for proving accuracy results for deferred corrections," submitted to *SIAM J. Numer. Anal.*

42. H. Stetter, "Asymptotic expansions for the error of discretization algorithms for nonlinear functional equations," *Numer. Math.* **7** (1965), 18–31.

43. _____, *Analysis of Discretization Methods for ODE's*, Springer-Verlag, Berlin, 1973.

44. _____, "The defect correction principle and discretization methods," *Numer. Math.* **29** (1978), 425–443.

45. A. B. White, "On selection of equidistributing meshes for 2PBVP's," *SIAM J. Numer. Anal.* **16** (1979), 472–502.

46. P. E. Zadunaisky, "A method for the estimation of errors propagated in the numerical solution of a system of ODE's," *Proc. Astronomical Union*, *Symp. 25*, Academic Press, New York, (1966).

47. _____, "On the estimation of errors propagated in the numerical integration of ODE's," *Numer. Math* **27** (1976), 21–39.

48. E. G. Sewell, "Automatic generation of triangulations for piecewise polynomial approximation," Ph.D. Thesis, Purdue Univ., Lafayette, Ind. (1972).

49. Ir. J. P. Roos, "Some problems solved with the aid of SYSSOL/PASVAR," Akzo Res. Lab. Arnhem, Holland. Personal communication (1975).

50. M. Lentini and H. B. Keller, "Computation of Kármán swirling flows," *Lect. Notes Comp. Sc.* **76** (1979), 89–100.

51. M. Lentini, "Boundary value problems over semi-infinite intervals," Ph.D. Thesis, Caltech (1978).

52. H. B. Keller, personal communication (1972).

53. R. Mejía, J. L. Stephenson, and R. J. Le Veque, "A test problem for kidney models," *Math. Biosc.* **50** (1980), 129–131.

54. J. Stoer, and R. Bulirsch, *Introduction to Numerical Analysis*, Springer-Verlag (1980).

55. U. Ascher, J. Christiansen, and R. D. Russell, "Collocation software for boundary value ODE's," *ACM Trans. Math. Software* (1979).

56. M. L. Scott and H. A. Watts "SUPORT—a computer code for 2PBVP's via orthonormalization," SAND 75–0198, Sandia Labs., Albuquerque (1975).

57. B. Childs, M. Scott, J. W. Daniel, E. Demman, and P. Nelson, (editors), *Codes for BVP in ODE's*, *Lect. Notes Comp. Sc.* **76** (1979).

58. R. Bulirsch, J. Stoer, and P. Deuflhard "Numerical Solution of nonlinear two-point boundary value problems I" *Num. Math.*, *Handbook Series App.* (in preparation).

59. R. D. Russell, *Numerical solution of Boundary Value Problems*, Pub. 79–06, Esc. Comp. Univ. Central de Venezuela (1979).

60. H. B. Keller, "Numerical solution of boundary value problems for ordinary differential equations: survey and some recent results on difference methods," *In Numerical Solution of BVP for ODE's* (ed. A. K. Aziz), Academic Press, New York, 1975, 27–88.

61. M. Urabe, "An existence theorem for multipoint boundary value problems," *Funkcial. Ekvac.* **9** (1966), 43–60.

62. K. Böhmer, "Discrete Newton methods and iterated defect corrections," *Numer. Math.* **37** (1981), 167–192.

MULTIGRID METHODS FOR PARTIAL DIFFERENTIAL EQUATIONS

Dennis C. Jespersen

Apologia

This article is an attempt to describe multigrid methods in a manner suitable for novices. No previous exposure to multigrid methods is necessary in order to read it. Some readers may feel I am too informal and spend too much time introducing some topics. I ask them to remember this is not a research article and that the audience will include (I hope) people outside the numerical analysis community. I have in mind as reader the proverbial "mathematically mature" person who is not a specialist in numerical analysis and who might appreciate a little less terseness than is the norm in journal articles nowadays. Readers who have a numerical analysis background are invited to skim Section 1.

1. INTRODUCTION

How can one best solve very large systems of equations, say systems of equations with many thousands of unknowns? This is an important question, since such systems of equations regularly turn up in numerical solution of multidimensional heat conduction and fluid dynamics problems, for example. There is no general answer to the question; if you are faced with a very large set of equations to solve there may be little to do except try some general-purpose method and be prepared to spend a lot of time and money on computer solution.

If you know more about the problem, then more specific methods, which take advantage of some knowledge of the properties of the problem or the solution, can be used. In particular, if the large system of equations comes from discretization of some partial differential equation or system of partial differential equations, then the very fact that there is a continuum problem which is modeled by the discrete one may help in constructing efficient solution techniques. This is the essence of *multigrid* methods, so-called because they use a family of discrete problems (defined on several grids), all approximating the same continuous problem, to help in the solution of the large system of equations. The objectives of this article are to give a simple introduction to multigrid methods, to explain the key features of multigrid methods, and to try to give some insight as to why these methods may be "optimal" (in some sense) for solving certain classes of equations. This is primarily a descriptive article; a few key references will be given but no attempt will be made to provide a complete bibliography. The attention of numerical analysts has been drawn to multigrid methods by the work of Achi Brandt, and many of the ideas presented in this article are due to him.

The term "multigrid methods" is hard to define. There are several algorithms which can be thought of as falling into the class of multigrid methods, and adaptive methods (algorithms in which the computer automatically adapts the solution process to the solution) are sometimes considered to be a component of multigrid methods. The general principle is to keep the continuum origin of the problem in mind when constructing a numerical algorithm for its solution.

Before beginning with the description of multigrid methods, it is well to fix the framework for the ideas that follow. Assume we wish to solve some problem in which the variables of interest to us, such as temperature or velocities, are "given" as solutions of some partial differential equation or set of partial differential equations. For example, one might consider the steady-state heat-conduction equation (this is Poisson's equation, the topic of the next two chapters)

$$\Delta u = f \quad \text{in a region } \Omega$$

$$u = g \quad \text{on the boundary of } \Omega$$

or the steady-state Navier-Stokes equations

$$\mathbf{u} \cdot \nabla \mathbf{u} = (\text{Re})^{-1} \Delta \mathbf{u} + \nabla p \quad \text{in } \Omega \left(\mathbf{u} = (u_1, u_2)^T \text{ or } (u_1, u_2, u_3)^T \right)$$

$$\nabla \cdot \mathbf{u} = 0 \quad \text{in } \Omega$$

(plus suitable boundary conditions), where $\Delta u = \sum \dfrac{\partial^2 u}{\partial x_i^2}$ and $\nabla \cdot \mathbf{u} = \sum \dfrac{\partial u_i}{\partial x_i}$.

These equations can be solved exactly only in special cases. Thus one is led to consider numerical solution techniques. Standard finite difference or finite element techniques often lead (for a linear partial differential equation) to a set of linear equations. The task then is to solve the linear system.

There are two fundamentally different approaches to the solution of the linear system. One approach, the direct method, is simply to solve the set of linear equations to machine accuracy using as solution technique Gaussian elimination or some variant thereof. This approach has the undeniable appeal that there is no uncertainty as to what the computed solution represents: it is the solution (modulo rounding errors) of the discrete equations. The disadvantages of this method are practical in nature, having to do with storage requirements and computing time required. Problems in 2 and especially 3 dimensions can quickly grow to enormous size. For example, a finite-difference method for the steady-state Navier-Stokes equations in 3 dimensions on a $20 \times 20 \times 20$ grid gives a system of equations in 32000 unknowns (8000 grid points times 4 unknowns per grid point), leading to a matrix with more than 1,000,000,000 entries. Now, most of these entries are zeros and need be neither computed nor stored; the matrix is very "sparse." However, when a direct method such as Gaussian elimination is applied to the matrix, the phenomenon of "fill-in" occurs, in which nonzero entries are created by the addition of multiples of one row to another. A quick estimate shows that Gaussian elimination applied to such a matrix would require about 50,000,000 words of storage, so large a number as to create massive data management problems and enormously increase the time and expense of the calculation, and, in fact, the calculation would be extremely difficult in practice. For certain problems with regular geometries and "nice" operators (such as the Laplacian: linear, constant coefficients), direct methods have been devised that solve very large systems of equations much faster than Gaussian elimination. Methods utilizing the fast Fourier transform and cyclic reduction are examples of such fast direct methods. The requirement of certain geometries and operators is not too restrictive in practice, since such problems occur frequently. Fast direct methods are very valuable in such cases. Direct methods are fine when applicable, but their limited applicability restricts their usefulness.

The other class is the iterative methods; multigrid methods are in this class. Methods in this category do not attempt to solve the equations "exactly" (although they could produce a solution accurate to machine precision if allowed to run long enough). Rather, all that is desired is that a solution accurate to some tolerance ε be produced. This notion of an approximate solution is especially appropriate to finite difference methods for partial differential equations, since the exact solution of the discrete equations is (generally) in error itself; we cannot hope to capture the exact behavior of the solution of the differential equation on a discrete grid. Given that, it seems reasonable not to ask for the exact solution of the discrete equations; rather, one could ask for an approximate solution of the discrete equations, say, an approximate solution that differs from the solution of the differential equation by about the same amount that the exact solution of the difference equations differs from the continuum solution. Such a solution is all we can expect from the discrete equations anyway.

The class of iterative methods includes the Jacobi method, the Gauss-Seidel method, successive overrelaxation (SOR), alternating-direction implicit (ADI) methods, Chebyshev semi-iterative methods, conjugate direction methods, and multigrid methods. Iterative methods have the advantage of minimal storage, requiring typically only the nonzero entries of the matrix and a few vectors of corresponding length. Sometimes even the storage of the matrix is not required, all that may be needed is an *algorithm* which will produce the matrix-vector product $A\mathbf{x}$ given the vector $\mathbf{x}$. Iterative methods are also typically easier to program than direct methods. The main disadvantages of iterative methods are their sometimes slow convergence and the problem of deciding when to terminate the iterations. For some problems practical considerations dominate, and iterative methods are the only possible candidates for solution algorithms. For other problems, iterative methods may be able to reach an acceptable solution in less time than direct methods even when direct methods are applicable. Multigrid methods may be described as a class of iterative methods which attempt to take advantage of the existence of an underlying continuous problem in reaching a discrete solution.

In the next section the basic multigrid idea is introduced, beginning with a model problem in one space·dimension. The concepts of smoothing, restriction, and interpolation are discussed. Methods for analysis of multigrid methods are considered, and the section concludes with a set of exercises illustrating one means of analysis, the invariant subspace analysis. Section 3 discusses the full multigrid method, nonlinear problems, adaptive methods, and multigrid-finite element techniques (the latter two topics very briefly). In Section 4 work estimates for multigrid methods are derived and compared with work estimates for other solution techniques. Finally, Section 5 surveys some published applications of multigrid methods.

2. BASIC IDEAS

In this section we will illustrate the multigrid idea by applying the method to two model problems, the first in one space dimension and the second in two dimensions. The power of multigrid methods does not make itself felt in one dimension, but one dimension is a good place to start in understanding how the methods work.

So, consider the differential equation

$$u''(x) = f(x), \qquad 0 < x < 1$$

$$u(0) = 0, \qquad u(1) = 0. \tag{2.1}$$

Let us lay down a uniform grid on $[0,1]$ with mesh spacing $h := 1/(M+1)$ (the notation $a := b$ means a is defined as b) for some odd integer M and define grid points $x_j := jh$, $0 \leq j \leq M+1$. A Taylor series expansion shows that if u has four continuous derivatives, then

$$u(x_{j+1}) - 2u(x_j) + u(x_{j-1}) = h^2 u''(x_j) + O(h^4),$$

If $u(x)$ has four continuous derivatives and satisfies (2.1) we find, therefore,

$$h^{-2}\big(u(x_{j+1}) - 2u(x_j) + u(x_{j-1})\big) = f(x_j) + O(h^2), \qquad 1 \leq j \leq M.$$
$$\tag{2.2}$$

Let us now define a grid function $\mathbf{u}^1$ (think of $\mathbf{u}^1$ as an M-vector $(u_1^1, u_2^1, \ldots, u_M^1)^T$) by the set of equations

$$h^{-2}\big(u_{j-1}^1 - 2u_j^1 + u_{j+1}^1\big) = f(x_j), \qquad 1 \leq j \leq M,$$

where $u_0^1 := 0$ and $u_{M+1}^1 := 0$. This is the discrete analog to (2.2). We can write this system of equations as a matrix equation

$$A^1 \mathbf{u}^1 = \mathbf{f}^1, \tag{2.3}$$

where the superscript 1 refers to the grid, A^1 is the M-by-M matrix

$$h^{-2} \begin{pmatrix} -2 & 1 & & & & \\ 1 & -2 & 1 & & & \\ & 1 & -2 & 1 & & \\ & & & \ddots & & \\ & & & 1 & -2 & 1 \\ & & & & 1 & -2 \end{pmatrix}$$

and $\mathbf{f}^1$ is the M-vector $(f(x_1), f(x_2),\ldots, f(x_M))^T$. It is a standard result that the system (2.3) has a unique solution $\mathbf{u}^1_*$ and that $\max_j |u^1_{*j} - u(x_j)| \leqslant Ch^2$ as $h \to 0$, where C is some positive constant independent of h. Thus the discrete solution $\mathbf{u}^1_*$ approximates the continuous solution $u(x)$ at the grid points increasingly accurately as h decreases.

The system (2.3) can be easily solved by standard methods. Multigrid methods are not necessary to solve this problem. This problem is, however, a simple setting in which to consider the application of the basic multigrid ideas.

The simplest part of a multigrid method is called a "coarse grid correction" algorithm. The name comes from the use of a coarse grid (with mesh spacing $2h$, say) to correct an approximate solution on the fine (mesh spacing h) grid. Here is a precise description of a coarse grid correction algorithm that uses only one auxiliary grid.

Let $G^1(\cdot, \cdot)$ denote a relaxation process on the fine grid (examples for this and the other processes will be given soon). Let R^2_1 ("restriction") denote a mapping from grid functions on the fine grid to grid functions on the coarse grid. Let I^1_2 ("interpolation") denote a mapping from functions on the coarse grid to functions on the fine grid. Let A^2 be the matrix which is obtained by discretizing the differential equation on the coarse grid. The steps in the one-auxiliary grid process would be:

1. Choose an initial guess $\mathbf{u}^1$.
2. (Do ν relaxation sweeps): for $k = 1$ to ν do $\mathbf{u}^1 \leftarrow G^1(\mathbf{u}^1, \mathbf{f}^1)$. (Here the left arrow denotes replacement, so $a \leftarrow b$ means replace a by b.)
3. (Transfer residual to coarse grid): let $\mathbf{r}^1 := \mathbf{f}^1 - A^1\mathbf{u}^1$ (the residual), and let $\mathbf{r}^2 := R^2_1\mathbf{r}^1$.
4. (Solution process on coarse grid): Solve $A^2\mathbf{e}^2 = \mathbf{r}^2$.
5. (Interpolation to fine grid and solution update): Do $\mathbf{u}^1 \leftarrow \mathbf{u}^1 + I^1_2\mathbf{e}^2$.
6. (Stopping test): If some stopping criterion is satisfied, halt; else go to step 2.

A fuller explanation of each phase of the process follows. First, the relaxation process. We begin with some initial guess $\mathbf{u}^1$. The relaxation process for multigrid methods typically consists of a few steps of some convergent iterative process on the fine grid. Such a process might be the Jacobi method (sometimes called the method of simultaneous displacements) defined (for this model problem (2.1)) by $\mathbf{u}^{1,0}$ given; for $n = 0,1,\ldots$ do:

for $j = 1$ to M do:

$$u_j^{1,\,n+1} := \tfrac{1}{2}\Big[u_{j-1}^{1,\,n} + u_{j+1}^{1,\,n} - h^2 f(x_j)\Big] \tag{2.4}$$

(here the first superscript refers to the grid while the second superscript is an iteration counter). We will see later that the Jacobi method may not be a good choice of relaxation method in conjunction with multigrid methods, but an underrelaxed Jacobi method has some appeal. Another process is the Gauss-Seidel method (method of successive displacements), defined by $\mathbf{u}^{1,0}$ given; for $n = 0,1,\ldots$ do:

for $j = 1$ to M do:

$$u_j^{1,\,n+1} := \tfrac{1}{2}\Big[u_{j-1}^{1,\,n+1} + u_{j+1}^{1,\,n} - h^2 f(x_j)\Big]. \tag{2.5}$$

A more general relaxation algorithm might have the form

$$\mathbf{u}^{1,\,n+1} := \mathbf{u}^{1,\,n} + C^{-1}\big(A^1\mathbf{u}^{1,\,n} - \mathbf{f}^1\big)$$

where C is nonsingular; in this case the relaxation operator G is given by $G^1(\mathbf{u}^1,\mathbf{f}^1) = (I + C^{-1}A^1)\mathbf{u}^1 - C^{-1}\mathbf{f}^1$. Such methods are designed to be convergent processes and will, if iterated long enough, produce an acceptably accurate numerical solution. The multigrid idea is to take only a few steps of such a method, however, and then to proceed with phase 2. The rationale for this is that standard relaxation processes are very efficient at reducing high frequency components of the error, but inefficient at reducing low frequency error components. (We will look at this claim in more detail later.) Hence after a few steps (say one to three) of the relaxation method, the high frequencies of the error will be substantially reduced while the low frequencies might be only marginally reduced. At this stage the principle of relying on the existence of an underlying continuum problem is invoked, and a discrete problem on a coarser grid is introduced.

Phase 2 begins by defining the residual; if one denotes the error by

$$\mathbf{e}^1 := \mathbf{u}^1_* - \mathbf{u}^1$$

where $\mathbf{u}^1_*$ denotes the exact solution of the discrete equations, one sees that

$$A^1\mathbf{e}^1 = \mathbf{f}^1 - A^1\mathbf{u}^1 = \mathbf{r}^1$$

so that the original matrix problem (2.3) is solved if we solve the linear system

$$A^1\mathbf{e}^1 = \mathbf{r}^1. \tag{2.6}$$

The first step in "solving" this linear system is to transfer the residual to the coarser grid via some restriction operator R_1^2. Examples of possible restriction operators will be given below.

Phase 3 consists of the solution of a set of equations on the coarse grid. The coarse grid will have many fewer points than the fine grid (one-half as many in one dimension, one-fourth as many in two dimensions, and so on), thus the work required to solve the coarse grid problem should be much less than the work required to solve the fine grid problem. There is one possible problem here. Since there are fewer mesh points in the coarse grid, the solution of the coarse grid equations will not contain high frequencies; here "high" refers to rapid fluctuations on the scale of the fine grid. We hope that the relaxation process has smoothed out the error sufficiently well so that the high frequencies are (almost) gone, and hence the error and residual are well approximated on the coarse grid.

Phase 4 consists of the interpolation of the computed solution on the coarse grid to all points of the fine grid and the updating of the approximate solution on the fine grid. (An example of an interpolation process will be given below.) The hope is that the interpolated

$I_2^1 \mathbf{e}^2$ is sufficiently close to the exact error $\mathbf{e}^1$ that the updating process on the fine grid significantly improves the quality of the solution. After phase 4, phase 1 can begin again, with the improved $\mathbf{u}^1$ as the starting guess. The entire process is to be continued until some convergence criterion is met.

A further phase could be added here, namely, ν' cycles of some relaxation process (not necessarily the same as in phase 1). This "postrelaxation" would be intended to reduce high frequency components of the error which may have been excited in the transfer from the coarse grid to the fine grid.

One of the key ideas, and the source of the name "multigrid", is that the problem to be solved in phase 3 is a problem of the same type as the original problem. Therefore the multigrid process can be applied to it: we take some starting guess on the coarse grid, do a few relaxation sweeps, then transfer the problem to a still coarser grid, etc. The transferring to coarser grids might stop when so few grid points remain that the cost of the exact solution of the problem on that level is negligible. After solving the problem exactly on that level, the interpolation process begins, with transfers from coarser to finer levels until the finest level is reached. The whole algorithm can be written in an elegant recursive form, where the multigrid process for the solution on grid 1 calls the multigrid process for the solution on grid 2, which calls the multigrid process for solution on grid 3, etc. Here is such an algorithm, in a quasi-algorithmic language. "CGC" stands for "coarse grid correction."

procedure CGC(f, u, A, level, pre-relax, post-relax, mgcycles)
    {comment: level tells what level we're on (1 = finest level)
    A is the matrix
    f is the right-hand side
    u is the solution variable
    pre-relax tells how many relaxation steps to take
        before going to the next coarsest level
    post-relax tells how many relaxation steps to take
        after coming from the next coarsest level
    mgcycles defines the depth of the iterative procedure
    begin by calling CGC with level = finest level}

```
begin
  if (level = coarsest level) then
                    u := A⁻¹f;
  else
    begin
    for m := 1 to mgcycles do
      begin
      relax pre-relax times;
      residual: = restriction of f - Au to the next coarsest grid;
      call CGC (residual, v, A, level + 1, pre-relax, post-relax);
      u := u + interpolation of v from the next coarsest grid;
      relax post-relax times;
      end;
    end;
end.
```

If mgcycles = 1 the algorithm is called a "V-cycle", while if mgcycles = 2 the name "W-cycle" has been applied. These algorithms can be represented pictorially as follows (four levels has been assumed):

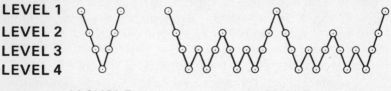

LEVEL 1
LEVEL 2
LEVEL 3
LEVEL 4

V-CYCLE   **W-CYCLE**

The reduction in the size of the problems that are to be solved on each successive level is exceedingly rapid in two and three dimensions; the number of grid points decreases by factors of 4 and 8, respectively, as each coarser level is reached. Thus a two-dimensional problem for which the finest grid was 256 by 256 would be reduced in 8 levels of recursion to a problem with just 1 grid point.

A host of questions suggest themselves. What are good relaxation processes? How many relaxation sweeps should be taken? How should one define the restriction and interpolation operators? Is the problem on the coarse grid really a problem of the same type as that on the fine grid, as asserted above? Will the whole process converge, and, if so, how well? How does one handle nonlinear problems? How should one manage the data base (i.e., the arrays of solution values at the various levels)? If the process converges, what is the order of magnitude of the number of arithmetic operations required? Some of these questions will be touched on in the remainder of this article.

Let us first consider some motivation for the coarse-grid-correction algorithm. The remark was made above that traditional relaxation processes are good at eliminating short wavelength errors, but are slow to remove long wavelength errors. A precise analysis, using Fourier methods, can be given for model problems, and this analysis and a means of approximately analyzing this phenomenon will be given later. If we grant the proposition concerning short vs. long wavelength error components, it follows that after a few steps of our relaxation process we will reach a point of diminishing returns. If the error is now "smooth" on the fine grid, both it and the residual should be well-representable on the coarse grid, so that transferring the residual to the coarse grid can be done without much loss of information. Now, an error component of low frequency on the fine grid appears on the coarse grid as a higher frequency component, hence it can be more efficiently removed by relaxation. (The relaxation process on the fine grid is often referred to as "smoothing," since the high-frequency content of the error on the fine grid has been reduced by the relaxation.)

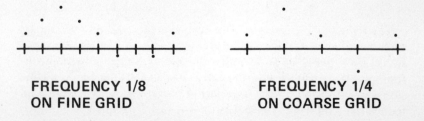

FREQUENCY 1/8   FREQUENCY 1/4
ON FINE GRID   ON COARSE GRID

The "smooth" structure of the error and residual leads one to consider very simple restriction and interpolation processes. For the one-dimensional problem, two restriction processes come quickly to mind. First is the simple transfer of even-subscripted residual values to the coarse grid:

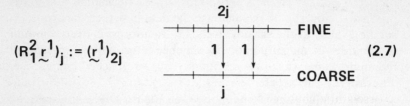

$$(R_1^2 \underset{\sim}{r}^1)_j := (\underset{\sim}{r}^1)_{2j} \tag{2.7}$$

Second is a weighted transfer of residual values to the coarse grid, perhaps:

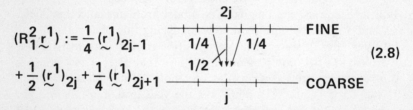

$$(R_1^2 \underset{\sim}{r}^1) := \frac{1}{4} (\underset{\sim}{r}^1)_{2j-1} + \frac{1}{2} (\underset{\sim}{r}^1)_{2j} + \frac{1}{4} (\underset{\sim}{r}^1)_{2j+1} \tag{2.8}$$

Both of these seem eminently reasonable. For the interpolation process, linear interpolation immediately suggests itself:

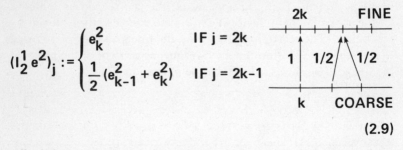

$$(I_2^1 e^2)_j := \begin{cases} e_k^2 & \text{IF } j = 2k \\ \frac{1}{2} (e_{k-1}^2 + e_k^2) & \text{IF } j = 2k-1 \end{cases} \tag{2.9}$$

All of these processes work well when one considers low-frequency oscillations: not much information is lost in the transfer from fine to coarse grids, and the interpolation process produces a function on the fine grid which consists primarily of low frequencies. In fact, these simple processes (and their extensions to two dimensions) are very well suited for use with multigrid methods.

A precise means of analyzing the two-grid process can be given in terms of standard matrix iterative methods. The one-auxiliary grid algorithm outlined above is a stationary linear iterative method, hence the error obeys the recursion

$$\mathbf{e}^{1,1} = T\mathbf{e}^{1,0}, \tag{2.10}$$

where T is some matrix describing the total process. Here the first superscript identifies the grid while the second superscript counts the number of steps of the iterative process. The matrix T is constructible from the relaxation, restriction, interpolation, and coarse-grid operators. To show how this construction proceeds, we begin with the relaxation process on the coarse grid. This is usually some stationary iterative process of the form

$$\mathbf{u}^{1,n+1} = \mathbf{u}^{1,n} + C^{-1}(A^1\mathbf{u}^{1,n} - \mathbf{f}^1), \tag{2.11}$$

where C is some nonsingular matrix such that C^{-1} is easily formed (or that linear systems of the form $Cx = b$ are easily solved). This immediately implies that the error is governed by the recursion

$$\mathbf{e}^{1,n+1} = (I + C^{-1}A^1)\mathbf{e}^{1,n}, \tag{2.12}$$

(*Exercise.* Show this!) and hence that the error after ν relaxation steps on the fine grid is given by

$$\mathbf{e}^{1,\nu} = (I + C^{-1}A^1)^\nu\mathbf{e}^{1,0} =: G^\nu\mathbf{e}^{1,0}. \tag{2.13}$$

Then the residual, which satisfies $\mathbf{r}^1 = A^1\mathbf{e}^1$ (see (2.6)), is transferred to the coarser grid via R_1^2, the matrix $(A^2)^{-1}$ is applied, the result is transferred back to the fine grid via the interpolation operator I_2^1, and this quantity is then added to $\mathbf{u}^1$. The result of combining all these steps is that the error satisfies the relation

$$\mathbf{e}^{1,\nu+1} = \left(I - I_2^1(A^2)^{-1}R_1^2A^1\right)G^\nu\mathbf{e}^{1,0}, \tag{2.14}$$

(*Exercise.* Verify this!) and so the matrix T of the total process is given by

$$T = \left(I - I_2^1(A^2)^{-1}R_1^2A^1\right)G^\nu. \tag{2.15}$$

One can then analyze the effect of different relaxation and interpolation processes by considering their effect on the matrix T. The desire is to make the spectral radius of T as small as possible. (This is because of a standard result that a stationary iterative process with iteration matrix T converges if and only if the spectral radius of T is less than 1, and, furthermore, the process asymptotically converges faster the smaller is the spectral radius of the iteration matrix.) The main drawback in this approach is that the matrix T is not generally conveniently given in a closed form, hence numerical techniques must generally be employed in the analysis, and these in turn limit one to studying model problems. When this analysis can be carried out it gives satisfactory results; for example, for the model problem (2.1) in one dimension with Jacobi underrelaxation for the smoothing and the linear operators (2.7) and (2.9) for the transfers, the spectral radius of the matrix T is on the order of $1/2$ for $\nu = 1$, on the order of $1/4$ for $\nu = 2$, and tends to 0 as $\nu \to \infty$, independent of h (see Hackbusch [7]). This allows one to conclude that the iterative process will produce a solution accurate to within a tolerance ε in $O(|\log\varepsilon|)$ steps, independent of h. Proofs of convergence for two-dimensional problems can be constructed using this idea as the basis for the analysis (see Hackbusch [8] and Wesseling [15], for example). The proofs are theoretically satisfying but, unfortunately, not useful in practice due to the loss of precision in estimates within the proof. For example, Wesseling proves the convergence of a multigrid method for rather general two-dimensional elliptic problems in the unit square, but the estimates are such that the validity of the proof requires more than 20,000,000 smoothing steps on the fine grid!

A technique called *local mode analysis* is advocated by Brandt [3] as a means of analyzing the relaxation process and speed of convergence of the overall multigrid method. It is claimed that this technique always gives reliable estimates of the overall convergence rate of the multigrid method. The main idea of the technique is to use Fourier analysis and look at the effect of the relaxation scheme on the high frequencies of the error. (A more refined version of this analysis considers the effect of the relaxation process, the restriction process, and the interpolation process, but for now let's just look at the effect of the relaxation scheme.) As applied to our model problem (2.1), the analysis might begin like this. Extend the error vector $\mathbf{e}^1$ to all grid points jh, $-\infty < j < \infty$, by first extending $\mathbf{e}^1$ as an odd function about $j = 0$ (this defines e_j^1 for $-M \leqslant j < 0$) and then extending $\mathbf{e}^1$ by periodicity, so $e_{j+2M+2}^1 = e_j^1$ for all j. Then we can write the Fourier series (it turns out to be convenient to work with complex Fourier series) for $\mathbf{e}^1$ as

$$e_k^1 = \sum_{j=-1/h+1}^{1/h} a_j e^{i\pi jkh}, \qquad -M \leqslant k \leqslant M+1, \qquad (2.16)$$

where

$$a_j = \frac{1}{2M+2} \sum_{k=-1/h+1}^{1/h} e_k^1 e^{-i\pi kjh}.$$

Now, consider, for example, the Jacobi overrelaxation method with parameter ω, defined by (2.4) with the last line of (2.4) replaced by

$$u_j^{1,n+1} := u_j^{1,n} + \omega h^2 \left(h^{-2} \left(u_j^{1,n} - 2u_j^{1,n} + u_j^{1,n} \right) - f(x_j) \right).$$

We see that the error satisfies the homogeneous difference equation

$$e_j^{1,n+1} = e_j^{1,n} + \omega \left(e_{j-1}^{1,n} - 2e_j^{1,n} + e_{j+1}^{1,n} \right). \qquad (2.17)$$

If we write a_j' for the Fourier coefficients of $\mathbf{e}^1$ after one step of the relaxation process, we see that

$$e_k^{1,1} = \sum_j a_j' e^{i\pi jkh} = \sum_j \left(a_j + \omega \left(a_j e^{-i\pi jh} - 2a_j + a_j e^{i\pi jh} \right) \right) e^{i\pi jkh}$$

and hence

$$a_j' = a_j (1 + \omega (2\cos \pi jh - 2)) = a_j (1 - 4\omega \sin^2 \pi jh/2). \qquad (2.18)$$

Of the modes $e^{i\pi jkh}$ in the sums above, the ones which are faithfully represented on the coarse grid are those with $|jh| \leqslant 1/2$.

For example, for $jh = 1/2$ we have the mode $e^{i\pi k/2}$, which is alternately $+1$ and -1 at the coarse grid points, the highest possible frequency on the coarse grid. On the other hand, modes with $|jh| > 1/2$ will "alias" when transferred to the coarse grid. (For example, the mode with $jh = 1$ is $e^{i\pi k}$ which is a high frequency mode on the fine grid but which is identically 1 on the coarse grid if we consider the restriction operator (2.7). It is thus a low frequency mode; aliasing has occurred.) So, call the modes with $|jh| > 1/2$ the *high frequency* modes and judge the performance of the relaxation scheme by the *smoothing rate*

$$\mu := \max_{1/2 < |jh| \leqslant 1} |a'_j/a_j|. \tag{2.19}$$

This measures the worst possible high frequency error reduction. For the example above,

$$\mu = \max_{1/2 < |jh| \leqslant 1} |1 - 4\omega \sin^2 \pi jh/2|.$$

Treating jh as a continuous variable now, we see

$$\mu = \max(|1 - 2\omega|, |1 - 4\omega|).$$

Thus it is clear that the optimal ω is $1/3$, which gives a smoothing rate of $1/3$. (*Exercise*: What is the smoothing rate of the Jacobi method (2.4)? The answer to this exercise indicates why the Jacobi method without parameters does not have much appeal in conjunction with multigrid algorithms.)

In two dimensions, a Fourier expansion would look like

$$\sum_{j, m} a_{jm} e^{i\pi jkh} e^{i\pi jmh}$$

Now a high frequency is something which oscillates rapidly in *at least one* direction, hence, when the smoothing rate is calculated, the maximum is taken over the range $1/2 < \max(|jh|, |mh|) \leqslant 1$. More on this later in this section.

For another example, consider Gauss-Seidel as the relaxation process. For this example we have

$$e_j^{1,\,n+1} = \tfrac{1}{2}\Big(e_{j-1}^{1,\,n+1} + e_{j+1}^{1,\,n}\Big), \tag{2.20}$$

and hence a'_j is given by

$$a'_j = a_j \cdot \frac{e^{i\pi jh}}{2 - e^{-i\pi jh}}. \tag{2.21}$$

Now the smoothing factor is given by

$$\mu = \max_{1/2 < |jh| \leqslant 1} \left| \frac{e^{i\pi jh}}{2 - e^{-i\pi jh}} \right| \tag{2.22}$$

and it is easy to see that this is $5^{-1/2} \sim .447$. Each of the high frequencies is damped by at least this factor on each relaxation sweep. It is interesting that according to this analysis, the Gauss-Seidel method is not as favorable as the underrelaxed Jacobi method, contrary to the usual ranking of the two methods, which takes into account their behavior on all frequencies. The Gauss-Seidel method indeed damps the lower frequencies faster than the Jacobi method, but in the multigrid analysis one only requires that the relaxation method be efficient in removing high frequency components of the error. *Exercise*: Show that the smoothing rate of SOR (for any over-relaxation parameter) is not as good as that of Jacobi underrelaxation with $\omega = 1/3$. SOR for (2.1) is obtained by replacing the last line of (2.5) by

$$u_j^{1,\,n+1} = \tfrac{1}{2}\omega\Big(u_{j-1}^{1,\,n+1} + u_{j+1}^{1,\,n}\Big) - \omega h^2 f(x_j) + (1-\omega)u_j^{1,\,n}.$$

The local mode analysis is not rigorous. It assumes periodic boundary conditions, and can handle variable coefficient problems only by "freezing" the coefficients. Practitioners of local mode analysis apply it to problems with variable coefficients and any boundary conditions by freezing the coefficients at some local values and ignoring the boundaries. Heuristically, one might say that local mode analysis is concerned only with short wavelength oscillations, and for short wavelengths the boundary is unimportant (unless one is adjacent to the boundary) and coefficients do not vary much. Hence one expects the local mode analysis to be a valuable tool in the design and analysis of multigrid algorithms. Indeed, Brandt claims that the analysis always predicts the correct rate of convergence of the overall multigrid process, and if the rate of convergence observed in practice does not agree with the rate predicted by the local mode analysis, then there is a bug in the computer program!

We close this section with a description of the coarse-grid correction algorithm applied to Poisson's equation in 2 dimensions,

$$\Delta u = f \text{ in } \Omega := (0,1) \times (0,1),$$

$$u = 0 \text{ on the boundary of } \Omega. \tag{2.23}$$

This is the most studied (and overused!) example problem for elliptic equation solvers; the ability to solve this problem efficiently is *sine qua non* for any method which is claimed to be a good method for numerically solving elliptic equations.

Lay down a uniform grid with mesh width $h := 1/(M+1)$ in both directions, where M is an odd integer. For finite difference equations, use the standard 5-point discretization of the Laplacian,

$$\left(A^1 \mathbf{u}^1\right)_{jk} := h^{-2}\left(u^1_{j+1,k} + u^1_{j-1,k} + u^1_{j,k+1} + u^1_{j,k-1} - 4u^1_{jk}\right).$$

$$\tag{2.24}$$

(This is the natural extension to two dimensions of the one-dimensional discretization discussed above.) The system of equations is thus

$$\left(A^1 \mathbf{u}^1\right)_{jk} = f(x_j, y_k), \qquad 1 \leqslant j, k \leqslant M$$

$\left(\text{with } u_{jk} := 0 \quad \text{if } j = 0 \text{ or } k = 0 \text{ or } j = M+1 \text{ or } k = M+1\right).$

The proposed coarse-grid correction algorithm for this problem is the same as for the one-dimensional problem given above: relaxation on fine grid, transfer of residual to coarse grid, solution process on coarse grid, transfer back to fine grid, and update. The restriction and interpolation processes are the natural extensions to two dimensions of the one-dimensional processes in (2.7)—(2.9). For example, the linear interpolation operator is given by

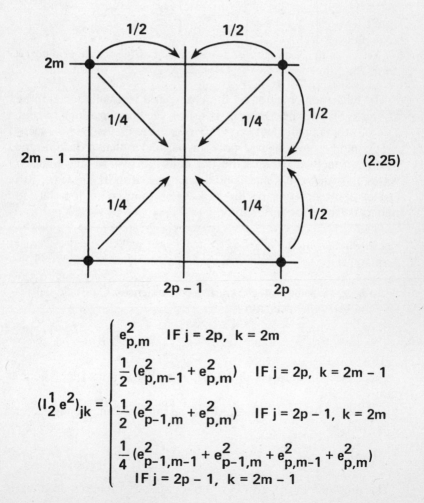

(2.25)

$$
(I_2^1 e^2)_{jk} = \begin{cases}
e^2_{p,m} & \text{IF } j = 2p, \ k = 2m \\[2ex]
\dfrac{1}{2}(e^2_{p,m-1} + e^2_{p,m}) & \text{IF } j = 2p, \ k = 2m-1 \\[2ex]
\dfrac{1}{2}(e^2_{p-1,m} + e^2_{p,m}) & \text{IF } j = 2p-1, \ k = 2m \\[2ex]
\dfrac{1}{4}(e^2_{p-1,m-1} + e^2_{p-1,m} + e^2_{p,m-1} + e^2_{p,m}) \\
\qquad \text{IF } j = 2p-1, \ k = 2m-1
\end{cases}
$$

and the restriction operator R_1^2 can be given by simple injection, analogous to (2.7), or by a weighted injection such as

$$(R_1^2 r^1)_{j,k} = \frac{1}{2} r^1_{2j,2k} + \frac{1}{8} (r^1_{2j-1,2k} + r^1_{2j+1,2k} + r^1_{2j,2k-1} + r^1_{2j,2k+1}) \qquad (2.26)$$

The relaxation operator on the coarse grid is usually taken to be of Gauss-Seidel type. One can consider Jacobi relaxations but for ease of programming and economy of storage Gauss-Seidel sweeps are in more common use; note in particular that a Jacobi-type relaxation method requires the use of two arrays for the solution vector **u**, while with Gauss-Seidel only one array is needed. With lexicographic ordering of the unknowns (bottom to top, left to right) and the error vector $e^1(x, y)$

$$e^1(x, y) = \sum_{|jh|, |kh| \leqslant 1} a_{jk} e^{ij\pi x} e^{ik\pi y}$$

(writing x, y instead of jh, kh for convenience), Gauss-Seidel relaxation gives the recursion

$$e^{1,n+1}(x, y) = \tfrac{1}{4}\big(e^{1,n+1}(x - h, y) + e^{1,n+1}(x, y - h)$$

$$+ e^{1,n}(x + h, y) + e^{1,n}(x, y + h)\big).$$

Local mode analysis then would indicate that the smoothing factor μ is given by

$$\mu = \max_{1/2 < \max(|jh|, |kh|) \leqslant 1} \left| \frac{e^{ij\pi h} + e^{ik\pi h}}{4 - e^{-ij\pi h} - e^{-ik\pi h}} \right|.$$

It is easy but not trivial to show that this maximum occurs at $\pi(x, y) = (\pi/2, \arccos(4/5))$, which yields $\mu = 1/2$. (*Exercise*: For the Jacobi underrelaxation with parameter ω, show that the optimal value of ω is $2/5$, giving a smoothing factor of $2/5$. This relaxation scheme probably would not be used in practice because of its extra storage requirement.)

It becomes clear (after working out a few smoothing factors by hand) that the job could get very tedious. In fact, a computer program called SMORATE has been written which takes information about the relaxation scheme as input and produces the smoothing rate μ as output. It is clear that such a program could be of great help in designing and optimizing multigrid algorithms. The task of choosing a suitable relaxation scheme (usually the most crucial component of a multigrid process) is greatly aided by automatic calculation of the smoothing factor.

EXERCISES ON INVARIANT SUBSPACES

The matrix iterative analysis is very cumbersome (virtually undoable in two and more dimensions), and the local mode analysis is not rigorous (except in the case of periodic boundary conditions). In some cases the total process can be analyzed in a way which is exact and which can be carried out in two or more dimensions. These cases are those in which each basic process (relaxation, restriction, interpolation) leaves certain subspaces invariant. These exercises will introduce these ideas. The first to notice and exploit these invariant subspaces was apparently Hackbusch [7].

First consider the two-level process applied to the one-dimensional problem $u'' = f$ on $[0,1]$. The discrete problem is $A^1 \mathbf{u}^1 = \mathbf{f}^1$. See (2.3). For the error analysis, we may consider the homogeneous case $\mathbf{f}^1 = 0$.

1. Show that the eigenvalues of A^1 are $\lambda^1_m := \frac{-4}{h^2} \sin^2 \pi m h /2$ $(1 \leqslant m \leqslant M)$ and that the eigenvector $\mathbf{v}^1_m$ associated with λ^1_m has components $(\mathbf{v}^1_m)_j = \sin(jm\pi h), 1 \leqslant j, m \leqslant M$.
2. (a) Show that the Jacobi iterative process $\mathbf{u}^1 \to \mathbf{u}^1 + \omega h^2 A^1 \mathbf{u}^1$ preserves the eigenvectors $\mathbf{v}^1_m$, in fact $\mathbf{v}^1_m \to (1 + \omega h^2 \lambda^1_m) \mathbf{v}^1_m$.

(b) *Definition.*   A mesh point jh is *red* if j is even, *black* if j is odd.

Definition.   *Red-black Gauss-Seidel* is the iterative process which proceeds as follows. First, for all red points do $u_j \leftarrow (u_{j-1} + u_{j+1})/2$. Then, for all black points do $u_j \leftarrow (u_{j-1} + u_{j+1})/2$. (This is just Gauss-Seidel with a different ordering of the unknowns.)

For $1 \leqslant m < (M+1)/2$, write $m' := M+1-m$.

Show that red-black Gauss-Seidel leaves the two-dimensional subspace V_m spanned by $\mathbf{v}_m^1$ and $\mathbf{v}_{m'}^1$ invariant; in fact, show that one step of red-black Gauss-Seidel sends $\mathbf{v}_m^1$ to

$$\cos m\pi h \cos^2 m\pi h / 2 \left(\mathbf{v}_m^1 + \mathbf{v}_{m'}^1 \right).$$

Hence, on V_m, red-black Gauss-Seidel has the matrix representation

$$\begin{pmatrix} c_{2m}c_m^2 & c_{2m}c_m^2 \\ -c_{2m}s_m^2 & -c_{2m}s_m^2 \end{pmatrix} \qquad \begin{aligned} c_{2m} &:= \cos m\pi h, \\ s_m &:= \sin m\pi h/2. \\ c_m &:= \cos m\pi h/2, \end{aligned}$$

Also, for $m = (M+1)/2$, $\mathbf{v}_m^1 \rightarrow 0$ under red-black Gauss-Seidel.

3.   Show that forming the residual sends $\mathbf{v}_m^1$ to $-\lambda_m^1 \mathbf{v}_m^1$.

4.   (a) Consider the unweighted restriction process (2.7). Show that under this restriction operator,

$$\mathbf{v}_m^1 \rightarrow \begin{cases} \mathbf{v}_m^2 & \text{if} \quad 1 \leqslant m < (M+1)/2 \\ 0 & \text{if} \quad m = (M+1)/2 \\ -\mathbf{v}_{M+1-m}^2 & \text{if} \quad (M+1)/2 < m \leqslant M \end{cases}$$

where $\mathbf{v}_m^2$ is defined by $(\mathbf{v}_m^2)_j = \sin(2h\pi mj)$, $1 \leqslant j, m \leqslant (M-1)/2$ and $A^2 \mathbf{v}_m^2 = \lambda_m^2 \mathbf{v}_m^2$. Hence $(\mathbf{v}_m^1, \mathbf{v}_{m'}^1) \rightarrow (\mathbf{v}_m^2, -\mathbf{v}_m^2)$ for $1 \leqslant m \leqslant (M-1)/2$, i.e., on V_m the restriction operator (2.7) has the matrix representation $[1, -1]$ (a 1-by-2 matrix).

(b) Show that the weighted restriction operator (2.8) has the matrix representation $[\cos^2 m\pi h/2, -\sin^2 m\pi h/2]$ on V_m.

5. Show that the exact solution process on the coarser grid has the matrix representation $[1/\lambda_m^2]$ on the space spanned by v_m^2. (Here the superscript denotes the grid.)

6. Show that under the interpolation process (2.9) $v_m^2 \to c_m^2 v_m^1 - s_m^2 v_{m'}^1, 1 \leqslant m \leqslant (M-1)/2$. (Note that a high frequency has been excited. This indicates that relaxation after update on the fine grid may be beneficial.) Hence the interpolation process has matrix representation

$$\begin{pmatrix} c_m^2 \\ -s_m^2 \end{pmatrix}.$$

7. Show that the total process leaves V_m invariant. On this subspace the matrix representation of the total process is

$$T_m = \left\{ \begin{pmatrix} 1 & 0 \\ 0 & 1 \end{pmatrix} - \begin{pmatrix} c_m^2 \\ -s_m^2 \end{pmatrix} \cdot \frac{1}{\lambda_m^2} \cdot R \cdot \begin{pmatrix} \lambda_m^1 & 0 \\ 0 & \lambda_{m'}^1 \end{pmatrix} \right\} \cdot G^\nu$$

where $R = [1, -1]$ or $[c_m^2, -s_m^2]$, ν is the number of relaxation steps, and

$$G = \begin{pmatrix} 1 + \omega h^2 \lambda_m^1 & 0 \\ 0 & 1 + \omega h^2 \lambda_{m'}^1 \end{pmatrix} \text{ or } G = c_{2m} \begin{pmatrix} c_m^2 & c_m^2 \\ -s_m^2 & -s_m^2 \end{pmatrix}.$$

8. *Example.* Consider Jacobi relaxation, $\omega = 1/4$, $\nu > 0$, and $R = [1, -1]$. Show that

$$T_m = \begin{pmatrix} 0 & \cot^2 m\pi h/2 \\ \tan^2 m\pi h/2 & 0 \end{pmatrix} \cdot \begin{pmatrix} c_m^{2\nu} & 0 \\ 0 & s_m^{2\nu} \end{pmatrix}$$

while for $m = (M+1)/2$, T_m is the 1-by-1 matrix $[2^{-\nu}]$, and hence $\rho(T) = 2^{-\nu}$. Here T is the matrix of the total process and $\rho(T)$ denotes the spectral radius of T. What smoothing rate is predicted by local mode analysis in this case?

Remark. The spectral radius of T decreases exponentially with ν, but this is not the whole story. The norm of T is more important in the early stages of a relaxation process, and one can show that $\|T\|_2 \sim 1/(e\nu)$ as $\nu \to \infty$.

9. *Example.* Show that red-black Gauss-Seidel with $\nu = 1$ and $R = [c_m^2, -s_m^2]$ produces a direct method, i.e.,

$$T_m = \begin{pmatrix} 0 & 0 \\ 0 & 0 \end{pmatrix} \text{ for all } m.$$

What smoothing rate is predicted by local mode analysis in this case?

10. Show that red-black Gauss-Seidel with $\nu = 1$ and $R = [1, -1]$ gives a matrix T_m with eigenvalues 0 and $-c_{2m}^2$. Hence $\rho(T) = \cos^2 \pi h$, so this process is not very good. Notice that local mode analysis won't distinguish between this case and the previous one.

11. Show that this analysis can be extended to two space dimensions. Consider the standard 5-point difference operator as in (2.24). Show that the eigenvalues and corresponding eigenvectors are

$$\lambda_{mn}^1 = -\frac{4}{h^2}\left(\sin^2 m\pi h/2 + \sin^2 n\pi h/2\right), \quad 1 \leq m, n \leq M$$

$$\left(v_{mn}^1\right)_{jk} = \sin m\pi jh \sin n\pi kh, \qquad\qquad 1 \leq m, n, j, k \leq M.$$

Let V_{mn} denote the subspace spanned by

$$v_{mn}^1, v_{m'n'}^1, v_{m'n}^1, \text{ and } v_{mn'}^1, 1 \leq m, n \leq (M-1)/2.$$

Definition. A grid point is *red* if $j + k$ is even, *black* if $j + k$ is odd.

Definition. *Red-black Gauss-Seidel* is the relaxation process wherein first all the mesh function values at the red points are adjusted so that the difference equations are satisfied at the red points, then all the mesh function values at the black points are adjusted so that the difference equations are satisfied at the black points.

Show that V_{mn} is invariant under the total process, provided that the relaxation operator is either (underrelaxed) Jacobi or red-black Gauss-Seidel, the restriction operator is either the operator (2.26) or the unweighted restriction operator mentioned above (2.26), and the interpolation operator is the bilinear interpolation (2.25). Hence the spectral radius of the total process can be calculated by computing eigenvalues of 4-by-4 matrices.

12. Show that all these ideas extend to three space dimensions (this is an open-ended problem).

3. EXTENSIONS

In this section we will describe some extensions and variants of the coarse-grid correction algorithm, and also consider the application of multigrid methods to finite element equations.

The first extension will justify the "multi" in the name multigrid method. Most of the discussion in the previous section was phrased in terms of only two grids. It turns out that the recursive nature of the multigrid process makes it possible to analyze the properties of a *multigrid* process in terms of a two-grid process. Thinking back to the algorithm *procedure CGC* of Section 2, let's construct the total process matrix which describes the behavior of the error under one step of the algorithm. Consider a sequence of grids $\Gamma^1, \Gamma^2, \ldots, \Gamma^L$ where Γ^1 is the finest grid and Γ^L is the coarsest. Write γ for the depth of the recursion (so if $\gamma = 1$ we get the V-cycle algorithm, for example), ν for the number of relaxation steps before proceeding to the next coarsest grid, and assume that no relaxation sweeps are carried out after returning from the next coarsest grid. Write R_k^{k+1} for the restriction map from level k to level $k+1$ and write I_{k+1}^k for the interpolation map from level $k+1$ to level k. Write G_k for the relaxation matrix on level k, so under one relaxation sweep on level k an error $\mathbf{e}^k$ is mapped to $G_k\mathbf{e}^k$. Write $T_{k,j}$ for the total process matrix starting on level k (finer grid) and going down to level j (coarser grid). We know from the previous section that $T_{k,k+1} = (I - I_{k+1}^k(A^{k+1})^{-1}R_k^{k+1}A^k)G_k^\nu$. Our objective is to determine $T_{k,j}$. This can be done recursively.

Proposition. The matrices $T_{k,j}$ are given recursively by

$$T_{k,k+1} = \left(I - I_{k+1}^k(A^{k+1})^{-1}R_k^{k+1}A^k\right)G_k^\nu, \qquad 1 \leqslant k < L$$

and

$$T_{k,j} = T_{k,k+1} + I_{k+1}^k T_{k+1,j}^\gamma (A^{k+1})^{-1}R_k^{k+1}A^kG_k^\nu, \qquad 1 \leqslant k < j \leqslant L.$$

Proof. Let's first work out the matrix $T_{1,3}$; this will show how the calculation goes. We begin with some error $\mathbf{e}^{1,0}$ on level 1. After ν relaxation steps we have the error $\mathbf{e}^{1,\nu}$. Then the right-hand side on level 2 is defined as $\mathbf{f}^2 = R_1^2A^1\mathbf{e}^{1,\nu}$. Since the initial guess on level 2 is taken to be zero, the initial error on level 2, $\mathbf{e}^{2,0}$, is $(A^2)^{-1}\mathbf{f}^2$. This maps to an error $T_{2,3}^\gamma\mathbf{e}^{2,0}$ under γ steps of the multigrid method (which is actually now a two-grid method) starting on level 2. Since approximate solution = true solution − error, the approximate solution on level 2 is now $(A^2)^{-1}\mathbf{f}^2 - T_{2,3}^\gamma\mathbf{e}^{2,0}$. Now the solution on level 1 is updated, which gives a new expression for the error,

$$\mathbf{e}^{1,\nu+1} = \mathbf{e}^{1,\nu} - I_2^1\left[(A^2)^{-1}\mathbf{f}^2 - T_{2,3}^\gamma\mathbf{e}^{2,0}\right]$$

$$= T_{1,2}\mathbf{e}^{1,0} + I_2^1T_{2,3}^\gamma(A^2)^{-1}R_1^2A^1G_1^\nu\mathbf{e}^{1,0},$$

and hence the matrix $T_{1,3}$ is given by

$$T_{1,3} = T_{1,2} + I_2^1T_{2,3}^\gamma(A^2)^{-1}R_1^2A^1G_1^\nu.$$

For the case of general k and j, the same arguments apply, *mutatis mutandis*. The only change is that an initial error $\mathbf{e}^{k+1,0}$ on level $k+1$ is mapped to $T_{k+1,j}^{\gamma}\mathbf{e}^{k+1,0}$ by the multigrid process between levels $k+1$ and j. This completes the proof.

The recursive expression for the error operators enables one to prove convergence, under certain hypotheses. Suppose that

1. $\|I_{k+1}^{k}\|\|(A^{k+1})^{-1}R_{k}^{k+1}A^{k}G_{k}^{\nu}\| \leqslant C$ for $1 \leqslant k < L$, where C is some positive constant;
2. $\|T_{k,k+1}\| \leqslant \varepsilon \leqslant \min\{1/4, 1/(4C)\}$;
3. $\gamma \geqslant 2$.

Then we have $\|T_{k,j}\| \leqslant 2\varepsilon \leqslant 1/2$ for $2 \leqslant k < j \leqslant L$. The proof is by induction on the difference $j - k$. For $j - k = 1$, the result holds by assumption. For $j - k > 1$, we have from the equation for $T_{k,j}$ that

$$\begin{aligned}
\|T_{k,j}\| &\leqslant \varepsilon + C\|T_{k+1,j}^{\gamma}\| \\
&\leqslant \varepsilon + C(2\varepsilon)^{\gamma} && \text{by the inductive hypothesis} \\
&\leqslant \varepsilon + C4\varepsilon^{2} && \text{since } \gamma \geqslant 2 \\
&\leqslant 2\varepsilon.
\end{aligned}$$

The weakness of this result lies in the hypotheses, which are very difficult to verify. Hackbusch [9] has made progress in this direction, but it would be desirable to have simpler means of verifying the hypotheses. Another difficulty is in getting realistic estimates of the constant C. If C is estimated to be too large, then an unrealistically large number of relaxation steps will have to be taken in order that assumption 2 above is satisfied (it is this difficulty that caused the estimate of 20,000,000 relaxation steps mentioned in Section 2). Notice that $\gamma = 1$ will also lead to convergence, provided that $C \leqslant 1/2$. Experience has shown that the multigrid method performs much better in practice than the proofs indicate; there is a wide gap between theory and practical experience, and it would be desirable to close this gap.

The second extension of the coarse-grid correction algorithm has to do with the problem of getting a good starting guess. The efficiency of an iterative method is often crucially dependent on the quality of the initial guess; the better the initial guess the sooner the iteration can be terminated. Thus one would like to specify an intelligent initial guess for the coarse-grid correction process. This is in general a difficult job. Then the thought comes to mind, why not specify an initial guess on the fine grid by interpolating some approximate solution from the coarse grid? Indeed, it is eminently sensible to specify the initial guess u^1 as $I_2^1 u^2$, where u^2 has been obtained by (approximately) solving the coarse grid problem. How does one solve the coarse-grid problem? By multigrid methods, naturally. How does one get an initial guess for the method started on the coarse grid? By interpolating a solution from a still coarser grid! Thus one is led to look at a coarse-to-fine process as the basis for a multigrid algorithm, turning the algorithms of the previous section upside down, as it were. This idea has been called the *full multigrid* method. An algorithm might be outlined as follows:

1. Solve the coarsest-grid problem directly.
2. Interpolate the solution to the next finest grid.
3. Perform the coarse-grid correction algorithm beginning on this level. If this is the finest grid, stop; else go to 2.

This algorithm automatically produces a good initial guess on the finest level because it uses the interpolated solution from the next-to-finest level, and the problem on this level has been solved accurately. This algorithm also has the attractive feature of producing good solutions at all immediate levels. In particular, the solution at the next-to-finest level could be compared with the solution at the finest level to get an error estimate.

In adaptive multilevel methods, one allows for adaptive switching between levels, where the process moves from one level to another by deciding if certain criteria are fulfilled. For example, if the rate of convergence of the relaxation process slows down on a given level (this requires some means of judging the rate of convergence and some standard for judging if the rate is too slow) the algorithm should automatically switch to a coarser level. On the other hand, if the problem on a given level has been solved to within truncation error, a switch should be made to a finer level. Coding such an algorithm is quite tricky; almost all published multigrid programs are of coarse-grid correction type rather than full multigrid type. (Hemker [10] has published some ALGOL 68 algorithms for adaptive multilevel computations.) It would be very desirable to have robust general-purpose adaptive multigrid software available. The current state of multigrid software is much more primitive than, say, software for initial value problems for ordinary differential equations. For ordinary differential equations there are several extremely flexible, powerful, and robust codes which are widely available. It will be quite a while yet before software for multigrid methods attains this level of maturity.

The exposition so far has been restricted to linear problems. Many very important and difficult problems from applications areas are nonlinear. For example, the Navier-Stokes equations mentioned in the first section are a nonlinear system. Heat conduction problems in which the conductivity depends on the temperature (as is the case in many applications) are also examples of nonlinear problems. It is very important to be able to handle nonlinear problems.

One possibility for treating nonlinearities is as follows. Write the differential equation as $L(u) = 0$. Discretize this equation somehow, giving the finite-dimensional set of nonlinear equations

$$L^1(\mathbf{u}^1) = 0.$$

Then consider Newton's method as a solution algorithm for this finite-dimensional system:

$$\mathbf{u}^{1,0} \text{ given; for } n = 0, 1, \ldots \text{ do:}$$

$$\text{solve } DL^1(\mathbf{u}^{1,n}) \Delta\mathbf{u}^{1,n} = -L^1(\mathbf{u}^{1,n});$$

$$\text{put } \mathbf{u}^{1,n+1} := \mathbf{u}^{1,n} + \Delta\mathbf{u}^{1,n}$$

where DL^1 is the Jacobian matrix of the mapping L^1. Thus at each Newton step we have to solve a linear problem, and these linear problems are candidates for solution by our previous multigrid algorithms. Newton's method is an excellent method, when it

works. The classic Kantorovich theorem gives a precise description
of the convergence behavior of the method, including the ultimately
quadratic convergence of the method (under certain hypotheses)
and bounds on how close to the solution the initial guess must be.
The basic problem with Newton's method, known to all solvers of
nonlinear equations, is the problem of finding a good initial guess.
In fact, the unmodified Newton method is unsuitable as a general-
purpose equation solver because of the possibly small domain of
convergence of the method. Some modifications must be made,
such as going to a damped Newton's method, or using a globally
convergent method and switching to Newton's method when indi-
cations are that Newton's method will succeed. The coarse-to-fine
viewpoint of multigrid mentioned above is a natural candidate for
obviating this problem. Instead of seeking a good initial guess on
the fine grid, one solves the nonlinear problem directly on the
coarsest grid (where a good solution can be obtained at low cost)
and then begins the process of coarse-to-fine transfers outlined in
the algorithm above. This idea is very natural in the multigrid
context and is liable to produce a guess on the finest grid that is
sufficiently accurate that Newton's method will converge without
difficulty.

There is an objection to the use of Newton's method that is
partly a matter of philosophy and partly a practical matter. Why,
one may ask, should you spend all that effort to get a solution to
the linear equations at each step of Newton's method when what
you really want is a solution of the nonlinear system of equations?
Would it not be more natural to attack the nonlinear equations
directly? This can indeed be done. An algorithm called the Full-
Approximation Storage (FAS, also referred to as Full-Approxima-
tion Scheme) has been outlined by Brandt. Suppose the problem on
the fine grid is written as

$$L^1(\mathbf{u}^1) = \mathbf{f}^1.$$

As in the linear problem, we carry out a few steps of some
relaxation process on the fine grid. (This may entail solving some
nonlinear equations. For example, a Gauss-Seidel process applied
to a system of nonlinear equations would proceed by adjusting the

jth unknown so that the jth equation was satisfied, where j runs through all the unknowns. If the jth equation is nonlinear in the jth unknown, we have to solve a scalar nonlinear equation. Fortunately, scalar nonlinear equations are much easier to solve than systems of nonlinear equations.) After the relaxation process we have some improved solution $\mathbf{u}^1$; form the residual $\mathbf{r}^1 := \mathbf{f}^1 - L^1(\mathbf{u}^1)$. We hope the residual and the error are slowly fluctuating on the scale of the fine grid. If this is the case, then the equation

$$L^1(\mathbf{u}^1_*) - L^1(\mathbf{u}^1) = \mathbf{r}^1$$

can be well approximated on the coarse grid, so take as coarse-grid problem (assuming that L^2 is defined somehow)

$$L^2(\mathbf{u}^2) - L^2(R_1^2 \mathbf{u}^1) = R_1^2 \mathbf{r}^1,$$

which would be rewritten as

$$L^2(\mathbf{u}^2) = R_1^2 \mathbf{r}^1 + L^2(R_1^2 \mathbf{u}^1).$$

After solving the coarse-grid problem, update the solution on the fine grid; to do this, realize that the approximation to the error is $\mathbf{u}^2 - R_1^2 \mathbf{u}^1$, hence update via

$$\mathbf{u}^1 \leftarrow \mathbf{u}^1 + I_2^1(\mathbf{u}^2 - R_1^2 \mathbf{u}^1).$$

This scheme is equivalent to the scheme outlined in Section 2 if the operator L^1 is linear. (*Exercise.* Show this!) It has the advantage of not requiring any linearizations of nonlinear operators and also of working with the same equation at each level, so that only one relaxation process need be programmed. The disadvantage of this scheme is the increased work requirement; the approximate solution $\mathbf{u}^1$ must be restricted to a solution on the coarser grid, leading to extra work. As to a comparison between the Newton-multigrid approach and the FAS scheme, there seems to be no clear evidence as of this writing as to which approach is superior.

All the description above has been for finite difference approximations of partial differential equations. An extremely successful method for numerical solution of boundary value problems is the finite element method. Time and space do not permit an extensive discussion of the finite element method here. Here it will suffice to remark that the finite element method (applied to linear problems) leads to large systems of linear equations, and much effort has been expended by those in the field on deriving efficient solution techniques. The matrices are sparse but not well-structured, so a great deal of attention has been paid to techniques for Gaussian elimination suited for sparse matrices. The linear systems arising in the finite element method are quite often solved by Gaussian elimination, and implementing Gaussian elimination requires a great deal of attention to matters of data management.

There is no reason why multigrid methods cannot be applied to the systems of equations arising from finite element methods. Indeed, in line with the coarse-to-fine viewpoint outlined above, triangulations (common in the finite element method) are easy to refine: and the interpolation map is easy to define; since the finite element space on the coarse grid is a subspace of the finite element space on the fine grid, the interpolation mapping can be taken as the natural inclusion.

There is a powerful and sophisticated mathematical theory which can be brought to bear on finite element formulations. Using this theory, proofs of convergence and work estimates of optimal order (more about this notion later) have been given by Bank and Dupont [1]. One important point should be mentioned, however.

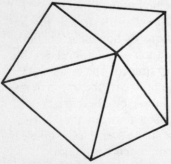

**COARSE
TRIANGULATION**

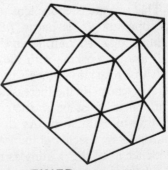

**FINER
TRIANGULATION**

Usually in finite element methods the cost of problem setup is a substantial portion of the overall cost. Hence the payoff from use of multigrid with finite elements might not be as substantial as with finite differences.

4. WORK ESTIMATES

Multigrid methods have attracted a great deal of attention lately and are being studied intensively; why is this so? One reason some workers are very interested in multigrid methods is that these methods may be of optimal order of computational complexity, as was mentioned above. In this section a brief explanation of this notion will be given and multigrid methods will be compared with other methods in terms of computational complexity.

The notion of computational complexity is important in the study of algorithms and thus has been extensively studied by computer scientists. Two basic questions in the theory of computational complexity have to do with operations counts: how many operations are required to solve a *problem*? and, how many operations does a particular *algorithm* take? (Naturally, the second question is easier than the first.) In numerical analysis an "operation" is often defined as a multiplication. This is partly for historical reasons and partly for reasons of convenience when it comes to doing operations counts for particular algorithms. First, computers historically took longer to perform multiplications than additions (this is not necessarily the case any more), and so multiplications tended to be the dominant factor in determining the running time of an algorithm. Second, in many algorithms (particularly in numerical linear algebra) the number of additions is (almost) equal to the number of multiplications and thus if one has counted the multiplications one automatically has the number of additions. (For purposes of counting operations, a division is counted as a multiplication and a subtraction as an addition.)

How many operations does it take (theoretically) to numerically solve a partial differential equation? As it stands, the question is very vague. Let us try to tighten it somewhat. Consider a partial differential equation in two independent variables in a region Ω.

Suppose a grid of N^2 points is laid down on Ω and we consider finite difference methods for the numerical solution technique. Thus, after discretization the problem consists of solving N^2 equations in as many unknowns. How many operations are needed for this? Since the input consists of $O(N^2)$ pieces of information and the output consists of another $O(N^2)$ pieces of information, it should be the case that the theoretical minimum number of operations required to compute the solution is $O(N^2)$. (This is not a proof, it is only a heuristic argument.)

Let us now consider the order of magnitude of arithmetic operations required by various finite difference algorithms. To fix ideas, let us take Ω as the unit square, the differential equation as the Poisson equation, and the finite difference approximation as the usual five-point operator (see (2.24)). Then the following table can be constructed for various solution techniques. For the direct methods, we count the number of operations required for the solution, for the iterative methods we count the number of operations required to reduce the initial error by a factor of h^2 (this is to ensure that the error in the iterative solution is of the same order of magnitude as the error in the exact solution of the difference scheme).

METHOD	OPERATIONS	COMMENTS
Gaussian elimination, natural ordering	$O(N^4)$	Banded matrix, dimension N^2, bandwidth $O(N)$, storage $O(N^3)$
Gaussian elimination, nested dissection ordering	$O(N^3)$	Storage is $O(N^2 \log N)$
Jacobi relaxation	$O(N^4 \log N)$	Spectral radius of the iteration matrix is $1 - ch^2$. Need $O(h^2 \lvert \log h \rvert)$ iterations, each iteration costs $O(N^2)$ operations.
Gauss-Seidel relaxation	$O(N^4 \log N)$	Spectral radius of the iteration matrix is $1 - 2ch^2$ (same c as for Jacobi method).

SOR with optimal ω (SOR = Successive Over-Relaxation)	$O(N^3 \log N)$	Spectral radius of the iteration matrix is $1 - O(h)$. Need $O(N^2)$ operations per sweep.
SSOR with optimal parameters (SSOR = Symmetric SOR)	$O(N^{2.5} \log N)$	Spectral radius of the iteration matrix is $1 - O(h^{.5})$.
ADI with optimal parameters (ADI = Alternating Direction Implicit)	$O(N^2 \|\log N\|^2)$	Spectral radius of the iteration matrix is $1 - O(1/\|\log h\|^2)$. Each step requires $2N$ tridiagonal solves.
FFT (Fast Fourier Transform)	$O(N^2 \log_2 N)$	Direct method; requires regular geometry.
Cyclic reduction	$O(N^2 \log_2 N)$	Direct method; preprocessing phase with work $O(N^3)$ required. Requires regular geometry.
FACR (Fourier Analysis + Cyclic Reduction)	$O(N^2 \log_2 \log_2 N)$	Combination of FFT and cyclic reduction. Requires regular geometry.
Multigrid	?	

What is the operations count for multigrid? Well, an estimate might proceed as follows. Suppose that the iteration matrix T for a multigrid cycle (fine to coarse) has norm $\|T\| \leqslant g < 1$, where the upper bound is independent of h. Then the number of iterations to reduce the initial error by a factor of $O(h^2)$ is $O(\|\log h\|)$. How much work is required for each iteration? Let us suppose we have L levels, with level 1 = finest level and level L = coarsest level. Suppose we do t coarse-grid correction cycles starting from each grid (in the notation of the algorithm in Section 2, mgcycles = t). Write $f(k) := $ work required for one complete down-and-back cycle starting from level k. Suppose the number of relaxation sweeps on each level is bounded by a constant p, independent of the finest mesh size and of the number of levels. Also suppose that the work required for each relaxation sweep is bounded by dN^2, a constant times the number of mesh points. Then the work required for the total process starting from level 1 is given by

$$f(1) \leqslant c \cdot N + t \cdot f(2) \leqslant c \cdot N + t \big(c \cdot (N/2)^2 + t \cdot f(3) \big) \leqslant \cdots$$

$$\leqslant c \cdot N^2 \big(1 + t/4 + (t/4)^2 + \cdots \big) + t^{L-1} \cdot f(L)$$

(where c is a constant depending only on p and d). If now $t < 4$, this last expression is bounded by

$$c \cdot N^2 \cdot 4/(4-t) + t^{L-1} \cdot f(L).$$

Finally, if $t^{L-1} f(L)$ is $O(N^2)$, we have the final result that the work of one complete multigrid iteration from fine to coarse and back is $O(N^2)$. Since $O(|\log h|)$ iterations are required to reduce the initial error by a factor of h^2, the total work to reduce the initial error by a factor h^2 is $O(N^2 \log N)$.

This is not yet the theoretical minimum. In order to obtain a slightly better bound let us start from the coarse grid and work toward the fine grid. This is the "full multigrid" approach (see Bank and Dupont [1]). Suppose that each iteration of the process beginning on level k reduces the error by a factor of $g < 1$. (In practice g is on the order of .5 to .1.) Let us write u_k for the exact solution of the level k equations, and $u(k)$ for the exact solution of the partial differential equation restricted to grid k. In the following argument a set of discrete norms will be used, one for each level. For the sake of notational ease we will not distinguish between the norms by subscripts. Write I_{k+1}^k for the interpolation from grid $k+1$ to k, coarse to fine. Let the grids have grid sizes $h_L > h_{L-1} > \cdots > h_1$ with $h_{k+1} = 2h_k$. Make the following assumptions.

1. The interpolation maps are uniformly bounded: there exists a constant C such that $\|I_{k+1}^k\| \leqslant C$ for $k = 1, \ldots, L-1$.
2. $\|u(k) - u_k\| \leqslant Ch_k^2$ for $k = 1, 2, \ldots, L$, where C is independent of k.
3. The level L equations are solved exactly.

4. Starting from level k, a total of s down-and-back sweeps are taken, where s is chosen such that $g^s < 1/4$.
5. $\|u(k) - I_{k+1}^k u(k+1)\| \leqslant Ch_{k+1}^2$, where C is independent of k.

Write $\tilde{u}_k$ for the computed solution at level k. Then we have the following lemma.

LEMMA. $\|u_k - \tilde{u}_k\| = O(h_k^2)$ for all k. The constant depends on g and the constants in 1–5 above but is independent of L and N.

Proof. Write $e_k := u_k - \tilde{u}_k$. We have $e_L = 0$ by assumption. For $k < L$,

$$\|e_k\| \leqslant g^s \|u_k - I_{k+1}^k \tilde{u}_{k+1}\|$$

$$\leqslant g^s \{ \|u_k - u(k)\| + \|u(k) - I_{k+1}^k u(k+1)\|$$

$$+ \|I_{k+1}^k (u(k+1) - u_{k+1})\| + \|I_{k+1}^k (u_{k+1} - \tilde{u}_{k+1})\| \}$$

$$\leqslant g^s \{ Ch_k^2 + Ch_{k+1}^2 + Ch_{k+1}^2 + C\|e_{k+1}\| \}$$

$$= 9Cg^s h_k^2 + Cg^s \|e_{k+1}\|.$$

A standard argument now shows that $\|e_k\| \leqslant \delta_k$, where δ_k is the solution of the majorizing difference equation

$$\delta_k = 9Cg^s h_k^2 + Cg^s \delta_{k+1}, \qquad k < L, \qquad \delta_L := 0.$$

Solving this difference equation, we find that $\|e_k\|$ satisfies

$$\|e_k\| \leqslant \frac{9Cg^s}{1 - 4g^s} h_k^2.$$

Thus a solution accurate to within truncation error can be computed starting on the coarse grid and proceeding to the finest grid.

Now for the work estimate. Let $F(k) := $ the cost of one full cycle starting from level k. From arguments above, $F(k) = O(h_k^2)$; this uses the assumption that $t \leqslant 3$. Then the total work is bounded by

$$F(L) + s \sum_{k=L-1}^{1} F(k) \leqslant F(L) + s \sum_{k=L-1}^{1} C4^{(L-k)} h_L^{-2}$$

$$= F(L) + sC(1/3)(4^L - 4) h_L^{-2}$$

$$\leqslant F(L) + (1/3) sC h_1^{-2}$$

$$= O(h_1^{-2}),$$

if $F(L) = O(h_1^{-2})$. (The essence of the proof resides in the fact that the initial error on grid k need only be reduced by a factor of 4, and this can be done by a fixed number of cycles, independent of the number of grid points.)

Notice that this argument depends on the bound for the norm of the iteration matrix of a complete cycle. There are few proofs of convergence for multigrid methods. The proofs that have been given essentially bound the spectral radius (or the norm) of the iteration matrix. In these cases it has indeed been proved that the rate of convergence is independent of h, and hence that the multigrid method is of optimal order of computational complexity (if one accepts the arguments above concerning $O(N^2)$ as the optimal order).

One should note that there is a constant hidden in the $O(\cdot)$ notation, and the constant can be very important for the comparison of two methods. For example, suppose that the constant in the estimate for the FFT method above were 6; since the N would usually in practice be at most 256, the work estimate for the FFT would be $48N^2$. This means that in order for an $O(N^2)$ method to be superior to FFT in practice, the constant in the work estimate could be no bigger than 48. In any event, a careful comparison of different methods requires more than just an order of magnitude estimate. Order of magnitude estimates are asymptotically decisive, but may be misleading in practice. To emphasize this point, the constant in the FACR work estimate is 3 (Swarztrauber [14]), so for $N \leqslant 256$, the work estimate for FACR is $\leqslant 9N^2$! The fast direct methods have limited applicability, however; they are restricted to problems in which the region Ω is a square or rectangle. Multigrid methods are applicable in general regions. Even for the unit square, where fast direct methods are applicable, recent work has shown that the best multigrid method will usually produce an acceptable solution faster than the best direct method (Foerster and Witsch [5]).

One detailed investigation (Hackbusch [7]) of a particular multigrid program yielded a work estimate for one complete fine-to-coarse and back cycle of

$$41 \cdot A(\Omega) \cdot h^{-2} + L(\partial\Omega) \cdot O(h^{-1}),$$

where $A(\Omega)$ is the area of Ω (it is assumed Ω is contained in the unit square) and $L(\partial\Omega)$ is the length of the boundary of Ω. The important point about this estimate is that the constant is given and is of a reasonable size. (In fact, for the five-point difference formula the 41 can be replaced by 27.6; the 41 was for a more general nine-point difference formula.)

One point has been glossed over so far. Think back to the coarse-grid correction algorithm of Section 2. It required a matrix A^2 on the coarse grid. How is this matrix defined? In practice, two possibilities have been used. The first is to choose A^2 as the matrix that arises from the differencing scheme which produced A^1 on the fine grid. This has the undeniable advantage of offering simplicity in programming. The second alternative is to use the restriction and interpolation operators to help define A^2: $A^2 := R_1^2 A^1 I_2^1$. This has the disadvantage that some preprocessing is usually required to determine the entries of A^2 (suppose the differential operator has variable coefficients, for example), and A^2 may turn out to have a different stencil than A^1. For example, if A^1 is the five-point second-order approximation of the Laplacian in two dimensions, and if the restriction and interpolation operators are linear, then the operator A^2 is a nine-point difference operator. To get an idea of possible advantages of this choice of A^2, suppose we take

$$R_1^2 := (\text{constant}) \left(I_2^1 \right)^T \tag{4.1}$$

$$A^2 := R_1^2 A^1 I_2^1, \tag{4.2}$$

so once A^1 and I_2^1 are chosen, R_1^2 and A^2 are determined (up to a constant).

Recall the update step in the coarse-grid correction algorithm:

$$\mathbf{u}^1 \leftarrow \mathbf{u}^1 + I_2^1 \mathbf{e}^2. \qquad (4.3)$$

We see the correction lies in the range of I_2^1 which equals, by virtue of (4.1), the orthogonal complement of the null space of the restriction operator. What is the null space of the restriction operator? The claim is that these vectors are all highly oscillatory. Accept this for the moment. Further accept the claim that standard relaxation processes are efficient at diminishing highly oscillatory components of the error. Since the correction lies in the orthogonal complement of these highly oscillatory vectors, the relaxation process is not disturbed by the updating procedure (it is conceivable in general that the updating step reintroduces rapidly oscillating components into the error, thus negating some of the work done by the relaxation process).

We haven't yet looked at the possible advantage of choosing A^2 by the formula (4.2). Recall from (2.15) that the matrix of the total process involves the matrix representing the coarse-grid correction

$$I - I_2^1 (A^2)^{-1} R_1^2 A^1.$$

With A^2 and R_1^2 defined by (4.1) and (4.2), this becomes

$$I - I_2 \left((cI_2^1)^T A^1 I_2^1 \right)^{-1} (cI_2^1)^T A^1.$$

EXERCISE.   Show that this matrix can be written as

$$I - P^1 (A^1)^{-1} P^1 A^1,$$

where $P^1 := I_2^1 (I_2^1)^\dagger$ and $\dagger$ denotes the Moore-Penrose generalized inverse. Show further that P^1 is a symmetric projection with range $(P^1) = [\text{null space}(R_1^2)]^\perp$.

Define $W^1 := \text{null space } (R_1^2)$ (the oscillatory vectors). Write $\mathbf{e}^1 = \mathbf{e}_1^1 + \mathbf{e}_2^1$ where $\mathbf{e}_2^1 \in W^1$ and $\mathbf{e}_1^1 \in (W^1)^\perp$.

EXERCISE.   Show that the updated error $\tilde{e}^1$ has the decomposition

$$\tilde{e}^1 = \tilde{e}_1^1 + \tilde{e}_2^1$$

where

$$\tilde{e}_2^1 = e_2^1, \qquad \tilde{e}_1^1 = - P^1(A^1)^{-1}P^1A^1e_2^1.$$

You may assume P^1A^1 restricted to $(W^1)^\perp$ is nonsingular.

The implication of these results is that with the definitions (4.1) and (4.2), one coarse grid correction cycle eliminates the old smooth part of the error, while the old oscillatory part feeds into the new smooth part. One can see from the invariant subspace analysis that these properties need not hold if (4.1) and (4.2) do not hold. It is entirely possible for the old smooth part of the error to excite a new oscillatory part of the error. On the other hand, if (4.1) and (4.2) hold, the relaxation procedure and coarse-grid correction procedure do not interact unfavorably with one another. These ideas were first presented by McCormick [12].

It is not clear at this writing which method of defining A^2 is to be preferred. Only time will tell.

EXERCISE.   Let R_1^2 be the restriction operator given by (2.8). Show that a basis for the null space of R_1^2 is $\{v_1, \ldots, v_{(M+1)/2}\}$, where $v_1 := (2, -1, 0, \ldots, 0)^T$, $v_2 := (0, -1, 2, -1, 0, \ldots, 0)^T$, $v_{(M+1)/2} := (0, \ldots 0, -1, 2)^T$. Convince yourself that any linear combination of these vectors is "oscillatory".

5. APPLICATIONS

There has been a great preponderance of theoretical investigations of multigrid methods as compared to practical applications of multigrid techniques. The multigrid idea is still so new that "best" methods are not known in many cases, and in some sense the time is not ripe for significant applications of multigrid. Another limiting factor is that multigrid methods are quite tricky to program compared with other algorithms for solving the systems of linear equations resulting from finite difference approximations to partial differential equations. Efforts are underway to produce multigrid software for general-purpose use, and much progress has been made, but there is much yet to be done. In this section we will take a brief look at some published multigrid applications.

Hackbusch [7] wrote and presented a program to solve the Helmholtz equation with Dirichlet boundary conditions

$$- \Delta u + cu = f \text{ in } \Omega, \ c = \text{nonnegative constant on } \partial \Omega$$

$$u = g \tag{5.1}$$

in an arbitrary region in R^2. The relaxation scheme used was Gauss-Seidel with a special ordering of the grid points: defining a point (jh, kh) to be (odd, odd) if j and k are both odd, to be (even, odd) if j is even and k is odd, etc., the points were ordered (odd, odd), (odd, even), (even, odd), (even, even) for the Gauss-Seidel sweeps. Two differencing schemes were used, one of second order and one of fourth order, with suitable adjustments if needed at mesh points near an irregular boundary. The restriction mapping was defined as the transpose of the interpolation mapping, and the coarse-grid matrix was defined by the restriction, interpolation, and fine-grid matrices as in (4.2):

$$A^2 := R_1^2 A^1 I_2^1 = \left(I_2^1 \right)^T A^1 I_2^1. \tag{5.2}$$

This definition requires a preprocessing phase to compute the matrices on the various coarser grids because of the possible complications due to irregular geometry. The results of tests with various domains showed an average spectral radius of about .05 to .17, depending on the differencing scheme used (second or fourth order), the region in question (a disk with a slit showed a slower convergence rate than a disk, for example), and depending somewhat on the value of h. Values of h used were in the range $1/8$ to $1/128$. If c was allowed to be negative and close to $- \lambda_1$, the first eigenvalue of the Laplacian operator, the convergence slowed dramatically.

A group of researchers in Germany has written a suite of programs for fairly general linear scalar elliptic equations in two-dimensional domains (Foerster and Witsch [5]). The boundary conditions can be of the Dirichlet or Neumann type. The relaxation schemes used are checkerboard Gauss-Seidel (for pointwise relaxation) or zebra-line Gauss-Seidel (for pointwise relaxation). (Line Gauss-Seidel consists of simultaneously updating unknowns along, say, a vertical line so as to satisfy the difference equations at all the points on the line. Zebra-line Gauss-Seidel is line Gauss-Seidel with the lines ordered in "zebra" fashion: first the even-numbered lines, then the odd-numbered lines.) The restriction and interpolation operators are linear but are given by use of intermediate grids. Some of their results indicate exceptional speed for the solution process.

Mol [13] presents a method for elliptic two-dimensional boundary value problems with applications to the Navier-Stokes equations. The region is taken to have regular (i.e., square or rectangular) geometry. One system considered was the two-dimensional Navier-Stokes equations in stream function-vorticity formulation:

$$\Delta \psi = \omega$$

$$\Delta \omega = Re \left(\frac{\partial \omega}{\partial x} \frac{\partial \psi}{\partial y} - \frac{\partial \omega}{\partial y} \frac{\partial \psi}{\partial x} \right) \tag{5.3}$$

where Re is the Reynolds number, a nondimensional fluid dynamics parameter which gives the ratio of convective to diffusive transport. Here a 7-point scheme was used for the differencing operator, and the restriction was taken as the transpose of the interpolation operator. The relaxation process was taken to be given by an incomplete LU decomposition. (Incomplete LU decomposition goes as follows: to produce an iterative method for $Ax = b$, find a lower triangular matrix $\tilde{L}$ and an upper triangular matrix $\tilde{U}$ such that $\tilde{L}\tilde{U} \sim A$, and apply a standard iterative method to the equivalent linear system $(\tilde{L}\tilde{U})^{-1}Ax = (\tilde{L}\tilde{U})^{-1}b$.) As an application, the driven cavity problem was solved. (The driven cavity problem has become a standard test problem for two-dimensional Navier-Stokes solvers. The physical problem is this: A viscous fluid is contained in a square box. The bottom, left and right walls of the box are fixed, while the top wall of the box moves to the right with a constant velocity. The problem is to find the steady-state motion of the fluid inside the box. Physically, the moving top wall drags some fluid with it and sets up a circulation inside the cavity. Depending on the Reynolds number, there may be countercirculation regions near the corners of the cavity. This is a difficult problem for large Reynolds number, say $Re > 100$. The system of differential equations is the Navier-Stokes equations (5.3) with boundary conditions $\psi = 0$ on all walls and $\partial \psi / \partial n = 0$ on the left, bottom, and right walls, and $\partial \psi / \partial n = 1$ on the top wall). This leads to a nonlinear system of equations. The method of solution was to apply Newton's method to the nonlinear system, and solve the linear equations at each step of Newton's method by the multigrid technique. Results were presented for Reynolds numbers of 50, 100, and 150, and a finest mesh of $h = 1/34$. In all cases, the error was reduced by a factor of more than 10 in each multigrid cycle.

Wesseling and Sonneveld [16] also solve the driven-cavity problem. They again consider the system (5.3). Central differences were used for the partial derivatives, linear interpolation for the coarse-to-fine transfers, the restriction operator was defined as the transpose of the interpolation operator, and the coarse-grid matrices were defined using the interpolation, restriction, and fine-grid matrices as in (4.2). This again requires a preprocessing phase to determine the coefficients of the coarse-grid matrices. The method for handling the nonlinear equations was Newton's method, solving the inner linear systems via multigrid. The relaxation method was an incomplete LU decomposition. No relaxation iterations were applied on the fine grid before descending to the coarse grid; the incomplete LU was used as relaxation after returning to the fine grid. An operation count is presented for the linear part of the algorithm: the preprocessing phase requires $728\ h^{-2}$ operations (an operation is here an add, multiply, divide, or square root), while a complete coarse-grid correction cycle costs $259\ h^{-2}$ operations.

Brandt and Dinar [3] study steady-state fluid dynamics problems, starting with the generalized Cauchy-Riemann equations

$$\frac{\partial u}{\partial x} + \frac{\partial v}{\partial y} = f,$$
$$\frac{\partial u}{\partial y} - \frac{\partial v}{\partial x} = g \qquad \text{in } \Omega. \qquad (5.4)$$

This is an elliptic system, so the multigrid ideas should extend successfully from the scalar case. For the numerical solution central differences are used on a staggered grid. A special relaxation scheme is invented, a "Distributive Gauss-Seidel" scheme. (In the usual Gauss-Seidel scheme, one sweeps through the unknowns, adjusting the jth unknown so that the jth equation is satisfied. In the "Distributive Gauss-Seidel" scheme, one sweeps through the unknowns, adjusting the jth unknown and some of its neighbors so that the jth equation is satisfied.) This turns out to give a smoothing rate of .5, the same as for the Poisson equation (cf. Section 2). Also considered are the steady Stokes equations

$$-\Delta \mathbf{u} + \nabla p = \mathbf{f}, \qquad \left(\mathbf{u} = (u, v)^T\right) \qquad (5.5)$$
$$\operatorname{div} \mathbf{u} = 0$$

and the incompressible Navier-Stokes equations in primitive variables

$$-\frac{1}{Re}\Delta \mathbf{u} + \mathbf{u} \cdot \nabla \mathbf{u} + \nabla p = \mathbf{f}, \qquad \left(\mathbf{u} = (u, v)^T\right). \qquad (5.6)$$
$$\operatorname{div} \mathbf{u} = 0$$

In both cases staggered grids are again used and Distributive Gauss-Seidel is the relaxation scheme (in the Navier-Stokes equations upstream differencing is used on the first-order terms if $h \cdot Re$ is large). Again the smoothing rate turns out to be .5, the same as for the Poisson and Cauchy-Riemann equations.

Brandt and Cryer [4] considered the solution by multigrid techniques of variational inequalities:

$$Lu \leqslant f \quad \text{in} \quad \Omega,$$

$$u \geqslant 0 \quad \text{in} \quad \Omega,$$

$$u(Lu - f) = 0 \quad \text{in} \quad \Omega,$$

$$u = g \quad \text{on} \quad \partial\Omega, \qquad (5.7)$$

where L is a self-adjoint, negative definite elliptic operator. This is a nonlinear problem because of the unknown position of the interface $\{(x, y): u(x, y) = 0\}$. Linear interpolation and unweighted restriction are used. The relaxation scheme used is "projected Gauss-Seidel" (carry out the Gauss-Seidel algorithm with the additional feature that if any u_j turns out to be negative, set it to 0 in view of the restriction $u \geqslant 0$, and carry on). A program is presented which implements the full multigrid algorithm, starting on the coarsest grid and working up to the finest level. (This program is one of the few published multigrid programs. The lack of published programs is both a testament to the trickiness of multigrid programming and an impediment to rapid progress. Publication of programs should be encouraged so various researchers need not keep "inventing the wheel.") A problem of seepage through a porous dam is presented, discretized on a 128×192 grid. The spectral radius of the overall cycle turns out to be about .65. A total of 5.41 work units (1 work unit is defined as the work of one relaxation sweep on the finest level) are used to solve the problem to an acceptable accuracy using 5 levels. (This beats an SOR method by a factor of about 6.)

Jameson [11] attacked the two-dimensional transonic full-potential equation

$$\left(\rho\phi_x\right)_x + \left(\rho\phi_y\right)_y = 0. \tag{5.8}$$

Here $\rho = \rho(|\nabla\phi|)$ and the problem is highly nonlinear. Not only that, but in the case of transonic flow the character of the equation changes from elliptic where the flow is subsonic to hyperbolic where the flow is supersonic. This has important implications for the numerical techniques used. The problem considered was flow over a two-dimensional airfoil. The region exterior to the airfoil was mapped conformally onto the exterior of a circle to simplify the geometry. Central differencing was used with artificial viscosity in the supersonic region to bias the differencing in the upwind direction there. The nonlinear equations were preconditioned by a one-sided difference operator and a generalized ADI scheme was invented and used for the relaxation. A fixed (nonadaptive) strategy was used, with one relaxation sweep before and after each cycle (in the notation of the algorithm in Section 2, $\nu = \nu' = $ mgcycles $= 1$ for all k). The nonlinear version (Full Approximation Scheme) of the multigrid method was used. The finest grid was 64×192, and up to 6 grids were employed. The solution of the equation is not smooth (the first derivatives are discontinuous—there is a shock). The multigrid ideas were developed with elliptic equations and smooth solutions in mind, so this problem presents a difficult and challenging test to the multigrid concept. With 5 grids, it was found that the error decreased 8 orders of magnitude in 50 work units; this is equivalent to a spectral radius of .67. It is likely that as of this writing this is the most difficult applied problem that has been solved by multigrid methods (but this may be controversial, and other researchers are entitled to other opinions!).

On surveying the published multigrid applications, one is struck by their small number. It is perhaps worth mentioning that the first conference devoted to multigrid methods took place in October, 1981. More applications of multigrid methods were presented at that conference, but it is clear that there is much to do toward making multigrid usable and effective.

BIBLIOGRAPHY

1. R. Bank and T. Dupont, "An optimal order process for solving finite element equations," *Math. Comp.*, **36** (1981), 35–51.
2. A. Brandt, "Multi-level adaptive solutions to boundary-value problems," *Math. Comp.*, **31** (1977), 333–390.
3. A. Brandt and N. Dinar, "Multigrid solutions to elliptic flow problems," *Numerical Methods for Partial Differential Equations*, ed. S. Parter, Academic Press, 1979.
4. A. Brandt and C. Cryer, "Multigrid algorithms for the solution of linear complementarity problems arising from free boundary problems," MRC Technical Summary Report 2131, Mathematics Research Center, University of Wisconsin-Madison, 1980. Also *Siam J. Sci. Stat. Comp.*, **4** (1983), 655–684.
5. H. Foerster and K. Witsch, "On efficient multigrid software for elliptic problems on rectangular domains," preprint no. 458, Bonn University, 1981.
6. H. Foerster, K. Stuben, and U. Trottenberg, "Non-standard Multigrid Techniques using Checkered Relaxation and Intermediate Grids," *Elliptic Problem Solvers*, ed. M. Schultz, Academic Press, to appear.
7. W. Hackbusch, "On the multi-grid method applied to difference equations," *Computing*, **20** (1978), 291–306.
8. _____, "Convergence of multi-grid iterations applied to difference equations," *Math. Comp.*, **34** (1980), 425–440.

9. _____, "On the convergence of multi-grid iterations," *Beitrage zur Numer. Math.*, **9** (1981), 213–239.

10. P. W. Hemker, "On the structure of an adaptive multi-level algorithm," *BIT*, **20** (1980), 289–301.

11. A. Jameson, "Acceleration of transonic potential flow calculations on arbitrary meshes by the multiple grid method," *Proc. of AIAA 4th Computational Fluid Dynamics Conference* (1979), 122–146.

12. S. McCormick, "An algebraic interpretation of multigrid methods," *SIAM J. Numerical Analysis*, **19** (1982), 548–560.

13. W. J. A. Mol, "Numerical solution of the Navier-Stokes equations by means of a multigrid method and Newton-iteration," *Seventh International Conference on Numerical Methods in Fluid Dynamics*, ed. W. C. Reynolds and R. W. MacCormack, *Lecture Notes in Physics*, **141** (1981), Springer-Verlag.

14. P. Swarztrauber, "The Methods of cyclic reduction, Fourier analysis, and the FACR algorithm for the discrete solution of Poisson's equation in a rectangle," *SIAM Review*, **19** (1977), 490–501.

15. P. Wesseling, "The Rate of convergence of a multiple grid method," *Numerical Analysis Proceedings, Dundee 1979*, ed. G. A. Watson, *Lecture Notes in Mathematics*, **773** (1980), Springer-Verlag.

16. P. Wesseling and P. Sonneveld, "Numerical experiments with a multiple grid and a preconditioned Lanczos method," *Approximation Problems for Navier-Stokes Equations*, ed. R. Rautmann, *Lecture Notes in Mathematics*, **771** (1980), Springer-Verlag.

FAST POISSON SOLVERS

Paul N. Swarztrauber

1. INTRODUCTION

The purpose of this chapter is to introduce the reader to a particular class of efficient computer-aided methods for solving Poisson's equation (1.1). Given a function $f(x, y)$ defined on a rectangle $a < x < b$; $c < y < d$, then we wish to find a function $u(x, y)$ such that

$$u_{xx}(x, y) + u_{yy}(x, y) = f(x, y). \qquad (1.1)$$

In general there will be many functions that satisfy (1.1) and one must specify boundary conditions to obtain a unique solution. For example, if $u(x, y)$ is specified on the boundary of the rectangle and satisfies (1.1) on the rectangle, then the solution is unique. In this chapter we will solve Poisson's equation subject to several standard boundary conditions.

There are two main reasons for interest in the efficient solution of Poisson's equation. First, the equation arises in many areas of science and engineering and, second, its solution can require a substantial amount of computing time. As we will show, a single solution can be obtained with only a modest amount of computing time; however, in many applications, repeated solutions are necessary, which can require a significant amount of computing time.

For example, in models of atmospheric dynamics, Poisson's equation is imbedded in a system of time-dependent partial differential equations. This system is solved computationally by numerical integration in which the solution is tabulated at a sequence of time levels, t_n. A single solution of such a system may require many hours of computer time in which Poisson's equation is solved, possibly several thousand times. In addition, the system of equations will be solved many times in order to determine the effect of changing model parameters, i.e., to simulate different atmospheric conditions.

Similar calculations are performed in many other disciplines including oceanography, stellar physics, electrostatics and magnetics, reactor design and oil reservoir simulation, to name just a few. Indicative of the great diversity of applications is the work of Baum and Rehm [14]. Using model calculations much like those discussed above, they have been able to analyze the propagation of a fire in a room.

With the advent of the fast Poisson solvers, the time required to compute solutions has dropped by at least an order of magnitude and up to a factor of 50. This has resulted in a marked improvement in the overall performance of the large model calculations. Prior to the fast Poisson solvers, at least 50% of the computing time for a model was used in the repeated solution of Poisson's equation using an iterative method such as successive overrelaxation (SOR). In the same models, but with a fast Poisson solver, only about 10% of the total computing time is spent solving Poisson's equation.

When a problem of this type is encountered, usually the first goal is to find a closed-form solution, i.e., a formula into which we can plug the coordinates of any point in the rectangle and out of which comes the exact solution $u(x, y)$. More often than not, such closed-form solutions cannot be found, but even when they are, their use for tabulating the solution may be inefficient when compared to the methods that will be discussed in this chapter.

A quite satisfactory alternative to a closed-form solution is an approximate solution of (1.1) in the form of a tabulation. Given integers M and N we first define a grid of points (x_i, y_j) for

$$x_i = a + i\delta x \qquad i = 0,\ldots, M \qquad (1.2)$$

$$y_j = c + j\delta y \qquad j = 0,\ldots, N, \qquad (1.3)$$

where $\delta x = \dfrac{b-a}{M}$ and $\delta y = \dfrac{d-c}{N}$. The goal is to compute an approximate value $u_{i,j}$ of the solution $u(x_i, y_j)$ at each of the grid points. To this end we require the $u_{i,j}$ to satisfy a finite difference approximation of (1.1), which is developed in the next section. The fact that $u_{i,j}$ is only an approximate solution is of little concern since it is known (for example, see [19]) that we can make it as accurate as we like by increasing M and N in order to make δx and δy sufficiently small. Of course, this is limited by the available computing resource, but for most applications one can obtain a result that is accurate to three or four decimal digits with only a moderate amount of computing, particularly since the advent of the fast Poisson solvers.

Fast Poisson solvers came into being in 1965 when Hockney [15] used Fourier analysis to solve Poisson's equation. This approach was superior to any other at that time and it continues to be one of the most successful of the fast Poisson solvers. The method was developed further and in more detail in [16], which is also an important paper on the subject. Since then a great deal has been written about the method, and we will refer to some of this literature throughout the chapter.

In both [15] and [16], Hockney discusses a method, developed in collaboration with Golub, called cyclic reduction. In its original form, cyclic reduction was quite unstable. However this fact was initially not noticed since cyclic reduction was combined with the Fourier method in a manner that reduced both error growth and computing time. This combination is called the FACR(ℓ) algorithm. Nevertheless, cyclic reduction by itself applies to a larger class of problems than the Fourier method and hence Buneman [5] made a significant contribution when he stabilized the cyclic reduction method, making it a viable "fast Poisson solver" with all the attributes that had distinguished the Fourier method. The method of cyclic reduction, including Buneman's algorithm and variants, is discussed in detail by Buzbee et al. [6]. This is one of the most important papers on cyclic reduction and is recommended for those who wish to continue reading about this subject. The Fourier and cyclic reduction methods are much superior to the iterative or relaxation methods that were previously used to solve Poisson's equation.

a) They require much less computing time; the speed-up was comparable to the speed-up achieved by using the fast Fourier transform over the slow Fourier transform. There is a vast amount of literature about the fast Fourier transform. For an account of recent work and a list of references see Swarztrauber [29].

b) They require half the storage of the previous methods. The computations can be done in place, i.e., the solution can be stored over the right hand side $f_{i,j} = f(x_i, y_j)$. This is an important advantage for these problems, which characteristically require a large amount of storage.

c) The Fourier and cyclic reduction methods are classified as direct methods. Using a direct method, the exact solution is obtained in a finite number of operations. This is in contrast with the iterative methods that theoretically require an infinite number of operations to obtain an exact solution. Of course, in practice, the iteration stops once the error has been reduced to an acceptable level called the error tolerance. For many problems the selection of the error tolerance is a nontrivial matter and a large amount of computing time can be wasted if it is improperly specified. The point of this is that an error tolerance is not needed when using a direct method. It is an important but little publicized advantage of the direct methods.

d) Because of the speed and reliability of the fast Poisson solvers, they are attractive candidates for implementation in software packages. Indeed, several packages exist, including that of Swarztrauber and Sweet [30] which is specifically designed for use by scientists. Machura and Sweet [21] have recently given a comprehensive description of software that is available for partial differential equations.

It is important to note that, although the fast Poisson solvers are clearly superior to the iterative methods for problems that both methods can solve, the iterative methods can be used on a much larger class of problems and still provide the most effective means of solving many elliptic partial differential equations. Brandt [4] has developed an efficient iterative method called the multigrid method that can be applied to a wide class of problems. The conjugate gradient method is also an effective method which, along with multigrid, is currently receiving much attention (see Concus, Golub and O'Leary [9]).

To a large extent the Fourier and cyclic reduction methods complement one another. The fast Fourier transform is very inefficient for prime M or N. On the other hand, as a result of the work of Sweet [31], cyclic reduction is quite efficient for any N. This point may not be too important since N or M can usually be selected as non prime, or at least highly composite, in which case the Fourier method has been found to be somewhat faster than cyclic reduction. The difference is not great and depends on a number of factors including the machine, compiler, and program [17], [28].

Although in practice the Fourier method is faster than cyclic reduction, it is important to note that cyclic reduction can be used to solve a larger class of problems. Schumann and Sweet [25] have developed variants of cyclic reduction for "staggered" grids and Swarztrauber [26] has developed a generalized cyclic reduction algorithm for solving an arbitrary separable elliptic equation. The degree to which the Fourier and cyclic reduction methods complement one another is exemplified by the fact that they can be combined into a third method, called the FACR method, which is faster than either the Fourier or cyclic reduction methods used separately. More on this later.

Prior to the seventies, the subject of fast Poisson solvers was limited to the Fourier and cyclic reduction methods. Dorr [12] reviews the status of these methods at that time as well as a number of other "not so fast" direct methods. In the past decade, a number of important fast Poisson solvers have been developed, and consequently, for those individuals who intend to pursue the subject, we will devote the remainder of this section to a brief review of some of the more recent results.

In order to quantify the relevance of these developments we need some measure of the computing resource that is required by a method. If we define an operation as either a multiplication, division, addition, or subtraction then it is common to use the number of operations, or operation count, as a measure of the resource that is required. For an N by N grid the total operation count consists of an expression which has several terms, each of which is a function of N. The *asymptotic* operation count is defined as the term in the operation count that dominates the expression as N gets very large. We will ignore any proportionality constants. For example, the asymptotic operation count of SOR is proportional to $N^3 \log N$, which is expressed by $O(N^3 \log N)$. The alternating-direction implicit (ADI) method [19], which is also an iterative method, has an asymptotic operation count $O(N^2 \log^2 N)$. The asymptotic operation count for both the Fourier and cyclic reduction methods is $O(N^2 \log N)$ but as mentioned above, the Fourier method is somewhat faster in practice because it has a smaller proportionality constant. These counts are compared in detail in [28] and [32].

The fastest known numerically stable method for solving Poisson's equation combines both the Fourier and cyclic reduction methods. Hockney [16] introduced the FACR(ℓ) algorithm in which ℓ steps of cyclic reduction were combined with the Fourier method. He also observed experimentally that an optimum value of ℓ existed. Swarztrauber [28] showed that ℓ proportional to $\log N$ is optimal and that with this value of ℓ the asymptotic operation count is $O(N^2 \log \log N)$. For a 64 by 64 grid the FACR method is only about 20% faster than the Fourier method since there is not that much difference between $\log N$ and $\log \log N$. It is interesting to note that, for all practical purposes, $\log \log N$ is constant. For an enormous grid, say, $N = 10^6$, $\log_2 \log_2 N \approx 4.3$; hence for grids that are within current or even anticipated capabilities of computers, $\log_2 \log_2 N$ is bounded by 5.

In spite of this observation, a question very naturally arises, namely, does a method exist with an asymptotic operation count that is proportional to N^2? The answer to this question is a qualified yes. Actually, there are several methods but they are not as accurate as the methods that we have already mentioned. Dorr [13] describes the shooting or marching algorithm which is $O(N^2)$

and he observes that it has been around for a number of years. The difficulty with this method is that it is extremely unstable, i.e., the error as a function of N grows rapidly and the method cannot be used in practice except for quite small grids where N is less than about 10, depending on the accuracy of the computer.

Roughly speaking, algorithms are classified as stable or unstable depending on whether the asymptotic error grows as a monomial or exponential function of N, respectively. On the other hand, the utility of a particular algorithm depends on the proportionality constant and the size of N. Nevertheless, this definition of stability seems quite satisfactory from a practical standpoint since the algorithms that are classified as unstable are generally not found to be useful—until they are stabilized like cyclic reduction!

Bank [1] and Bank and Rose [2], [3] develop a multiple marching method in which the marching and Fourier methods are combined into a new method that also has an asymptotic operation count that is proportional to N^2. Although it is not as accurate as the Fourier method, it is sufficiently accurate to make it useful in practice. Working independently, Lorenz [20] also developed a multiple marching method that he implemented in a FORTRAN program on the CDC 7600. He solved Poisson's equation on a 64 by 64 grid in .0195 seconds, which is equivalent to *two* SOR iterations. This is the fastest known solution of this problem (on the 7600). For the same problem the Fourier method took .062 seconds and cyclic reduction took .073 seconds. Each of these times are at least an order of magnitude less than the time required by SOR.

The multiple marching method gains its speed at some expense of accuracy. Lorenz's solution was accurate to 8 decimal digits compared to 12 digits for the Fourier method. On the other hand 8 digits is certainly sufficient for most applications and so it seems reasonable to use the multiple marching method and take advantage of the speed. But one should also note that if the multiple-shooting method were run on a less accurate computer, say, one with 8 digits compared with the 15 digits on the CDC 7600, then only one or possibly none of the digits would be accurate. As mentioned before, the best method depends on many factors. Hockney has examined the adaption of fast Poisson solvers to vector and parallel computers in [18].

There are other interesting $O(N^2)$ methods, including the total reduction method of Schroeder et al. [24] and the point-cyclic reduction method of Detyna [10]. Each of these methods has its own particular advantages but again none are as accurate as the Fourier method. Important work remains to be done in this area; in particular the question remains, does an $O(N^2)$ method exist that is as accurate as the Fourier method? If so, then it is asymptotically the best, since there are N^2 values of $u_{i,j}$.

The fast Poisson solvers, at least as they were originally developed, could be used only if the problem was defined on a rectangle. Much has changed since then and indeed now the solution can be obtained on irregular regions with very little sacrifice in performance. Buzbee et al. [7] use the capacitance matrix method with the fast Poisson solver to compute the solution on an irregular region. Proskurowski and Widlund [23] have developed the method further and Proskurowski [22] has implemented it in software. Fast Poisson solvers can also be used to solve more complex elliptic equations using a method called D'yakonov iteration, which is the approach taken by Concus and Golub [8].

Since this chapter is intended as an introduction to fast Poisson solvers, we will limit our discussion in succeeding sections to the Fourier and cyclic reduction methods. Because of this limitation we can describe these methods in some detail with the hope that the reader can, in fact, gain a working knowledge of the methods. The exposition proceeds in terms of examples in which four problems (A through D) are solved. In the next section, these problems are defined and Poisson's equation is solved in each, but subject to different boundary conditions. A large system of linear equations for the approximate solution of each problem is also derived in Section 2. These systems are solved using the Fourier method in Section 3 and the cyclic reduction method in Section 4. In Section 5 we deal with several computational problems that must be solved before the methods are useful in practice. We show that certain problems can only be solved in the least squares sense. We also provide an efficient method for computing the matrix polynomials that are associated with cyclic reduction. Finally, at the end of the section, we present Buneman's stable cyclic reduction algorithm.

2. FINITE DIFFERENCE APPROXIMATIONS

In this section we will define and discretize four sample problems by replacing the continuous differential problem with a large system of equations for the approximate solution. It would be ideal to obtain an exact closed-form solution of Poisson's equation; however, as noted in the introduction, this is not possible in most cases and hence we look for an approximate solution. The goal is to compute an approximate value $u_{i,j}$ of the solution $u(x_i, y_j)$ at each of the grid points defined in (1.1) and (1.2).

If the $u_{i,j}$ are solutions to an approximate Poisson equation then it seems reasonable that they will approximate the solution of the Poisson equation. This intuitive observation is, in fact, correct under most circumstances, as shown by formal mathematical development given in a number of places, including [19]. Therefore, we will require the $u_{i,j}$ to satisfy a finite difference approximation to the Poisson equation that is obtained from (1.1) by replacing the second partial derivatives by centered second-difference quotients

$$\frac{u_{i+1,j} - 2u_{i,j} + u_{i-1,j}}{\delta x^2} + \frac{u_{i,j+1} - 2u_{i,j} + u_{i,j-1}}{\delta y^2} = f_{i,j}.$$

$$(2.1)$$

The problem now becomes one of finding the solution of a very large linear system of equations with right-hand side $f_{i,j} = f(x_i, y_j)$ and unknowns $u_{i,j}$. The system (2.1) is not yet complete and must be augmented with linear equations obtained from the boundary conditions. We will consider three possible boundary conditions that occur frequently when solving the Poisson equation.

a) **The value of the solution is specified at the boundary.**

If the solution is specified at $x = a$, for example, then we are given a function, say, $b_a(y)$, such that $u(a, y) = b_a(y)$ for $c \leqslant y \leqslant d$. In this case $u_{0,j} = b_a(y_j)$ is known and the index i of the unknowns begins with $i = 1$. The equation for $i = 1$ is obtained by substituting the boundary condition into (2.1)

$$\frac{u_{2,j} - 2u_{1,j}}{\delta x^2} + \frac{u_{1,j+1} - 2u_{1,j} + u_{1,j-1}}{\delta y^2} = f_{1,j} - \frac{b_a(y_j)}{\delta x^2}.$$

$$(2.2)$$

Hence the boundary conditions enter the linear system of equations through the right hand side.

b) **The derivative of the solution is specified at the boundary.**

If the derivative of the solution is specified at $x = a$ then a function, say, $b_a(y)$, is given such that $\frac{\partial u}{\partial x}(a, y) = b_a(y)$. To obtain the discrete form of this boundary condition, replace the partial derivative with a finite difference approximation centered at $x = a$ or $i = 0$:

$$\frac{u_{1,j} - u_{-1,j}}{2\delta x} = b_a(y_j).$$ (2.3)

Note that this introduces a point outside the boundary at $x_{-1} = a - \delta x$ and a corresponding unknown $u_{-1,j}$. If we evaluate (2.1) at $x = a$ or $i = 0$, then we obtain

$$\frac{u_{1,j} - 2u_{0,j} + u_{-1,j}}{\delta x^2} + \frac{u_{0,j+1} - 2u_{0,j} + u_{0,j-1}}{\delta y^2} = f_{0,j}.$$ (2.4)

The unknown $u_{-1,j}$ outside the boundary can now be eliminated between equations (2.3) and (2.4):

$$\frac{2u_{1,j} - 2u_{0,j}}{\delta x^2} + \frac{u_{0,j+1} - 2u_{0,j} - u_{0,j-1}}{\delta y^2} = f_{0,j} + \frac{2}{\delta x} b_a(y_j).$$

$$(2.5)$$

When a derivative boundary condition is specified, system (2.1) is augmented with (2.5), and the boundary condition enters the linear system through the right-hand side. Since the solution is unknown at $x = a$, the i index of the unknowns $u_{i,j}$ begins with $i = 0$.

The temporary introduction of a point at $x = a - \delta x$ is a common computational technique called the virtual point method. By using this method, all finite differences are centered with the result that all the approximations have the same accuracy. If the solution is sufficiently differentiable at the boundary, then this technique can be justified by the fact that the solution can be continued across the boundary.

c) The solution is periodic.

If $u(a + x, y) = u(b + x, y)$ for all x, then the solution is said to be periodic in the x direction. In terms of the discrete variables, this condition takes the form $u_{-1,j} = u_{M-1,j}$ and $u_{M,j} = u_{0,j}$. The set $u_{i,j}$ for $i = 0, \ldots, M - 1$ is the smallest set of nonredundant unknowns. If they are determined then $u_{i,j}$ can be determined everywhere.

Since the solution is periodic, we know that the Poisson equation and (2.1) are valid at all x and in particular at $x = a$ and $x = b$. If we evaluate (2.1) at $i = 0$ subject to the discrete periodic boundary condition, then we obtain

$$\frac{u_{1,j} - 2u_{0,j} + u_{M-1,j}}{\delta x^2} + \frac{u_{0,j+1} - 2u_{0,j} + u_{0,j-1}}{\delta y^2} = f_{0,j}.$$

$$(2.6)$$

Similarly, if we evaluate (2.1) at $i = M - 1$ subject to the discrete periodic boundary conditions, then we obtain

$$\frac{u_{0,j} - 2u_{M-1,j} + u_{M-2,j}}{\delta x^2} + \frac{u_{M-1,j+1} - 2u_{M-1,j} + u_{M-1,j-1}}{\delta y^2}$$

$$= f_{M-1,j}.$$

$$(2.7)$$

When a periodic boundary condition is specified, system (2.1) is augmented with both (2.6) and (2.7).

This development must be repeated for the remaining sides of the rectangle with their corresponding boundary conditions in order to obtain a complete set of equations for the unknowns $u_{i,j}$. Also, the equations at the corners contain contributions from both boundary conditions. There are a number of possible combinations of boundary conditions. Any of the conditions given above can occur in either the x or the y direction, for a total of 25 possible combinations. In the remainder of this section we will describe four of these combinations in some detail. A complete set of finite difference equations will be derived for each of these four problems, including the boundary and corner equations.

As we will see, the main difference between the problems occurs on the boundary. The problems were chosen to illustrate the solution of the Poisson equation subject to all possible combinations of the boundary conditions. Although only four problems are described, they can be used to develop the method for any combination of boundary conditions. For example, if the solution is specified at $x = a$ and $x = b$ and the derivative of the solution is specified at $y = c$ and $y = d$, then the discretization in the x direction would follow the development given in Problem A, which follows, and the discretization in the y direction would follow the development that is given in Problem B. Once the finite difference equations are derived, the problem reduces to solving a very large system of equations. It *is* therefore quite reasonable to ask why Gaussian elimination is *not* used to solve the equations. The answer is that Gaussian elimination is *less* efficient than the direct methods that we will discuss. The asymptotic operation count for Gaussian elimination is $O(N^4)$ compared with $O(N^2 \log N)$ for the direct methods that are developed in the subsequent sections.

Problem A. Given functions $b_a(y)$, $b_b(y)$, $b_c(x)$ and, $b_d(x)$ defined on the boundary and $f(x, y)$ defined in the interior of the rectangle, we wish to determine a function $u(x, y)$ such that

$$u(a, y) = b_a(y), \qquad u(b, y) = b_b(y),$$
$$u(x, c) = b_c(x), \qquad u(x, d) = b_d(x)$$

and

$$u_{xx}(x, y) + u_{yy}(x, y) = f(x, y) \qquad (2.8)$$

are on the interior of the rectangle. This problem is known as Dirichlet's problem.

As we discussed above, the goal is to find an approximate solution of this problem, and to that end we proceed to derive the finite difference approximations. There are a total of $(M-1)(N-1)$ unknowns $u_{i,j}$ on the interior of the region, and hence we must derive $(M-1)(N-1)$ equations for their solution. The equations on the interior of the region are not affected by the boundary condition, and, therefore, we can obtain the following $(M-3)(N-3)$ equations from (2.1) for $i = 2, \ldots, M-2$ and $j = 2, \ldots, N-2$:

$$\frac{u_{i+1,j} - 2u_{i,j} + u_{i-1,j}}{\delta x^2} + \frac{u_{i,j+1} - 2u_{i,j} + u_{i,j-1}}{\delta y^2} = f_{i,j} = g_{i,j}.$$

$$(2.9)$$

The grid function $g_{i,j}$ is introduced in order to facilitate the incorporation of the boundary conditions into the right-hand side of the equations. As we will show below, $g_{i,j}$ is equal to $f_{i,j}$ except on or near the boundary, where $g_{i,j}$ is equal to $f_{i,j}$ plus a term or terms that result from the boundary conditions.

In this example, we assume that the solution is specified on all sides of the rectangle. Thus we assume that we are given four functions, $b_a(y)$, $b_b(y)$, $b_c(x)$ and, $b_d(x)$ that specify the value of the solution on the boundaries $x = a$, $x = b$, $y = c$ and, $y = d$, respectively. If we repeat the discretization of boundary condition (a) at all four boundaries, then we obtain the following equations from (2.1), which is altered near the boundary. At $x = a + \delta x$ we obtain the following equations for $i = 1$ and $j = 2, \ldots, N-2$:

$$\frac{u_{2,j} - 2u_{1,j}}{\delta x^2} + \frac{u_{1,j+1} - 2u_{1,j} + u_{1,j-1}}{\delta y^2} = f_{1,j} - \frac{b_a(y_j)}{\delta x^2} = g_{1,j}.$$

$$(2.10)$$

At $x = b - \delta x$ we obtain the following equations for $i = M - 1$ and $j = 2, \ldots, N - 2$:

$$\frac{-2u_{M-1,j} + u_{M-2,j}}{\delta x^2} + \frac{u_{M-1,j+1} - 2u_{M-1,j} + u_{M-1,j-1}}{\delta y^2}$$

$$= f_{M-1,j} - \frac{b_b(y_j)}{\delta x^2} = g_{M-1,j}. \qquad (2.11)$$

At $y = c + \delta y$ we obtain the following equations for $j = 1$ and $i = 2, \ldots, M - 2$:

$$\frac{u_{i+1,1} - 2u_{i,1} + u_{i-1,1}}{\delta x^2} + \frac{u_{i,2} - 2u_{i,1}}{\delta y^2} = f_{i,1} - \frac{b_c(x_i)}{\delta y^2} = g_{i,1}.$$

$$(2.12)$$

at $y = d - \delta y$ we obtain the following equations for $j = N - 1$ and $i = 2, \ldots, M - 2$:

$$\frac{u_{i+1,N-1} - 2u_{i,N-1} + u_{i-1,N-1}}{\delta x^2} + \frac{-2u_{i,N-1} + u_{i,N-2}}{\delta y^2}$$

$$= f_{i,N-1} - \frac{b_d(x_i)}{\delta y^2} = g_{i,N-1}. \qquad (2.13)$$

Near the four corners of the rectangle, the right-hand side of (2.9) is modified to include contributions from two boundary conditions. The discrete Poisson equation (2.9) centered at $i = j = 1$ takes the form

$$\frac{u_{2,1} - 2u_{1,1}}{\delta x^2} + \frac{u_{1,2} - 2u_{1,1}}{\delta y^2} = f_{1,1} - \frac{b_a(y_1)}{\delta x^2} - \frac{b_c(x_1)}{\delta y^2} = g_{1,1}.$$

$$(2.14)$$

At $i = M-1$ and $j = 1$,

$$\frac{-2u_{M-1,1} + u_{M-2,1}}{\delta x^2} + \frac{u_{M-1,2} - 2u_{M-1,1}}{\delta y^2}$$

$$= f_{M-1,1} - \frac{b_b(y_1)}{\delta x^2} - \frac{b_c(x_{M-1})}{\delta y^2} = g_{M-1,1}. \qquad (2.15)$$

At $i = 1$ and $j = N-1$,

$$\frac{u_{2,N-1} - 2u_{1,N-1}}{\delta x^2} + \frac{-2u_{1,N-1} + u_{1,N-2}}{\delta y^2}$$

$$= f_{1,N-1} - \frac{b_a(y_{N-1})}{\delta x^2} - \frac{b_d(x_1)}{\delta y^2} = g_{1,N-1}. \qquad (2.16)$$

At $i = M-1$ and $j = N-1$,

$$\frac{-2u_{M-1,N-1} + u_{M-2,N-1}}{\delta x^2} + \frac{-2u_{M-1,N-1} + u_{M-1,N-2}}{\delta y^2}$$

$$= f_{M-1,N-1} - \frac{b_b(y_{N-1})}{\delta x^2} - \frac{b_d(x_{M-1})}{\delta y^2} = g_{M-1,N-1}. \qquad (2.17)$$

Equations (2.9) through (2.17) total $(M-1)(N-1)$ equations for the $(M-1)(N-1)$ unknowns $u_{i,j}$ in the interior of the rectangle.

Problem B. Given functions $b_a(y)$, $b_b(y)$, $b_c(x)$ and $b_d(x)$ defined on the boundary and $f(x, y)$ defined in the interior of the rectangle, we wish to determine a function $u(x, y)$ such that

$$\frac{\partial u}{\partial x}(a, y) = b_a(y) \qquad (2.18)$$

$$\frac{\partial u}{\partial x}(b, y) = b_b(y) \qquad (2.19)$$

$$\frac{\partial u}{\partial y}(x, c) = b_c(x) \qquad (2.20)$$

$$\frac{\partial u}{\partial y}(x, d) = b_d(x) \qquad (2.21)$$

and on the interior of the rectangle

$$u_{xx}(x, y) + u_{yy}(x, y) = f(x, y). \tag{2.22}$$

As before, we seek an approximate solution to this problem, which is called the Neumann problem. The finite difference equations can be obtained following the discretization of the derivative boundary condition (b) that was given at the beginning of this section. The solution is unknown on the boundaries and, hence, there are $(M+1)(N+1)$ unknowns, namely, $u_{i,j}$ for $i = 0,\ldots, M$ and $j = 0,\ldots, N$. Therefore we must determine a system of $(M+1)(N+1)$ equations in these unknowns. The equations on the interior of the region are not affected by the boundary condition and, hence, we obtain the following $(M-1)(N-1)$ equations from (2.1) for $i = 1,\ldots, M-1$ and $j = 1,\ldots, N-1$:

$$\frac{u_{i+1,j} - 2u_{i,j} + u_{i-1,j}}{\delta x^2} + \frac{u_{i,j+1} - 2u_{i,j} + u_{i,j-1}}{\delta y^2} = f_{i,j} = g_{i,j}.$$

$$\tag{2.23}$$

Following the discretization of boundary condition (b) we obtain the following equations for $i = 0$ and $j = 1,\ldots, N-1$:

$$\frac{2u_{1,j} - 2u_{0,j}}{\delta x^2} + \frac{u_{0,j+1} - 2u_{0,j} + u_{0,j-1}}{\delta y^2} = f_{0,j} + \frac{2}{\delta x} b_a(y_j) = g_{0,j}.$$

$$\tag{2.24}$$

At $x = b$, we obtain the following equations for $i = M$ and $j = 1,\ldots, N-1$

$$\frac{-2u_{M,j} + 2u_{M-1,j}}{\delta x^2} + \frac{u_{M,j+1} - 2u_{M,j} + u_{M,j-1}}{\delta y^2}$$

$$= f_{M,j} - \frac{2}{\delta x} b_b(y_j) = g_{M,j}. \tag{2.25}$$

At $y = c$ we obtain the following equations for $i = 1, \ldots, M-1$ and $j = 0$:

$$\frac{u_{i+1,0} - 2u_{i,0} + u_{i-1,0}}{\delta x^2} + \frac{2u_{i,1} - 2u_{i,0}}{\delta y^2} = f_{i,0} + \frac{2}{\delta y} b_c(x_i) = g_{i,0}.$$

$$(2.26)$$

At $y = d$ we obtain the following equations for $i = 1, \ldots, M-1$ and $j = N$:

$$\frac{u_{i+1,N} - 2u_{i,N} + u_{i-1,N}}{\delta x^2} + \frac{-2u_{i,N} + 2u_{i,N-1}}{\delta y^2} = f_{i,N} - \frac{2}{\delta y} b_d(x_i)$$

$$= g_{i,N}. \qquad (2.27)$$

The remaining equations are obtained at the corners of the rectangle. At $i = j = 0$ the right-hand side includes contributions from two boundary conditions,

$$\frac{2u_{1,0} - 2u_{0,0}}{\delta x^2} + \frac{2u_{0,1} - 2u_{0,0}}{\delta y^2} = f_{0,0} + \frac{2}{\delta x} b_a(c) + \frac{2}{\delta y} b_c(a) = g_{0,0}.$$

$$(2.28)$$

At $i = M$ and $j = 0$, we obtain

$$\frac{2u_{M-1,0} - 2u_{M,0}}{\delta x^2} + \frac{2u_{M,1} - 2u_{M,0}}{\delta y^2} = f_{M,0} - \frac{2}{\delta x} b_b(c) + \frac{2}{\delta y} b_c(b)$$

$$= g_{M,0}. \qquad (2.29)$$

At $i = 0$ and $j = N$, we obtain

$$\frac{2u_{1,N} - 2u_{0,N}}{\delta x^2} + \frac{2u_{0,N-1} - 2u_{0,N}}{\delta y^2} = f_{0,N} + \frac{2}{\delta x} b_a(d) - \frac{2}{\delta y} b_d(a)$$

$$= g_{0,N}. \qquad (2.30)$$

Finally, at $i = M$ and $j = N$ we obtain

$$\frac{2u_{M-1,N} - 2u_{M,N}}{\delta x^2} + \frac{2u_{M,N-1} - 2u_{M,N}}{\delta y^2}$$

$$= f_{M,N} - \frac{2}{\delta x} b_b(d) - \frac{2}{\delta y} b_d(b) = g_{M,N}. \tag{2.31}$$

Equations (2.23) through (2.31) total $(M+1)(N+1)$ equations for the $(M+1)(N+1)$ unknowns $u_{i,j}$ on the rectangle. Actually this system of equations is singular and its "solution" presents some special difficulties that are discussed in Section 5.

Problem C. Given functions $b_a(y)$, $b_b(y)$, $b_c(x)$ and, $b_d(x)$ defined on the boundary and $f(x, y)$ defined in the interior of the rectangle, we wish to determine a function $u(x, y)$ such that

$$u(a, y) = b_a(y) \tag{2.32}$$

$$\frac{\partial u}{\partial x}(b, y) = b_b(y) \tag{2.33}$$

$$u(x, c) = b_c(x) \tag{2.34}$$

$$\frac{\partial u}{\partial y}(x, d) = b_d(x) \tag{2.35}$$

and on the interior of the rectangle

$$u_{xx}(x, y) + u_{yy}(x, y) = f(x, y). \tag{2.36}$$

As before, we seek an approximate solution of the problem and to this end we will derive the finite difference equations using the discretizations that were given in examples (a) and (b) at the beginning of this section. The solution is given at $x = a$ and $y = c$; however, it is unknown at $x = b$ and $y = d$ and, hence, there are MN unknowns, namely, $u_{i,j}$ for $i = 1, \ldots, M$ and $j = 1, \ldots, N$. Therefore we must derive a system of MN equations in these unknowns. The equations on the interior of the region are not affected by the boundary condition and, therefore, we can obtain the following $(M-2)(N-2)$ equations from (2.1) for $i = 2,\ldots,$ $M-1$ and $j = 2,\ldots, N-1$:

$$\frac{u_{i+1,j} - 2u_{i,j} + u_{i-1,j}}{\delta x^2} + \frac{u_{i,j+1} - 2u_{i,j} + u_{i,j-1}}{\delta y^2} = f_{i,j} = g_{i,j}.$$

(2.37)

Following the discretization of boundary condition (a) in Section 2, we obtain the following equations for $i = 1$ and $j = 2,\ldots, N-1$:

$$\frac{u_{2,j} - 2u_{1,j}}{\delta x^2} + \frac{u_{1,j+1} - 2u_{1,j} + u_{1,j-1}}{\delta y^2} = f_{1,j} - \frac{1}{\delta x^2} b_a(y_j) = g_{1,j}.$$

(2.38)

At $x = b$ we obtain the following equations for $i = M$ and $j = 2,\ldots, N-1$:

$$\frac{-2u_{M,j} + 2u_{M-1,j}}{\delta x^2} + \frac{u_{M,j+1} - 2u_{M,j} + u_{M,j-1}}{\delta y^2}$$

$$= f_{M,j} - \frac{2}{\delta x} b_b(y_j) = g_{M,j}.$$

(2.39)

At $y = c + \delta y$ we obtain the following equations for $i = 2,\ldots, M-1$ and $j = 1$:

$$\frac{u_{i+1,1} - 2u_{i,1} + u_{i-1,1}}{\delta x^2} + \frac{u_{i,2} - 2u_{i,1}}{\delta y^2} = f_{i,1} - \frac{1}{\delta y^2} b_c(x_i) = g_{i,1}.$$

(2.40)

At $y = d$ we obtain the following equations for $i = 2,\ldots, M-1$ and $j = N$:

$$\frac{u_{i+1,N} - 2u_{i,N} + u_{i-1,N}}{\delta x^2} + \frac{-2u_{i,N} + 2u_{i,N-1}}{\delta y^2} = f_{i,N} - \frac{2}{\delta y} b_d(x_i)$$

$$= g_{i,N}.$$

(2.41)

The remaining equations are obtained at or near the corners of the rectangle. At $i=1$ and $j=1$ the right-hand side contains contributions from two boundary conditions

$$\frac{u_{2,1}-2u_{1,1}}{\delta x^2}+\frac{u_{1,2}-2u_{1,1}}{\delta y^2}$$

$$=f_{1,1}-\frac{1}{\delta x^2}b_a(\delta y)-\frac{1}{\delta y^2}b_c(\delta x)=g_{1,1}. \qquad (2.42)$$

At $i=M$ and $j=1$ we obtain

$$\frac{2u_{M-1,1}-2u_{M,1}}{\delta x^2}+\frac{u_{M,2}-2u_{M,1}}{\delta y^2}$$

$$=f_{M,1}-\frac{2}{\delta x}b_b(\delta y)-\frac{1}{\delta y^2}b_c(b)=g_{M,1}. \qquad (2.43)$$

At $i=1$ and $j=N$ we obtain

$$\frac{u_{2,N}-2u_{1,N}}{\delta x^2}+\frac{2u_{1,N-1}-2u_{1,N}}{\delta y^2}=f_{1,N}-\frac{1}{\delta x^2}b_a(d)-\frac{2}{\delta y}b_d(\delta x)$$

$$=g_{1,N}. \qquad (2.44)$$

Finally at $i=M$ and $j=N$ we obtain

$$\frac{2u_{M-1,N}-2u_{M,N}}{\delta x^2}+\frac{2u_{M,N-1}-2u_{M,N}}{\delta y^2}$$

$$=f_{M,N}-\frac{2}{\delta x}b_b(d)-\frac{2}{\delta y}b_d(b)=g_{M,N}. \qquad (2.45)$$

Equations (2.37) through (2.45) total MN equations for the MN unknowns $u_{i,j}$.

Problem D. Given a function $f(x, y)$, we seek a function $u(x, y)$ that is periodic in both x and y, i.e., for any x and y

$$u(a + x, y) = u(b + x, y) \quad \text{and} \quad u(x, c + y) = u(x, d + y)$$

$$(2.46)$$

and, on the interior of the rectangle,

$$u_{xx}(x, y) + u_{yy}(x, y) = f(x, y). \qquad (2.47)$$

The development of the finite difference equations for the periodic boundary conditions follows the development given for condition (c) at the beginning of the section. The solution at $x = b$ and $y = d$ is identical to that at $x = a$ and $y = d$, respectively, and, therefore, it is not necessary to compute the solution on either $x = b$ or $y = d$. However, the solution is unknown at all other points on the rectangle and hence there are MN unknowns; namely, $u_{i,j}$ for $i = 0, \ldots, M - 1$ and $j = 0, \ldots, N - 1$. Therefore we wish to derive a system of MN finite difference equations in these unknowns. The equations on the interior of the region are not affected by the boundary condition and hence we can obtain the following $(M - 2)(N - 2)$ equations from (2.1) for $i = 1, \ldots, M - 2$ and $j = 1, \ldots, N - 2$:

$$\frac{u_{i+1,j} - 2u_{i,j} + u_{i-1,j}}{\delta x^2} + \frac{u_{i,j+1} - 2u_{i,j} + u_{i,j-1}}{\delta y^2} = f_{i,j} = g_{i,j}.$$

$$(2.48)$$

Following the discretization of boundary condition (c) we obtain the following equations for $i = 0$ and $j = 1, \ldots, N - 2$:

$$\frac{u_{1,j} - 2u_{0,j} + u_{M-1,j}}{\delta x^2} + \frac{u_{0,j+1} - 2u_{0,j} + u_{0,j-1}}{\delta y^2} = f_{0,j} = g_{0,j}.$$

$$(2.49)$$

At $x = b - \delta x$ we obtain the following equations for $i = M - 1$ and $j = 1, \ldots, N - 2$:

$$\frac{u_{0,j} - 2u_{M-1,j} + u_{M-2,j}}{\delta x^2} + \frac{u_{M-1,j+1} - 2u_{M-1,j} + u_{M-1,j-1}}{\delta y^2}$$

$$= f_{M-1,j} = g_{M-1,j}. \tag{2.50}$$

At $y = c$ we obtain the following equations for $i = 1, \ldots, M - 2$ and $j = 0$:

$$\frac{u_{i+1,0} - 2u_{i,0} + u_{i-1,0}}{\delta x^2} + \frac{u_{i,1} - 2u_{i,0} + u_{i,N-1}}{\delta y^2} = f_{i,0} = g_{i,0}.$$

$$\tag{2.51}$$

At $y = d - \delta y$ we obtain the following equations for $i = 1, \ldots, M - 2$ and $j = N - 1$:

$$\frac{u_{i+1,N-1} - 2u_{i,N-1} + u_{i-1,N-1}}{\delta x^2} + \frac{u_{i,0} - 2u_{i,N-1} + u_{i,N-2}}{\delta y^2}$$

$$= f_{i,N-1} = g_{i,N-1}. \tag{2.52}$$

The remaining equations are obtained at or near the corners of the rectangle. At $i = 0$ and $j = 0$

$$\frac{u_{1,0} - 2u_{0,0} + u_{M-1,0}}{\delta x^2} + \frac{u_{0,1} - 2u_{0,0} + u_{0,N-1}}{\delta y^2} = f_{0,0} = g_{0,0}.$$

$$\tag{2.53}$$

At $i = M - 1$ and $j = 0$ we obtain

$$\frac{u_{0,0} - 2u_{M-1,0} + u_{M-2,0}}{\delta x^2} + \frac{u_{M-1,1} - 2u_{M-1,0} + u_{M-1,N-1}}{\delta y^2}$$

$$= f_{M-1,0} = g_{M-1,0}. \tag{2.54}$$

at $i = 0$ and $j = N - 1$ we obtain

$$\frac{u_{1,N-1} - 2u_{0,N-1} + u_{M-1,N-1}}{\delta x^2} + \frac{u_{0,0} + 2u_{0,N-1} - u_{0,N-2}}{\delta y^2}$$

$$= f_{0,N-1} = g_{0,N-1}. \tag{2.55}$$

Finally at $i = M - 1$ and $j = N - 1$ we obtain

$$\frac{u_{0,N-1} - 2u_{M-1,N-1} + u_{M-2,N-1}}{\delta x^2}$$

$$+ \frac{u_{M-1,0} - 2u_{M-1,N-1} + u_{M-1,N-2}}{\delta y^2} = f_{M-1,N-1} = g_{M-1,N-1}.$$

$$\tag{2.56}$$

Equations (2.48) through (2.56) total MN equations for the MN unknowns $u_{i,j}$. Actually this system of equations is singular and its "solution" presents some special difficulties that are discussed in Section 5. This completes the derivation of the finite difference equations for Problem D, which is the last sample problem that we will consider. In the remaining sections we will describe the Fourier and cyclic reduction methods for the rapid solution of these equations.

3. THE FOURIER METHOD

In this section we will use the Fourier method to solve the large systems of equations derived in the previous section. These equations will also be solved by cyclic reduction in the next section. Both methods are described as efficient direct methods in order to distinguish them from Gaussian elimination, which is also classified as a direct method.

The Fourier method can be divided into the following phases:

I.   Transform the equations from physical into Fourier space.
II. Compute the Fourier coefficients of the solution.
III. Transform the solution from Fourier back into physical space.

The fast Fourier transform (FFT) is the key to the efficiency of this method. It also simplifies the method considerably since software is available for the FFT, with the result that phases I and III can be completed by simply calling subprograms that compute the transforms. However, each of the problems A through D requires a different Fourier transform. For example, Problem A requires a sine transform and Problem B requires a cosine transform. Like the full-complex Fourier transform, these transforms also have a "fast" counterpart. All of these fast transforms are available from the National Center for Atmospheric Research in a software package called FFTPACK. The development of the "fast" versions of the various transforms is described in [29].

Like the previous section, this section will be divided into four parts in which the Fourier method will be used to solve each of the four problems that were defined in the previous section

Problem A. In this part we will use the Fourier method to solve the Dirichlet problem, in which the solution is specified on the boundary. In the first phase of the Fourier method, the equations are transformed into Fourier space, where they decouple and can be easily solved. We will show that a solution can be obtained in the form

$$u_{i,j} = \sum_{k=1}^{M-1} \hat{u}_{k,j} \sin ki\frac{\pi}{M}. \tag{3.1}$$

The reason for seeking a solution in this form is that the Fourier coefficients $\hat{u}_{k,j}$ can be determined more easily than $u_{i,j}$. Once the $\hat{u}_{i,j}$ are determined, the solution $u_{i,j}$ can be determined quite efficiently from (3.1) using the fast Fourier transform. To this end we will transform equations (2.9) through (2.17) into Fourier space in order to obtain a system of equations for the $\hat{u}_{k,j}$.

Since the $g_{i,j}$ are known, we can first compute

$$\hat{g}_{k,j} = \frac{2}{M} \sum_{i=1}^{M-1} g_{i,j} \sin ik\frac{\pi}{M}. \tag{3.2}$$

Since the $\sin ik\dfrac{\pi}{M}$ are orthogonal we know that $g_{i,j}$ can also be expressed in terms of $\hat{g}_{k,j}$ by the inverse transform

$$g_{i,j} = \sum_{k=1}^{M-1} \hat{g}_{k,j}\sin ik\frac{\pi}{M}. \qquad (3.3)$$

The transforms in (3.1) and (3.2) constitute phases III and I, respectively. Actually, (3.1) and (3.2) are not used directly since the FFT provides a much more efficient way to compute these transforms. After the tabulation of the trigonometric functions, the direct calculation of either (3.1) or (3.2) would require $O(N^2)$ operations, compared with $O(N\log N)$ operations for the FFT. We can ignore the operations that are required for the trigonometric functions since they can be computed and stored at the beginning and used repeatedly for all subsequent transforms. In this way they do not significantly contribute to the overall computing time. If we now substitute (3.1) and (3.3) into (2.9) we obtain

$$\frac{1}{\delta x^2} \sum_{k=1}^{M-1} \hat{u}_{k,j}\left[\sin(i+1)k\frac{\pi}{M} - 2\sin ik\frac{\pi}{M} + \sin(i-1)k\frac{\pi}{M}\right]$$

$$+ \sum_{k=1}^{M-1}\left(\frac{\hat{u}_{k,j+1} - 2\hat{u}_{k,j} + \hat{u}_{k,j-1}}{\delta y^2}\right)\sin ik\frac{\pi}{M}$$

$$= \sum_{k=1}^{M-1} \hat{g}_{k,j}\sin ik\frac{\pi}{M}. \qquad (3.4)$$

But the term in brackets has the form

$$\sin(i+1)k\frac{\pi}{M} - 2\sin ik\frac{\pi}{M} + \sin(i-1)k\frac{\pi}{M}$$

$$= \left(2\cos k\frac{\pi}{M} - 2\right)\sin ik\frac{\pi}{M} = -4\sin^2 k\frac{\pi}{2M}\sin ik\frac{\pi}{M}.$$

$$(3.5)$$

Therefore, (3.4) can be written

$$\sum_{k=1}^{M-1}\left[\hat{u}_{k,j+1}-\left(2+4\rho^2\sin^2 k\frac{\pi}{2M}\right)\hat{u}_{k,j}+\hat{u}_{k,j-1}\right]\sin ik\frac{\pi}{M}$$

$$=\delta y^2\sum_{k=1}^{M-1}\hat{g}_{k,j}\sin ik\frac{\pi}{M}, \tag{3.6}$$

where $\rho=\dfrac{\delta y}{\delta x}$.

Equation (3.6) is valid for $i=2,\ldots,M-2$ and $j=2,\ldots,N-2$. If we repeat this derivation starting with equations (2.10) and (2.11), then we obtain two equations just like (3.6) that hold for $i=1$ and $i=M-1$. Therefore (3.6) holds for $i=1,\ldots,M-1$. Finally, since the set $\sin ik\dfrac{\pi}{M}$ is orthogonal, we can equate coefficients on both sides of (3.6) with the result that

$$\hat{u}_{k,j+1}-\left(2+4\rho^2\sin^2 k\frac{\pi}{2M}\right)\hat{u}_{k,j}+\hat{u}_{k,j-1}=\delta y^2\hat{g}_{k,j}. \tag{3.7}$$

This provides most of the equations for the coefficients $\hat{u}_{k,j}$. The remaining equations are obtained from (2.12) through (2.17). If we transform equations (2.12), (2.14), and (2.15) in a manner analogous to the transformations of (2.9), (2.10), and (2.11), then we obtain

$$\hat{u}_{k,2}-\left(2+4\rho^2\sin^2 k\frac{\pi}{2M}\right)\hat{u}_{k,1}=\delta y^2\hat{g}_{k,2}. \tag{3.8}$$

If we transform equations (2.13), (2.16), and (2.17), then we obtain

$$-\left(2+4\rho^2\sin^2 k\frac{\pi}{2M}\right)\hat{u}_{k,N-1}+\hat{u}_{k,N-2}=\delta y^2\hat{g}_{k,N-1}. \tag{3.9}$$

The coefficients $\hat{u}_{k,j}$ are determined in phase II. For any k we can determine $\hat{u}_{k,j}$ for $j=1,\ldots,N-1$ from the system of $N-1$ equations consisting of (3.8), (3.7) with $j=2,\ldots,N-2$, and (3.9). If we examine the coefficient matrix for these equations, it is apparent that only nonzero elements occur on the diagonal as well as just above and below the diagonal. This system is called tridiagonal and it can be solved very efficiently using an algorithm that is described in a number of places, including [19]. Also, a number of programs are available for solving tridiagonal systems [11]. The coefficients $\hat{u}_{k,j}$ are determined by solving $M-1$ independent tridiagonal systems for $k=1,\ldots,M-1$, which completes the second phase of the Fourier method.

We end this part with a review of the Fourier method for solving Problem A, in which the solution is specified on the boundary.

   I.   Compute $\hat{g}_{k,j}$ from $g_{i,j}$ using the FFT to compute (3.2).
  II.   Compute $\hat{u}_{k,j}$ from $\hat{g}_{k,j}$ by solving the systems of tridiagonal equations (3.7), (3.8) and (3.9).
 III.   Compute $u_{i,j}$ from $\hat{u}_{k,j}$ using the FFT to compute (3.1).

Problem B.   In this part we will use the Fourier method to solve the Neumann problem, in which the derivative of the solution is specified on the boundary. We present the solution in less detail since the development is similar to that given for Problem A. We begin with the first phase of the Fourier method, in which the equations are transformed into Fourier space, where they decouple into $M+1$ independent tridiagonal systems of linear equations that can be solved quite easily.

We will show that a solution can be obtained in the form

$$u_{i,j} = \sum_{k=0}^{M} {}'' \hat{u}_{k,j}\cos ki\frac{\pi}{M}. \qquad (3.10)$$

where the double prime notation on the sum indicates that the first and last terms are multiplied by one half. We note that (3.10) is different than the form (3.1) used for Problem A, where the solution was specified on the boundary. This difference is due to the difference between equations (2.10) and (2.24). Whereas the transformation (3.1) will decouple (2.10), the transformation (3.10) is required to decouple (2.24).

We proceed now to the first phase of the Fourier method, in which the equations (2.23) through (2.31) are transformed into

Fourier space in order to obtain a system of equations for $\hat{u}_{k,j}$. First we compute

$$\hat{g}_{k,j} = \frac{2}{M} \sum_{i=0}^{M} {}'' g_{i,j} \cos ik \frac{\pi}{M}, \qquad (3.11)$$

which has the inverse

$$g_{i,j} = \sum_{k=0}^{M} {}'' \hat{g}_{k,j} \cos ik \frac{\pi}{M}. \qquad (3.12)$$

As in Problem A, the FFT is used to compute the transforms (3.10) and (3.11), which constitute phases III and I, respectively. If we transform (2.23), (2.24), and (2.25) following the development after (3.3) above, then for $k = 0, \ldots, M$ and $j = 1, \ldots, N-1$ we obtain

$$\hat{u}_{k,j+1} - \left(2 + 4\rho^2 \sin^2 k \frac{\pi}{2M}\right) \hat{u}_{k,j} + \hat{u}_{k,j-1} = \delta y^2 \hat{g}_{k,j}.$$

$$(3.13)$$

If we transform (2.26), (2.28), and (2.29), then for $k = 0, \ldots, M$ we obtain

$$2\hat{u}_{k,1} - \left(2 + 4\rho^2 \sin^2 k \frac{\pi}{2M}\right) \hat{u}_{k,0} = \delta y^2 \hat{g}_{k,0}. \qquad (3.14)$$

If we transform equations (2.27), (2.30), and (2.31), then for $k = 0, \ldots, M$ we obtain

$$-\left(2 + 4\rho^2 \sin^2 k \frac{\pi}{2M}\right) \hat{u}_{k,N} + 2\hat{u}_{k,N-1} = \delta y^2 \hat{g}_{k,N}. \qquad (3.15)$$

In the second phase we can determine the coefficients $\hat{u}_{k,j}$ from $M+1$ tridiagonal systems each with $N+1$ equations consisting of (3.13), (3.14) with $j = 1, \ldots, N-1$, and (3.15). There is an important extra step that must be taken for this problem that was not necessary for Problem A, in which the solution, rather than the derivative of the solution, was specified. This step is discussed in the first subsection of Section 5.

The third and final phase of the Fourier method consists of computing the solution $u_{i,j}$ from $\hat{u}_{k,j}$, using (3.10) and a modified version of the FFT [29].

Problem C. In this part we will use the Fourier method to solve the problem in which the solution is specified on $x = a$ and $y = c$ and the derivative of the solution is specified on $x = b$ and $y = d$. We begin with the first phase of the Fourier method, in which the equations are transformed into Fourier space, where they decouple into M independent tridiagonal systems of linear equations that are easily solved.

We will show that a solution can be obtained in the form

$$u_{i,j} = \sum_{k=1}^{M} \hat{u}_{k,j} \sin(2k-1) i \frac{\pi}{2M}. \tag{3.16}$$

We note that this form is different than the previous forms. It will decouple equations (2.37) as well as the boundary equations (2.38) and (2.39) when they are transformed into Fourier space.

In the first phase of the Fourier method, equations (2.37) through (2.45) are transformed into Fourier space in order to obtain a system of equations for $\hat{u}_{k,j}$. First we compute

$$\hat{g}_{k,j} = \frac{2}{M} \sum_{i=1}^{M}{}' g_{i,j} \sin(2k-1) i \frac{\pi}{2M}, \tag{3.17}$$

where the prime notation on the sum indicates that the last term is multiplied by one half. Equation (3.17) has the inverse

$$g_{i,j} = \sum_{k=1}^{M} \hat{g}_{k,j} \sin(2k-1) i \frac{\pi}{2M}. \tag{3.18}$$

As in the previous problems, the FFT is used to compute (3.16) and (3.17), which constitute phases III and I, respectively. Following the development after equation (3.3) above, we obtain

$$\hat{u}_{k,j+1} - \left(2 + 4\rho^2 \sin^2(2k-1)\frac{\pi}{4M}\right)\hat{u}_{k,j} + \hat{u}_{k,j-1} = \delta y^2 \hat{g}_{k,j}$$

(3.19)

for $k = 1, \ldots, M$ and $j = 2, \ldots, N-1$. If we transform (2.40), (2.42), and (2.43), then for $k = 1, \ldots, M$ we obtain

$$\hat{u}_{k,2} - \left(2 + 4\rho^2 \sin^2(2k-1)\frac{\pi}{4M}\right)\hat{u}_{k,1} = \delta y^2 \hat{g}_{k,1}. \qquad (3.20)$$

If we transform equations (2.41), (2.44), and (2.45), then for $k = 1, \ldots, M$ we obtain

$$-\left(2 + 4\rho^2 \sin^2(2k-1)\frac{\pi}{4M}\right)\hat{u}_{k,N} + 2\hat{u}_{k,N-1} = \delta y^2 \hat{g}_{k,N}.$$

(3.21)

In the second phase we can determine the coefficients $\hat{u}_{k,j}$ from M tridiagonal systems with N equations consisting of (3.19), (3.20) with $j = 2, \ldots, N-1$, and (3.21).

The third and final phase of the Fourier method consists of computing the solution $u_{i,j}$ from $\hat{u}_{k,j}$, using (3.16) and a modified version of the FFT [29].

Problem D. In this part we will use the Fourier method to solve the Poisson equation subject to periodic boundary conditions. For Problem D it is assumed that the solution is periodic in both x and y, i.e., for any x and y, the solution satisfies both $u(a+x, y) = u(b+x, y)$ and $u(x, c+y) = u(x, d+y)$. We begin with the first phase of the Fourier method, in which equations (2.48) through (2.56) are transformed into Fourier space, where they decouple into M independent tridiagonal systems of N linear equations in N unknowns that are easily solved.

If for the moment we assume that M is even, then a solution can be obtained in the form

$$u_{i,j} = \sum_{k=0}^{M/2} {}'' \left[\hat{u}_{k,j}^{(0)} \cos ik\frac{2\pi}{M} + \hat{u}_{k,j}^{(1)} \sin ik\frac{2\pi}{M} \right]. \qquad (3.22)$$

We note that this form is different than the previous forms. It will decouple equations (2.48) as well as the boundary equations (2.49) and (2.50) when they are transformed into Fourier space. We proceed now to transform (2.48) through (2.56) into Fourier space in order to obtain a system of equations for $\hat{u}_{k,j}^{(l)}$. First we compute

$$\hat{g}_{k,j}^{(0)} = \frac{2}{M} \sum_{i=0}^{M-1} {}'' g_{i,j} \cos ik \frac{2\pi}{M} \qquad (3.23)$$

and

$$\hat{g}_{k,j}^{(1)} = \frac{2}{M} \sum_{i=1}^{M-1} g_{i,j} \sin \frac{2\pi}{M}, \qquad (3.24)$$

which has the inverse

$$g_{i,j} = \sum_{k=0}^{M/2} {}'' \left[\hat{g}_{k,j}^{(0)} \cos ik \frac{2\pi}{M} + \hat{g}_{k,j}^{(1)} \sin ik \frac{2\pi}{M} \right]. \qquad (3.25)$$

As in the previous problems, the FFT is used to compute (3.24) and (3.25), which constitute phases III and I, respectively. Following the development after equation (3.3) we obtain

$$\hat{u}_{k,j+1}^{(l)} - \left(2 + 4\rho^2 \sin^2 k \frac{\pi}{M} \right) \hat{u}_{k,j}^{(l)} + \hat{u}_{k,j-1}^{(l)} = \delta y^2 \hat{g}_{k,j}^{(l)} \qquad (3.26)$$

for $k = 0, \ldots, M/2$; $j = 1, \ldots, N-2$ and $l = 0, 1$. If we transform (2.51), (2.53), and (2.54), then for $k = 0, \ldots, M/2$ and $l = 0, 1$ we obtain

$$\hat{u}_{k,1}^{(l)} - \left(2 + 4\rho^2 \sin^2 k \frac{\pi}{M} \right) \hat{u}_{k,0}^{(l)} + \hat{u}_{k,N-1}^{(l)} = \delta y^2 \hat{g}_{k,0}^{(l)}. \qquad (3.27)$$

If we transform equations (2.52), (2.55), and (2.56), then for $k = 0, \ldots, M/2$ and $l = 0, 1$ we obtain

$$\hat{u}_{k,0}^{(l)} - \left(2 + 4\rho^2 \sin^2 k \frac{\pi}{M} \right) \hat{u}_{k,N-1}^{(l)} + \hat{u}_{k,N-2}^{(l)} = \delta y^2 \hat{g}_{k,N-1}^{(l)}. \qquad (3.28)$$

In the second phase we determine the coefficients $\hat{u}_{k,j}^{(l)}$ from M periodic tridiagonal systems with N equations consisting of (3.26), (3.27) with $j = 1, \ldots, N - 2$, and (3.28). A method for solving periodic tridiagonal linear systems is presented at the end of Section 4. In Section 5 we discuss the "solution" of the singular tridiagonal system corresponding to $k = 0$. The third and final phase of the Fourier method consists of computing the solution $u_{i,j}$ from $\hat{u}_{k,j}^{(l)}$ using (3.22) and the real periodic FFT. If M is odd then a straightforward modification of the transforms (3.22) and (3.25) can be used.

This completes the description of the Fourier method. The implementation of this method can be greatly simplified by using the transforms that are available in the software package FFTPACK which is described in [29].

4.   CYCLIC REDUCTION

In this section we will use the cyclic reduction method to solve the large systems of equations that were derived in Section 2. Although the computational procedure is quite different than the Fourier method, both methods have the same asymptotic operation count and neither of them requires any additional storage. Each method has its own advantages and disadvantages. The method of cyclic reduction can be applied to a larger class of problems than the Fourier method, for example, see [25] and [26], but the Fourier method is usually faster when both methods are applicable.

Like the Fourier method, we will describe the method of cyclic reduction by using it to solve the problems described in the second section. We begin with the problem in which the solution is specified on all the boundaries.

Problem A.   The exposition is greatly simplified if we write the large system of equations in matrix form. To this end we combine all of the unknowns on the jth grid line into the single vector $\mathbf{u}_j^T = (u_{1,j}, u_{2,j}, \ldots, u_{M-1,j})$ with $M - 1$ components. This selection was arbitrary since we could also combine the components on the ith line, which would lead to an alternate description of cyclic reduction with the same computational properties. In practice, these vectors should be made as large as possible. That is, the grid should be orientated so that $N \leqslant M$, since the operation count is proportional to $MN \log N$.

If we multiply the equations on the jth grid line by δy^2, then (2.9) can be written in matrix form

$$\mathbf{u}_{j+1} + \mathbf{A}\mathbf{u}_j + \mathbf{u}_{j-1} = \mathbf{g}_j, \qquad (4.1)$$

where $\mathbf{g}_j^T = \delta y^2 (g_{1,j}, g_{2,j}, \ldots, g_{M-1,j})$ and $\mathbf{A}$ is a matrix with order $M-1$ given by

$$\mathbf{A} = \rho^2 \begin{bmatrix} -2\alpha & 1 & & & \\ 1 & -2\alpha & 1 & & \\ & 1 & \cdot & & \\ & & & \cdot & 1 \\ & & & 1 & -2\alpha \end{bmatrix} \qquad (4.2)$$

with $\alpha = 1 + \rho^{-2}$. For a sample case in which $N = 8$, the complete system of equations (2.9) through (2.17) has the matrix form

$$\begin{aligned}
\mathbf{A}\mathbf{u}_1 + \mathbf{u}_2 & = \mathbf{g}_1 \\
\mathbf{u}_1 + \mathbf{A}\mathbf{u}_2 + \mathbf{u}_3 & = \mathbf{g}_2 \\
\mathbf{u}_2 + \mathbf{A}\mathbf{u}_3 + \mathbf{u}_4 & = \mathbf{g}_3 \\
\mathbf{u}_3 + \mathbf{A}\mathbf{u}_4 + \mathbf{u}_5 & = \mathbf{g}_4 \\
\mathbf{u}_4 + \mathbf{A}\mathbf{u}_5 + \mathbf{u}_6 & = \mathbf{g}_5 \\
\mathbf{u}_5 + \mathbf{A}\mathbf{u}_6 + \mathbf{u}_7 & = \mathbf{g}_6 \\
\mathbf{u}_6 + \mathbf{A}\mathbf{u}_7 & = \mathbf{g}_7.
\end{aligned} \qquad (4.3)$$

The original cyclic reduction algorithm was highly *unstable*. Nevertheless, it will be presented here because it is quite intuitive and excellent for expository purposes. The method came into general use as a practical computational tool after a stable version was developed by Buneman [5], as discussed briefly in the third subsection of Section 5.

The method of cyclic reduction consists of three main phases: preprocessing, reduction and backsubstitution.

I. *Preprocessing.*

In the preprocessing phase we compute a sequence of matrices $A^{(r)}$ from the recurrence relation

$$A^{(r+1)} = 2I - (A^{(r)})^2 \qquad (4.4)$$

beginning with $A^{(0)} = A$. These matrices are used in the later phases and they need not be recomputed if additional solutions are computed. In certain applications, many solutions *are* computed that correspond to different right-hand sides g_j. Under these circumstances, $A^{(r)}$ *can* be computed once and for all in the preprocessing phase since they *are* independent of the g_j. Section 5 contains an important computational technique that enables us to avoid the explicit computation and storage of $A^{(r)}$.

II. *Reduction.*

The reduction phase consists of steps in which new systems of equations are generated. Each new system contains about half the number of equations that the previous system contains. Therefore if N has the form $N = 2^k$ for some integer k, then the reduction phase consists of $k - 1$ steps in which $k - 1$ systems are generated. In the examples that follow, we will set $N = 8$, with the result that two systems can be generated. The first system, (4.6), is generated from (4.3) by multiplying every other equation by $-A$ and adding the adjacent equations. For example, if we multiply the second equation in (4.3) by $-A$ and add the first and third equations, then we obtain the first equation in (4.6), where

$$g_2^{(1)} = g_1 - Ag_2 + g_3. \qquad (4.5)$$

The second equation in (4.6) is obtained in a similar manner by multiplying the fourth equation in (4.3) by $-A$ and adding the third and fifth. This process is again repeated in order to obtain the last equation in (4.6). Note that about half the unknown vectors u_j corresponding to odd j have been eliminated in the system (4.6).

$$A^{(1)}u_2 + u_4 = g_2^{(1)}$$

$$u_2 + A^{(1)}u_4 + u_6 = g_4^{(1)} \qquad (4.6)$$

$$u_4 + A^{(1)}u_6 = g_6^{(1)}.$$

In the second step of the reduction we generate a new system (4.7) from (4.6). Since the system (4.6) has the same form as (4.3) the second step is essentially like the first. We eliminate u_2 and u_6 by multiplying the second equation in (4.6) by $-A^{(1)}$ and adding both the first and third equations. The result is

$$A^{(2)}u_4 = g_4^{(2)}, \qquad (4.7)$$

where

$$g_4^{(2)} = g_2^{(1)} - A^{(1)}g_4^{(1)} + g_6^{(1)}. \qquad (4.8)$$

III. *Backsubstitution.*

The backsubstitution phase consists of steps in which the systems generated in the reduction phase are used in reverse order. The first step consists of solving the much reduced system (4.7) of $M-1$ linear equations in the $M-1$ unknowns for the components of the vector u_4. We note in passing that $A^{(2)}$ is not tridiagonal and hence (4.7) cannot be solved by the simple tridiagonal algorithm. However, in Section 5 we will show that it can be solved by the repeated application of the algorithm.

In the next step of backsubstitution, u_4 is substituted into the first and third equations in (4.6), which become

$$A^{(1)}u_2 = g_2^{(1)} - u_4$$
$$A^{(1)}u_6 = g_6^{(1)} - u_4. \qquad (4.9)$$

These equations are said to be uncoupled since they can be solved independently of one another for u_2 and u_6 in the same way that (4.7) can be solved for u_4. In the third and final step of the backsubstitution phase we substitute u_2, u_4 and u_6 into equations (4.3), which can then be written

$$Au_1 = g_1 - u_2$$
$$Au_3 = g_3 - u_2 - u_4$$
$$Au_5 = g_6 - u_4 - u_6$$
$$Au_7 = g_7 - u_6. \qquad (4.10)$$

Each of these equations can be solved independently for $\mathbf{u}_j$, where $j = 1, 3, 5$ and 7. This completes the backsubstitution phase and the solution of Problem A.

One might quite reasonably ask why this rather complicated effort is recommended when there exist other straightforward methods such as SOR. As mentioned in the Introduction, there are two main reasons why this approach is preferred. In addition to being much faster, cyclic reduction also requires less storage. Indeed, the solution is written over the right-hand side so that no additional storage is required. Furthermore, there is no convergence criteria associated with the method, which is a distinct advantage for many large problems that require the repeated solution of Poisson's equation.

In the example that was given above, we assumed that the number of vector equations $N - 1$ was 7. In general, N must have the form $N = 2^k$ for some integer k for this procedure to be defined. Sweet [31] has developed an algorithm that is efficient for arbitrary N, which can be useful if N must be a prime number since the Fourier method would be very inefficient. There are several important computational problems that must be solved before the procedure that was given above can be used in practice to solve Poisson's equation. These problems will be deferred until Section 5.

Problem B. Consider now the method of cyclic reduction for the Neumann problem, in which the derivative of the solution is specified on all boundaries. As before, we will describe the method by an example in which $N = 8$. If we define $\mathbf{u}_j^T = (u_{0,j}, \ldots, u_{M,j})$ and $\mathbf{g}_j^T = \delta y^2 (g_{0,j}, \ldots, g_{M,j})$, then equations (2.23) through (2.31) have the following matrix form:

$$
\begin{aligned}
\mathbf{A}\mathbf{u}_0 + 2\mathbf{u}_1 & & &= \mathbf{g}_0 \\
\mathbf{u}_0 + \mathbf{A}\mathbf{u}_1 + \mathbf{u}_2 & & &= \mathbf{g}_1 \\
\mathbf{u}_1 + \mathbf{A}\mathbf{u}_2 + \mathbf{u}_3 & & &= \mathbf{g}_2 \\
\mathbf{u}_2 + \mathbf{A}\mathbf{u}_3 + \mathbf{u}_4 & & &= \mathbf{g}_3 \\
\mathbf{u}_3 + \mathbf{A}\mathbf{u}_4 + \mathbf{u}_5 & & &= \mathbf{g}_4 \quad (4.11) \\
\mathbf{u}_4 + \mathbf{A}\mathbf{u}_5 + \mathbf{u}_6 & & &= \mathbf{g}_5 \\
\mathbf{u}_5 + \mathbf{A}\mathbf{u}_6 + \mathbf{u}_7 & & &= \mathbf{g}_6 \\
\mathbf{u}_6 + \mathbf{A}\mathbf{u}_7 + \mathbf{u}_8 &= \mathbf{g}_7 \\
2\mathbf{u}_7 + \mathbf{A}\mathbf{u}_8 &= \mathbf{g}_8
\end{aligned}
$$

where $\mathbf{A}$ is a matrix with order $M + 1$ given by

$$\mathbf{A} = \rho^2 \begin{bmatrix} -2\alpha & 2 & & & & \\ 1 & -2\alpha & 1 & & & \\ & 1 & \cdot & & & \\ & & & \cdot & & \\ & & & & \cdot & 1 \\ & & & & 2 & -2\alpha \end{bmatrix}. \qquad (4.12)$$

As before, the solution is obtained in three phases.

I. *Preprocessing.*

In the preprocessing phase we compute $\mathbf{A}^{(1)} = 2\mathbf{I} - \mathbf{A}^2$, $\mathbf{A}^{(2)} = 2\mathbf{I} - (\mathbf{A}^{(1)})^2$, and $\mathbf{A}^{(3)} = 4\mathbf{I} - (\mathbf{A}^{(2)})^2$. Note that the expression for $\mathbf{A}^{(3)}$ is different from the other two. The reason for this will be discussed below.

II. *Reduction.*

If we multiply the first equation in (4.11) by $-\mathbf{A}$ and add it to twice the second equation, then we obtain the first equation in (4.13) below, where $\mathbf{g}_0 = 2\mathbf{g}_1 - A\mathbf{g}_0$. If the last equation in (4.11) is multiplied by $-\mathbf{A}$ and added to twice the next to the last equation, then we obtain the last equation in (4.13). The remaining equations in (4.13) are obtained in the same way as equations (4.6), i.e., the second equation in (4.13) is obtained by multiplying the third equation in (4.11) by $-\mathbf{A}$ and adding the second and fourth equations:

$$\begin{aligned}
\mathbf{A}^{(1)}\mathbf{u}_0 &+ 2\mathbf{u}_2 & & & & = \mathbf{g}_0^{(1)} \\
\mathbf{u}_0 &+ \mathbf{A}^{(1)}\mathbf{u}_2 &+ \mathbf{u}_4 & & & = \mathbf{g}_2^{(1)} \\
& \mathbf{u}_2 &+ \mathbf{A}^{(1)}\mathbf{u}_4 &+ \mathbf{u}_6 & & = \mathbf{g}_4^{(1)} \quad (4.13) \\
& & \mathbf{u}_4 &+ \mathbf{A}^{(1)}\mathbf{u}_6 &+ \mathbf{u}_8 & = \mathbf{g}_6^{(1)} \\
& & & 2\mathbf{u}_6 &+ \mathbf{A}^{(1)}\mathbf{u}_8 & = \mathbf{g}_8^{(1)}.
\end{aligned}$$

Since this system has the same form as (4.11), we can repeat the procedure in order to derive the following equations:

$$
\begin{aligned}
\mathbf{A}^{(2)}\mathbf{u}_0 &\quad + 2\mathbf{u}_4 &\quad &= \mathbf{g}_0^{(2)} \\
\mathbf{u}_0 &\quad + \mathbf{A}^{(2)}\mathbf{u}_4 &\quad + \mathbf{u}_8 &= \mathbf{g}_4^{(2)} \\
&\quad 2\mathbf{u}_4 &\quad + \mathbf{A}^{(2)}\mathbf{u}_8 &= \mathbf{g}_8^{(2)}.
\end{aligned}
\tag{4.14}
$$

The next and final step of the reduction for this problem differs from the previous steps. Instead of further reducing system (4.14) to two vector equations in the vector unknowns $\mathbf{u}_0$ and $\mathbf{u}_8$, it is reduced to a single vector equation in the vector unknown $\mathbf{u}_4$ by multiplying the second equation by $-\mathbf{A}^{(2)}$ and adding the first and third equations. Since $\mathbf{A}^{(3)} = 4\mathbf{I} - (\mathbf{A}^{(2)})^2$, we obtain

$$
\mathbf{A}^{(3)}\mathbf{u}_4 = \mathbf{g}_4^{(3)}.
\tag{4.15}
$$

Although this last step appears to be the same as the earlier steps, it is, in fact, different. Unlike (4.15), each of the previous systems (4.11), (4.13), and (4.14) contain both $\mathbf{u}_0$ and $\mathbf{u}_8$. Also, the expression for $\mathbf{A}^{(3)}$ contains the term $4\mathbf{I}$ rather than $2\mathbf{I}$, which was contained in the expressions for $\mathbf{A}^{(1)}$ and $\mathbf{A}^{(2)}$.

III. *Backsubstitution.*

The backsubstitution begins with solving equation (4.15) for $\mathbf{u}_4$ which is then substituted into the first and last equations in (4.14) in order to determine $\mathbf{u}_0$ and $\mathbf{u}_8$, respectively. We can then determine $\mathbf{u}_2$ and $\mathbf{u}_6$ from the second and fourth equations in (4.13). Finally, with $\mathbf{u}_j$ known for even j, $\mathbf{u}_j$ for odd j can be determined from (4.16).

Problem C. Consider now the method of cyclic reduction for the third problem, in which the solution is specified on $x = a$ and $y = c$ and the derivative of the solution is specified on $x = b$ and $y = d$. As before, we will describe the method by an example in which $N = 8$. If we define $\mathbf{u}_j^T = (u_{1, j}, \ldots, u_{M, j})$ and $\mathbf{g}_j^T = \delta y^2(g_{1, j}, \ldots, g_{M, j})$, then equations (2.37) through (2.45) have the following matrix form:

$$
\begin{array}{rcl}
Au_1 + u_2 & = & g_1 \\
u_1 + Au_2 + u_3 & = & g_2 \\
u_2 + Au_3 + u_4 & = & g_3 \\
u_3 + Au_4 + u_5 & = & g_4 \\
u_4 + Au_5 + u_6 & = & g_5 \\
u_5 + Au_6 + u_7 & = & g_6 \\
u_6 + Au_7 + u_8 & = & g_7 \\
2u_7 + Au_8 & = & g_8,
\end{array}
\tag{4.16}
$$

where A is a matrix with order M given by

$$
A = \rho^2
\begin{bmatrix}
-2\alpha & 1 & & & \\
1 & -2\alpha & 1 & & \\
& 1 & \cdot & & \\
& & & \cdot & 1 \\
& & & 2 & -2\alpha
\end{bmatrix}.
\tag{4.17}
$$

As before, the solution is obtained in three phases.

I. *Preprocessing.*

The preprocessing phase consists of computing $A^{(1)} = 2I - A^2$, $A^{(2)} = 2I - (A^{(1)})^2$, and $A^{(3)} = 2I - (A^{(2)})^2$.

II. *Reduction.*

As in Problem B, the last equation in (4.18) below is obtained by multiplying the last equation in (4.16) by $-A$ and adding twice the next to the last equation. The remaining equations in (4.18) are obtained in the same way as equations (4.6). For example, the second equation in (4.18) is obtained by multiplying the fourth equation in (4.16) by $-A$ and adding the adjacent equations.

$$
\begin{array}{rcl}
A^{(1)}u_2 + u_4 & = & g_2^{(1)} \\
u_2 + A^{(1)}u_4 + u_6 & = & g_4^{(1)} \\
u_4 + A^{(1)}u_6 + u_8 & = & g_6^{(1)} \\
2u_6 + A^{(1)}u_8 & = & g_8^{(1)}.
\end{array}
\tag{4.18}
$$

From this system we derive the following system in the two unknown vectors $\mathbf{u}_4$ and $\mathbf{u}_8$:

$$\mathbf{A}^{(2)}\mathbf{u}_4 + \mathbf{u}_8 = \mathbf{g}_4^{(2)}$$

$$2\mathbf{u}_4 + \mathbf{A}^{(2)}\mathbf{u}_8 = \mathbf{g}_8^{(2)}. \tag{4.19}$$

Multiplying the second equation by $-\mathbf{A}^{(2)}$ and adding twice the first, we obtain

$$\mathbf{A}^{(3)}\mathbf{u}_8 = \mathbf{g}_8^{(3)}. \tag{4.20}$$

III. *Backsubstitution.*

The backsubstitution begins with solving equation (4.20) for $\mathbf{u}_8$, which is then substituted into the first equation in (4.19) in order to determine $\mathbf{u}_4$. As before, system (4.18) and then (4.16) are used to determine the remaining $\mathbf{u}_j$.

Problem D. The solution of the Poisson equation subject to periodic boundary conditions can also be obtained using the three phases of cyclic reduction, which is the approach that is taken in [6]. However, an alternate approach is given in [31] which is quite intuitive and makes use of the methods that have already been developed for Problems A and B. This approach can also be used to solve scalar linear systems when the coefficient matrix is periodic tridiagonal as in (4.22) below. As we will see, the solution of this system can be expressed in terms of solutions to two tridiagonal systems with $N/2$ equations.

As before, we will describe the method by an example in which $N = 8$. If we define $\mathbf{u}_j^T = (u_{0,j}, \ldots, u_{M-1,j})$ and $\mathbf{g}_j^T = \delta y^2 (g_{0,j}, \ldots, g_{M-1,j})$, then equations (2.48) through (2.56) have the following matrix form

$$
\begin{array}{llllllll}
\mathbf{A}\mathbf{u}_0 & +\mathbf{u}_1 & & & & & & +\mathbf{u}_7 = \mathbf{g}_0 \\
\mathbf{u}_0 + \mathbf{A}\mathbf{u}_1 & +\mathbf{u}_2 & & & & & & = \mathbf{g}_1 \\
& \mathbf{u}_1 + \mathbf{A}\mathbf{u}_2 & +\mathbf{u}_3 & & & & & = \mathbf{g}_2 \\
& & \mathbf{u}_2 + \mathbf{A}\mathbf{u}_3 & +\mathbf{u}_4 & & & & = \mathbf{g}_3 \\
& & & \mathbf{u}_3 + \mathbf{A}\mathbf{u}_4 & +\mathbf{u}_5 & & & = \mathbf{g}_4 \\
& & & & \mathbf{u}_4 + \mathbf{A}\mathbf{u}_5 & +\mathbf{u}_6 & & = \mathbf{g}_5 \\
& & & & & \mathbf{u}_5 + \mathbf{A}\mathbf{u}_6 & +\mathbf{u}_7 & = \mathbf{g}_6 \\
\mathbf{u}_0 & & & & & & +\mathbf{u}_6 + \mathbf{A}\mathbf{u}_7 & = \mathbf{g}_7.
\end{array} \tag{4.21}
$$

where A is a matrix with order M given by

$$A = \rho^2 \begin{bmatrix} -2\alpha & 1 & & & & 1 \\ 1 & -2\alpha & 1 & & & \\ & 1 & \cdot & & & \\ & & & \cdot & & 1 \\ 1 & & & & 1 & -2\alpha \end{bmatrix}. \quad (4.22)$$

We will show that equations (4.21) can be split into two smaller systems like those that we have already discussed. If we add the last equation in (4.21) to the second and the next to the last to the third and so forth, then we obtain the following system:

$$
\begin{array}{llll}
A(u_0 + u_0) + 2(u_1 + u_7) & & & = 2g_0 \\
(u_0 + u_0) + A(u_1 + u_7) & + (u_2 + u_6) & & = g_1 + g_7 \\
(u_1 + u_7) + A(u_2 + u_6) & + (u_3 + u_5) & & = g_2 + g_6) \quad (4.23) \\
(u_2 + u_6) + A(u_3 + u_5) & + (u_4 + u_4) & = g_3 + g_5 \\
2(u_3 + u_5) + A(u_4 + u_4) & = 2g_4.
\end{array}
$$

If we define new vectors

$$v_j = u_j + u_{8-j} \quad (4.24)$$

and

$$s_j = g_j + g_{8-j},$$

then (4.23) becomes

$$
\begin{array}{llll}
Av_0 + 2v_1 & & & = s_0 \\
v_0 + Av_1 & + v_2 & & = s_1 \\
& v_1 + Av_2 & + v_3 & = s_2 \quad (4.25) \\
& & v_2 + Av_3 & + v_4 = s_3 \\
& & & 2v_3 + Av_4 = s_4.
\end{array}
$$

But this system has the same form as (4.11) and, hence, we can use the procedure following (4.11) to solve (4.25) for v_j. However, we still need to determine u_j. To this end we define a second system of vector equations. If we subtract the last equation in (4.21) from the second, the next to the last from the third, etc., then we obtain the following system:

$$
\begin{aligned}
A(u_1 - u_7) + (u_2 - u_6) &&&= g_1 - g_7 \\
(u_1 - u_7) + A(u_2 - u_6) &&+ (u_3 - u_5) &= g_2 - g_6 \\
(u_2 - u_6) &&+ A(u_3 - u_5) &= g_3 - g_5.
\end{aligned}
\tag{4.26}
$$

If we define new vectors

$$
w_j = u_j - u_{8-j} \tag{4.27}
$$

and

$$
t_j = g_j - g_{8-j}
$$

then (4.26) becomes

$$
Aw_1 + w_2 \qquad = t_1
$$

$$
w_1 + Aw_2 + w_3 = t_2 \tag{4.28}
$$

$$
w_2 + Aw_3 = t_3.
$$

But this system has the same form as (4.3) and, hence, we can use the procedure following (4.3) to solve (4.28) for w_j. Having determined both v_j and w_j, we can now determine u_j from the first equations in (4.24) and (4.27):

$$
u_0 = \tfrac{1}{2} v_0
$$

$$
u_j = \tfrac{1}{2}(v_j + w_{8-j}) \qquad j = 1, 2, 3 \tag{4.29}
$$

$$
u_4 = \tfrac{1}{2} v_4
$$

$$
u_{8-j} = \tfrac{1}{2}(v_j - w_{8-j}) \qquad j = 1, 2, 3.
$$

5. COMPUTATIONAL NOTES

In this section we will discuss certain nuisance problems that must be dealt with before many of the earlier results are useful in practice. Three main problems will be discussed in this section. In sub-section I, we will show that certain problems may not have a solution unless the right-hand side is perturbed. We will show how to identify these problems and "solve" them in the least squares sense. In subsection II, we will present an efficient way of computing, storing, and working with the matrices $A^{(r)}$ that are used in cyclic reduction. Finally, in subsection III, we will present Buneman's stable cyclic reduction algorithm.

I.   *Least Squares Solutions.*

Many problems that occur in practice do not have a solution in the usual sense. Nevertheless, a solution can still be obtained if we are permitted to alter or perturb the problem somewhat. In general this is found to be quite acceptable, particularly if the perturbation is kept within the accepted level of error for the problem.

One does not have to contrive such problems; indeed both Problem B and Problem D may require a least squares solution. These problems are encountered in many applications and require special attention if important computational complications are to be avoided. This difficulty occurs when any combination of derivative and periodic boundary conditions are specified. An easy test can be made in order to determine if a least squares solution may be required; namely, if the solution plus a constant is also a solution, then the following discussion applies to the problem. In this part of the section we will show how to obtain a least squares solution, but we do not prove that it is a least squares solution. A proof of this fact appears in [30] as well as [27] for spherical geometry.

In order to illustrate the difficulty that can be encountered, we will examine Problem D, in which periodic boundary conditions are specified. For expository purposes, we will first examine the difficulty in the context of the continuous problem. Suppose, for the moment, that we wish to find a continuous function $u(x, y)$ that is periodic in both x and y and satisfies Poisson's equation

$$u_{xx}(x, y) + u_{yy}(x, y) = f(x, y) \qquad (5.1)$$

in the rectangle $a < x < b$ and $c < y < d$. If we assume that a solution exists then we can integrate both sides of (5.1) and obtain

$$\int_c^d \left(u_x(b, y) - u_x(a, y) \right) dy + \int_a^b \left(u_y(x, d) - u_y(x, c) \right) dx$$

$$= \int_a^b \int_c^d f(x, y) \, dx \, dy. \tag{5.2}$$

But by hypothesis the solution is periodic and hence $u_x(a, y) = u_x(b, y)$ and $u_y(x, c) = u_y(x, d)$, so that the left-hand side of (5.2) is zero. Therefore, we obtain the following constraint on $f(x, y)$, namely

$$\int_a^b \int_c^d f(x, y) \, dx \, dy = 0. \tag{5.3}$$

If $f(x, y)$ does not satisfy this constraint, then a solution of Problem D does not exist. In most cases this is not a problem since $f(x, y)$ can be perturbed so that the perturbed function does satisfy the constraint. For example, let

$$I_f = \int_a^b \int_c^d f(x, y) \, dx \, dy. \tag{5.4}$$

If we define

$$\hat{f}(x, y) = f(x, y) - \frac{I_f}{(b - a)(d - c)} \tag{5.5}$$

then $\hat{f}(x, y)$ satisfies the constraint, and the perturbed Poisson equation

$$u_{xx} + u_{yy} = \hat{f}(x, y) \tag{5.6}$$

has a solution. Furthermore, it can be shown that the solution of (5.6) is a least squares solution of (5.1). A word of caution: we can be comfortable with the solution of the perturbed problem (5.6) only if the second term on the right side of (5.5) is small relative to $f(x, y)$. Otherwise we may have, in fact, solved a problem that is very different from (5.1).

Similar results can be obtained for the discrete problem. If we sum all of the equations in (2.48) through (2.56), we find that the right side $g_{i,j}$ must satisfy the constraint

$$S_g = \sum_{i=0}^{M-1} \sum_{j=0}^{N-1} g_{i,j} = 0, \qquad (5.7)$$

which is the discrete analogue of (5.3). If $g_{i,j}$ does not satisfy (5.7), then a solution to equations (2.48) through (2.56) does not exist. In practice, as a result of computational or observational errors, (5.7) does not hold even though it may "almost" hold. Following the development for the continuous problem above, we can perturb $g_{i,j}$ by a constant so that the constraint does hold, i.e., if we define

$$\hat{g}_{i,j} = g_{i,j} - \frac{S_g}{MN} \qquad (5.8)$$

then $\hat{g}_{i,j}$ satisfies the constraint. If we replace $g_{i,j}$ by $\hat{g}_{i,j}$, a solution of equations (2.48) through (2.56) exists which is a least squares solution to the unperturbed equations.

The constraint for Problem B, where the derivative is specified on all the boundaries, is different from the constraint for the periodic boundary condition. Of course, there are problems other than those given as examples in which the solution may be specified to be periodic in one direction (say x) and the derivative of the solution with respect to y specified at $y = c$ and $y = d$. In general, for any problem that requires a least squares solution, the constraint can be written in the form

$$S_g = \sum_{i=0}^{M} \sum_{j=0}^{N} w_i w_j g_{i,j} = 0, \qquad (5.9)$$

where the weights w_i and w_j *are* given in Table 1 below. For example, if periodic boundary conditions *are* specified in the x direction, then w_i in (5.9) are selected from the row labeled "periodic" in the upper part of Table 1. If derivative boundary conditions *are* specified in the y direction, then w_j are selected from the row labeled "derivative" in the lower part of Table 1.

TABLE 1. Weights for Constraints on the right side of Poisson's equation.			
		i		
	0	$0 < i < M-1$	$M-1$	M
periodic	1	1	1	0
derivative	.5	1	1	.5
		j		
	0	$0 < j < N-1$	$N-1$	N
periodic	1	1	1	0
derivative	.5	1	1	.5

Once the weights are determined, we can perturb the right side by defining

$$\hat{g}_{i,j} = g_{i,j} - \frac{S_g}{\displaystyle\sum_{i=0}^{M}\sum_{j=0}^{N} w_i w_j}. \qquad (5.10)$$

When the right side has been perturbed, a solution can be obtained using either Fourier analysis or cyclic reduction even though the system of equations is singular. A slight modification of these methods is necessary in order to avoid a division by zero when solving the tridiagonal system. A simple IF test in FORTRAN can be used to test the last pivot when solving the tridiagonal systems. If the pivot is zero, then the corresponding unknown can be set to zero. Actually it can be set to any value since the solution is not unique. Usually the pivot is not identically zero, in which case the unknown will be set to some arbitrary value that depends on the roundoff error. In either case, a solution of the system will be found.

II. *Matrix Polynomials.*

In Section 4, the method of cyclic reduction made extensive use of certain matrices $A^{(r)}$. It turns out, as shown in [6], that these matrices do not have to be explicitly computed. This important observation results in considerable savings in both computer time and storage. Just prior to equations (4.9) it was stated that (4.8) could be solved by a repeated application of the Gauss algorithm for tridiagonal systems. We will show how this is done and at the same time develop an efficient way to represent the matrices $\mathbf{A}^{(r)}$.

From (4.4) we see that $\mathbf{A}^{(r)}$ is just a polynomial of degree 2^r in $\mathbf{A}$, which we will define as $p_{2^r}(\mathbf{A})$. If we let $p_{2^r}(x)$ denote its scalar analogue, then from (4.4) it can be seen that these scalar polynomials can be generated recursively starting with $p_1(x) = x$ and continuing with

$$p_{2^{r+1}}(x) = 2 - \left[p_{2^r}(x) \right]^2. \tag{5.11}$$

If we make the substitution $x = -2\cos\theta$ then

$$p_{2^r}(x) = -2\cos 2^r\theta \tag{5.12}$$

or

$$p_{2^r}(x) = -C_{2^r}(x), \tag{5.13}$$

where $C_{2^r}(x)$ is the Chebyshev polynomial with zeros

$$x_m^{(r)} = 2\cos(2m-1)\frac{\pi}{2^{r+1}} \qquad m = 1,\ldots,2^r. \tag{5.14}$$

Therefore $p_{2^r}(\mathbf{A})$ can be expressed in factored form

$$p_{2^r}(\mathbf{A}) = - \prod_{m=1}^{2^r} \left(\mathbf{A} - x_m^{(r)}I \right). \tag{5.15}$$

Hence, instead of storing the matrix $\mathbf{A}^{r)}$, we need only store the zeros $x_m^{(r)}$. Furthermore, all the matrix computations can be performed using the factored form (5.15). For example, consider the solution of equation (4.7), which can now be written in factored form

$$\left(\mathbf{A} - x_1^{(2)}\mathbf{I}\right)\left(\mathbf{A} - x_2^{(2)}\mathbf{I}\right)\left(\mathbf{A} - x_3^{(2)}\mathbf{I}\right)\left(\mathbf{A} - x_4^{(2)}\mathbf{I}\right)\mathbf{u}_4 = -\mathbf{g}_4^{(2)}.$$

$$\tag{5.16}$$

If we define $w_0 = -g_4^{(2)}$ and

$$w_1 = (A - x_2^{(2)}I)(A - x_3^{(2)}I)(A - x_4^{(2)}I)u_4, \qquad (5.17)$$

then (5.16) becomes

$$(A - x_1^{(2)}I)w_1 = w_0. \qquad (5.18)$$

But the matrix in parenthesis is tridiagonal, and therefore we can determine w_1 with the simple form of the Gauss algorithm. If we then define

$$w_2 = (A - x_3^{(2)}I)(A - x_4^{(2)}I)u_4, \qquad (5.19)$$

then (5.17) becomes

$$(A - x_2^{(2)}I)w_2 = w_1. \qquad (5.20)$$

Here the coefficient matrix is also tridiagonal, and we can determine w_2 using the Gauss algorithm. If we continue this process by defining

$$w_3 = (A - x_4^{(2)}I)u_4, \qquad (5.21)$$

then we can determine w_3 in the same manner, and finally from (5.21) we can determine u_4.

All of the matrix operations in cyclic reduction can be performed in this manner using factored forms and, therefore, it is unnecessary to explicitly compute and store the $A^{(r)}$. In the next part we will present Buneman's cyclic reduction algorithm in which matrix-vector multiplications are eliminated.

III. *Stable Cyclic Reduction.*

As mentioned earlier, the cyclic reduction method presented in the previous section is useful only for exposition and not for computation. The reason for this is that, in the course of computation, the roundoff errors accumulate to the extent that they dominate the results. The extent to which this happens depends on the size of

N and the accuracy of the computer; however, at the current state of the art, the loss of accuracy can be unacceptable for a grid size as small as $N = 10$.

If we examine the unstable algorithm closely, we can identify the source of the error. In Section 4, the method of cyclic reduction was divided into three phases: preprocessing, reduction, and backsubstitution. Beginning with the original system of equations, a sequence of linear systems is generated in the reduction phase. Each of these systems has about half the number of unknowns of the preceding system. By examining the sample reduction phase for Problem A in Section 4, we can determine the general form of the reduction phase. The right-hand sides of the $r + 1$st system are computed recursively from the right-hand sides of the rth system by

$$\mathbf{g}_j^{(r+1)} = \mathbf{g}_{j-2^r}^{(r)} - \mathbf{A}^{(r)}\mathbf{g}_j^{(r)} + \mathbf{g}_{j+2^r}^{(r)}. \tag{5.22}$$

In the backsubstitution phase, the linear systems that were generated in the reduction phase are used in reverse order. The unknown vectors $\mathbf{u}_j$ are computed as solutions of equation (5.23) below. The vectors $\mathbf{u}_{j-2^r}$ and $\mathbf{u}_{j+2^r}$ that appear on the right side are computed in a previous step of the backsubstitution.

$$\mathbf{A}^{(r)}\mathbf{u}_j = \mathbf{g}_j^{(r)} - \mathbf{u}_{j-2^r} - \mathbf{u}_{j+2^r}. \tag{5.23}$$

The source of error in the unstable algorithm is in the second term on the right side of (5.22). Any error that is present in $\mathbf{g}_j^{(r)}$ is greatly amplified when it is multiplied by $\mathbf{A}^{(r)}$. These multiplications do not occur in the backsubstitution phase. Indeed, the solution of (5.23) is theoretically obtained by multiplying the right side by the inverse of $\mathbf{A}^{(r)}$. The multiplications in the reduction phase can be eliminated by incorporating some of the inversions that occur in the backsubstitution phase into the reduction phase. Buneman [5] was able to do this with a simple transformation. If instead of computing $\mathbf{g}_j^{(r)}$ we can compute new vectors $\mathbf{p}_j^{(r)}$ and $\mathbf{q}_j^{(r)}$ such that

$$\mathbf{g}_j^{(r)} = \mathbf{A}^{(r)}\mathbf{p}_j^{(r)} + \mathbf{q}_j^{(r)}, \tag{5.24}$$

then the backsubstitution (5.23) takes the form

$$\mathbf{u}_j = \mathbf{p}_j^{(r)} + \left(\mathbf{A}^{(r)}\right)^{-1}\left(\mathbf{q}_j^{(r)} - \mathbf{u}_{j-2^r} - \mathbf{u}_{j+2^r}\right). \tag{5.25}$$

It remains to compute the vectors $\mathbf{p}_j^{(r)}$ and $\mathbf{q}_j^{(r)}$. If we substitute (5.24) into (5.22) and use the identity $[\mathbf{A}^{(r)}]^2 = 2\mathbf{I} - \mathbf{A}^{(r+1)}$, we obtain a single equation with a number of terms. If we combine terms that are multiplied by $\mathbf{A}^{(r+1)}$ then we obtain (5.26) below. The remaining terms correspond to (5.27):

$$\mathbf{p}_j^{(r+1)} = \mathbf{p}_j^{(r)} - \left(\mathbf{A}^{(r)}\right)^{-1}\left(\mathbf{p}_{j-2^r} + \mathbf{p}_{j+2^r} - \mathbf{q}_j^{(r)}\right) \tag{5.26}$$

$$\mathbf{q}_j^{(r+1)} = \mathbf{q}_{j-2^r} + \mathbf{q}_{j+2^r} - 2\mathbf{p}_j^{(r+1)}. \tag{5.27}$$

Note that the matrix-vector multiplications that occurred in the unstable reduction phase are no longer present.

The Buneman algorithm for stable cyclic reduction can now be summarized. Like the unstable algorithm it consists of three phases. The zeros of the matrix polynomials are computed in the pre-processing phase just as in the preprocessing phase of the unstable algorithm. The reduction phase of the Buneman algorithm consists of computing $\mathbf{p}_j^{(r)}$ and $\mathbf{q}_j^{(r)}$ from (5.26) and (5.27), beginning with $\mathbf{p}_j^{(0)} = 0$ and $\mathbf{q}_j^{(0)} = g_j$. Finally, the solution vectors $\mathbf{u}_j$ are computed in the backsubstitution phase using (5.25).

There is an important variant of the Buneman algorithm in which only the $\mathbf{q}_j^{(r)}$ are computed. It is presented in [6] together with a theoretical discussion on the stability of cyclic reduction.

REFERENCES

1. R. E. Bank, "Marching algorithms for elliptic boundary value problems. II: The non-constant coefficient case," *SIAM J. Numer. Anal.*, **14** (1977), 950–970.
2. R. E. Bank and D. J. Rose, "An $O(n^2)$ method for solving constant coefficient boundary value problems in two dimensions," *SIAM J. Numer. Anal.*, **12** (1975), 529–540.
3. _____, "Marching algorithms for elliptic boundary value problems. I: The constant coefficient case," *SIAM J. Numer. Anal.*, **14** (1977), 792–829.
4. A. Brandt, "Multi-level adaptive solutions to boundary-value problems," *Math. Comp.*, **31** (1977), 333–390.

5. O. Buneman, "A compact non-iterative Poisson solver," Rep. 294, Stanford University Institute for Plasma Research, Stanford, Calif., 1969.

6. B. L. Buzbee, G. H. Golub, and C. W. Nielson, "On direct methods for solving Poisson's equations," *SIAM J. Numer. Anal.*, 7 (1970), 627–656.

7. B. L. Buzbee, F. W. Dorr, J. A. George, and G. H. Golub, "The direct solution of the discrete Poisson equation on irregular regions," *SIAM J. Numer. Anal.*, 8 (1971), 722–736.

8. P. Concus and G. H. Golub, "Use of fast direct methods for the efficient numerical solution of nonseparable elliptic equations," *SIAM J. Numer. Anal.*, 10 (1973), 1103–1120.

9. P. Concus, G. H. Golub, and D. P. O'Leary, "A generalized conjugate gradient method for the numerical solution of elliptic partial differential equations," *Sparse Matrix Computations*, J. R. Bunch and D. J. Rose, eds., Academic Press, New York, 1976.

10. E. Detyna, "Point cyclic reductions for elliptic boundary-value problems. I. The constant coefficient case," *J. Comp. Phys.*, 33 (1979), 204–216.

11. J. J. Dongarra, C. B. Moler, J. R. Bunch, and G. W. Stewart, *LINPACK User's Guide*, SIAM publications, Philadelphia, 1979.

12. F. W. Dorr, "The direct solution of the discrete Poisson equation on a rectangle," *SIAM Rev.*, 12 (1970), 248–263.

13. _____, "The direct solution of the discrete Poisson equation in $O(N^2)$ operations," *SIAM Rev.*, 17 (1975), 412–415.

14. L. Garmon, "Science on fire," *Science News.*, 120, no. 14 (1981), 218.

15. R. W. Hockney, "A fast direct solution of Poisson's equation using Fourier analysis," *J. Assoc. Comput. Mach.*, 12 (1965), 95–113.

16. _____, "The potential calculation and some applications," *Methods of Computational Physics*, vol. 9, B. Adler, S. Fernbach and M. Rotenberg, eds., Academic Press, New York and London, 1969, 136–211.

17. _____, "Computers, compilers and Poisson-solvers," *Computers, Fast Elliptic Solvers and Applications*, U. Schumann, ed., Advance Publications, London, 1978.

18. R. W. Hockney and C. R. Jesshope, *Parallel Computers*, Adam Hilger, Bristol, 1981.

19. E. Isaacson and H. B. Keller, *Analysis of Numerical Methods*, John Wiley and Sons Inc., New York, 1966.

20. E. N. Lorenz, "A rapid procedure for inverting del-square with certain computers," *Mon. Wea. Rev.*, 104 (1976), 961–966.

21. M. Machura and R. A. Sweet, "A survey of software for partial differential equations," *ACM Trans. Math. Soft.*, 6 (1980), 461–488.

22. W. Proskurowski, "Numerical solution of Helmholtz's equation by implicit capacitance matrix methods," *ACM Trans. Math. Soft.*, 5 (1979), 36–49.

23. W. Proskurowski and O. Widlund, "On the numerical solution of Helmholtz's equation by the capacitance matrix method," *Math. Comp.*, 30 (1976), 433–468.

24. J. Schroeder, U. Trottenberg, and H. Reutersberg, "Reduktionsverfahren fuer differenzengleichungen bei randwertaufgaben," *Numer. Math.*, 26 (1976), 429–459.

25. U. Schumann and R. A. Sweet, "A direct method for the solution of Poisson's equation with Neumann boundary conditions on a staggered grid of arbitrary size," *J. Comp. Phys.*, **20** (1976), 171–182.
26. P. N. Swarztrauber, "A direct method for the discrete solution of separable elliptic equations," *SIAM J. Numer. Anal.*, **11** (1974), 1136–1150.
27. _____, "The direct solution of the discrete Poisson equation on the surface of the sphere," *J. Comp. Phys.*, **15** (1974), 46–54.
28. _____, "The methods of cyclic reduction, Fourier analysis and the FACR algorithm for the discrete solution of the Poisson equation on a rectangle," *SIAM, Rev.*, **19** (1977), 490–501.
29. _____, *Vectorizing the FFT's Parallel Computations*, G. Rodrigue, ed., Academic Press, New York, 1982.
30. P. N. Swarztrauber and R. A. Sweet, "Efficient FORTRAN subprograms for the solution of elliptic partial differential equations," *ACM Trans. Math. Soft.* **5** (1979), 352–364.
31. R. A. Sweet, "A cyclic reduction algorithm for solving block tridiagonal systems of arbitrary dimension," *SIAM J. Numer. Anal.*, **14** (1977), 706–720.
32. C. Temperton, "On the FACR(ℓ) algorithm for the discrete Poisson equation," *J. Comp. Phys.*, **34** (1980), 314–329.

POISSON'S EQUATION IN A HYPERCUBE: DISCRETE FOURIER METHODS, EIGENFUNCTION EXPANSIONS, PADE APPROXIMATION TO EIGENVALUES

By Peter Henrici

1. THE PROBLEM

Let the points of d-dimensional Euclidean space $\mathbb{R}^d$ be denoted by $\boldsymbol{\xi} = (\xi_1, \xi_2, \ldots, \xi_d)$, let $S := \{ \boldsymbol{\xi} : 0 < \xi_i < 1, \ i = 1, \ldots, d \}$ denote the open unit cube in R^d, and let S' denote its closure and ∂S its boundary. Let the function $f : S \rightarrow \mathbb{R}$ be, at the moment, continuous. We consider the problem of solving the Poisson's equation,

$$-\Delta u = f \tag{1.1}$$

$(\Delta := \sum_{i=1}^{d} \partial^2 / \partial \xi_i^2)$, for S under zero boundary conditions, i.e., to find the (unique) function $u : S \rightarrow \mathbb{R}$ that is continuous in S', zero

371

on ∂S, and whose second partial derivatives exist and satisfy (1.1) at every point $\xi \in S$. Physical interest is attached to our problem primarily in the cases $d = 2$ and (increasingly) $d = 3$, where the problem frequently arises in a larger context. If that context involves a dependence on time, or if an iteration is to be performed due to nonlinear features, (1.1) has to be solved many times, and the construction of fast and efficient solution algorithms for (1.1) therefore becomes a problem of great urgency.

Mathematically, the solution of our problem is a simple affair. Let $\gamma: S \times S \rightarrow \mathbb{R}$ denote Green's function for $-\Delta$ on S. Then, as is well known,

$$u(\xi) = \int_S \gamma(\xi, \eta) f(\eta) \, d\eta_1 \cdots d\eta_d. \tag{1.2}$$

Since γ may be said to be available in explicit form at least in the cases $d = 1, 2, 3$ (see Courant and Hilbert [3], p. 373, 378, 384), equation (1.2) even solves our problem constructively. However, since the evaluation of (1.2) for each ξ requires the numerical computation of a d-fold integral involving a singular integrand, direct use of (1.2) for $d \geq 2$ will be dismissed as unrealistic by most practicing applied mathematicians.

A survey of practical algorithms for solving our problem (so-called Fast Poisson Solvers) was given by Dorr [4]. In the present volume, Swarztrauber [16] presents a software-oriented discussion of some methods for solving (1.1) under more general boundary conditions. In this paper we discuss some additional approaches for solving the problem as stated above. These approaches, all well known in principle, include (i) general finite difference methods, including the Mehrstellenverfahren of Collatz; (ii) the finite element approach; (iii) expansion in series of eigenfunctions. Our main contribution consists in a transparent formalism which may serve to identify connections between various approaches. One such connection will be exploited in Section 8 to construct an algorithm for obtaining high-order finite difference and Mehrstellen operators for Poisson's equation. As a further application we elucidate the well-known fact that for problems with very smooth functions f, the Mehrstellen approach furnishes results that are more accurate than the finite element results obtained with comparable effort.

The formalism mentioned is a version of the d-dimensional discrete Fourier transform, an account of which is given in Section 2.

The author is indebted to Gene Golub and Martin Gutknecht for useful suggestions.

2. THE d-DIMENSIONAL DISCRETE FOURIER TRANSFORM

Here we summarize some facts from d-dimensional discrete Fourier analysis. Let N be an integer ≥ 1. We consider d-fold indexed sequences with complex elements,

$$x = \left\{ x_{k_1, k_2, \ldots, k_d} \right\},$$

which in each index are periodic with period N. Introducing the index vector

$$\mathbf{k} = (k_1, k_2, \ldots, k_d),$$

this means that

$$x_{\mathbf{k}+N\mathbf{q}} = x_{\mathbf{k}}$$

for any vector $\mathbf{q} \in \mathbb{Z}^d$ (i.e., having integer components). The linear space of these sequences will be denoted by Π_N, or, if it is necessary to indicate the dimension d, by $\Pi_N^{(d)}$.

In Π_N, the discrete Fourier operator $F_N = F_N^{(d)}: \Pi_N \to \Pi_N$ is defined as follows. Let

$$w = w_N := \exp \frac{2\pi i}{N},$$

and let Q_N denote the set of all index vectors $\mathbf{k} = (k_1, k_2, \ldots, k_d)$ such that $0 \leq k_i < N$, $i = 1, 2, \ldots, d$. Then

$$y = F_N x,$$

where

$$y_{\mathbf{m}} := N^{-d} \sum_{\mathbf{k} \in Q_N} x_{\mathbf{k}} w^{-(\mathbf{k}, \mathbf{m})}; \qquad (2.1)$$

here $(\mathbf{k}, \mathbf{m})$ denotes the scalar product in $\mathbb{R}^d$,

$$(\mathbf{k}, \mathbf{m}) := \sum_{i=1}^{d} k_i m_i.$$

It is well known (Henrici [9]) that y is again a sequence in Π_N; moreover, F_N maps Π_N *onto* Π_N, and the inverse operator is

$$F_N^{-1} = N^d \overline{F_N}, \tag{2.2}$$

where $\overline{F_N}$ is the operator F_N with the sign in the exponent of w reversed. It is fundamental in practical discrete Fourier analysis that there exist Fast Fourier Transform (FFT) algorithms that for $N = 2^{\ell}$ ($\ell = 1, 2, \ldots$) permit the evaluation of all elements of $F_N x$ in merely

$$(1 - 2^{-d}) N^d \mathrm{Log}_2 N$$

complex multiplications μ as opposed to the $N^{2d} \mu$ that are required when the formulas (2.1) are carried out literally. Under the name, "chirp z transform", there exist algorithms based on FFT that achieve the evaluation of $F_N x$ in $O(N^d \mathrm{Log}\, N)$ operations for all integers N, see Oppenheim and Schaefer [13].

If the Hadamard product $x \cdot y$ of two sequences in Π_N is defined by

$$(x \cdot y)_{\mathbf{k}} := x_{\mathbf{k}} y_{\mathbf{k}} \tag{2.3}$$

and the convolution product $x * y$ by

$$(x * y)_{\mathbf{k}} := \sum_{\mathbf{m} \in Q_N} x_{\mathbf{m}} y_{\mathbf{k} - \mathbf{m}}, \tag{2.4}$$

then there holds the convolution theorem,

$$F_N(x \cdot y) = F_N x * F_N y, \tag{2.5}$$

and its corollary,

$$F_N(x * y) = N^d F_N x \cdot F_N y. \tag{2.6}$$

In view of the formula

$$x * y = N^{2d} \overline{F_N} (F_N x \cdot F_N y), \qquad (2.7)$$

the convolution of two sequences, expensive if computed by carry-ing out (2.4) literally, is reduced to three discrete Fourier trans-forms and one Hadamard product, and can therefore be performed in $O(N^d \mathrm{Log}_2 N)\mu$.

3. FINITE DIFFERENCE METHODS

We return to our problem of solving Poisson's equation. Let $h = N^{-1}$ be the mesh constant of the discretization, where N is a positive integer. For any $\mathbf{j} \in \mathbb{Z}^d$ we write

$$\xi_{\mathbf{j}} := \mathbf{j}h$$

and put

$$f_{\mathbf{j}} := f(\xi_{\mathbf{j}}).$$

We denote by $u_{\mathbf{j}}$ an intended approximation to the value of the solution u at the mesh point $\xi_{\mathbf{j}}$.

Any standard finite difference approximation to $-\Delta u$ at the point $\xi_{\mathbf{j}}$ then can be written in the form

$$\sum_{\mathbf{i}} \delta_{\mathbf{i}} u_{\mathbf{j}-\mathbf{i}} \qquad (3.1)$$

where δ is a function defined on $\mathbb{Z}^d$ with generally very small support.

Example 3.1: If $d = 2$, then for the standard 5-point approxima-tion to $-\Delta$

$$h^2 \delta_{0,0} = 4,$$

$$h^2 \delta_{1,0} = h^2 \delta_{0,1} = h^2 \delta_{-1,0} = h^2 \delta_{0,-1} = -1, \qquad (3.2)$$

all other $\delta_i = 0$. Schematically we represent δ by the stencil

$$\frac{1}{h^2} \begin{array}{|c|c|c|} \hline & -1 & \\ \hline -1 & 4 & -1 \\ \hline & -1 & \\ \hline \end{array}$$

Example 3.2: The standard 9-point approximation to $-\Delta$ for $d = 2$ is given by the stencil

$$\frac{1}{6h^2} \begin{array}{|c|c|c|} \hline -1 & -4 & -1 \\ \hline -4 & 20 & -4 \\ \hline -1 & -4 & -1 \\ \hline \end{array}$$

Example 3.3: For $d = 2$, a more accurate approximation to $-\Delta$ is known to be given by the stencil

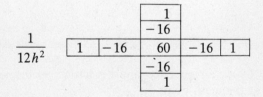

$$\frac{1}{12h^2} \begin{array}{|c|c|c|c|c|} \hline & & 1 & & \\ \hline & & -16 & & \\ \hline 1 & -16 & 60 & -16 & 1 \\ \hline & & -16 & & \\ \hline & & 1 & & \\ \hline \end{array}$$

(see Collatz [2], p. 542).

In all examples, the center square of the stencil corresponds to $i = 0$. We note that in all examples the sum of all δ_i adds up to 0, reflecting the fact that the finite difference approximation to $-\Delta u$ is exact if u is a constant. The condition

$$\sum_i \delta_i = 0$$

will be assumed for all finite difference approximations considered.

At each point ξ_j well in the interior of S, Poisson's equation is now simulated by a difference equation of the form

$$\sum_i \delta_i u_{j-i} = f_j. \tag{3.3}$$

It remains to take care of the boundary condition, and also to give a meaning to (3.3) in cases such as Example 3.3 where the stencil defining δ protrudes beyond S. Both problems are solved simultaneously by seeking a solution $u_{\mathbf{j}}$ of (3.3) in the space of sequences in Π_{2N} that are *odd* functions of each component of the index vector $\mathbf{j}$. Such sequences will briefly be called *odd*, and the space of all such sequences will be denoted by $\Pi_{2N}^{(o)}$. If $u \in \Pi_{2N}^{(o)}$, then $u_{\mathbf{j}} = 0$ not only whenever a component of $\mathbf{j}$ is zero, but also whenever a component of $\mathbf{j}$ equals N, for if

$$\mathbf{j} = (j_1, \ldots, N, \ldots, j_d)$$

and

$$\mathbf{j}' = (j_1, \ldots, -N, \ldots, j_d)$$

then on the one hand, because u has period $2N$,

$$u_{\mathbf{j}'} = u_{\mathbf{j}}$$

and on the other, because u is odd,

$$u_{\mathbf{j}'} = -u_{\mathbf{j}},$$

and both equations are compatible only if $u_{\mathbf{j}'} = u_{\mathbf{j}} = 0$. Thus any $u \in \Pi_{2N}^{(o)}$ simulates the boundary condition; moreover, no matter how big the stencil defining δ, the difference equation (3.4) can now be formed for any point $\boldsymbol{\xi}_{\mathbf{j}}$ in the interior of S.

The solution of the resulting system of $(N-1)^d$ equations for the $(N-1)^d$ nontrivial components of u is now easily found by discrete Fourier methods. As a by-product we obtain an explicit sufficient condition for the existence and uniqueness of the solution.

To this end we continue both δ and f (where defined) as periodic sequences into Π_{2N}. Then (3.3) is just the $\mathbf{j}$th component of a convolution equation between sequences in Π_{2N},

$$\delta * u = f. \tag{3.4}$$

So far, this has a meaning only for index vectors $\mathbf{j}$ such that $\xi_{\mathbf{j}}$ is congruent $\mathrm{mod}\, 2N$ to some interior point of S. However, assuming that a solution sequence $u \in \Pi_{2N}^{(o)}$ exists, we may use (3.4) to *define* f at the remaining points.

In order to state an explicit condition for the solubility of the system, we now assume that δ is an *even* sequence, i.e., that $\delta_{\mathbf{i}}$ is an even function of each component of the index vector $\mathbf{i}$. The space of all even sequences in Π_{2N} will be denoted by $\Pi_{2N}^{(e)}$. Because the convolution of an even and of an odd sequence is odd, it follows that if (3.4) has a solution u, then necessarily $f \in \Pi_{2N}^{(o)}$. Thus in order for (3.4) to have a meaning everywhere, f must be continued as an *odd* sequence in Π_{2N}.

Assume now there exists a $u \in \Pi_{2N}^{(o)}$ which satisfies (3.4). Applying F_{2N} to (3.4) and letting $\hat{f} := F_{2N}f$, we have

$$F_{2N}(\delta * u) = \hat{f}.$$

Letting $\hat{\delta} := F_{2N}\delta$, $\hat{u} := F_{2N}u$, this by virtue of the convolution theorem becomes

$$(2N)^d \hat{\delta} \cdot \hat{u} = \hat{f}. \tag{3.5}$$

Thus in the image space, the convolution (3.4) is replaced by a Hadamard multiplication, and if (3.5) is at all solvable for $\hat{u}$, it is solvable individually for each component $\hat{u}_{\mathbf{m}}$. A problem appears to arise for $\mathbf{m} = \mathbf{0}$, because by virtue of (3.2),

$$\hat{\delta}_{\mathbf{0}} = 0.$$

However, because both u and f are in $\Pi_{2N}^{(o)}$, there holds $\hat{u}_{\mathbf{0}} = \hat{f}_{\mathbf{0}} = 0$, and (3.5) is satisfied for $\mathbf{m} = \mathbf{0}$. A sufficient condition for the unique solvability of (3.5) thus is

$$\hat{\delta}_{\mathbf{m}} \neq 0, \qquad \mathbf{m} \neq \mathbf{0} \bmod 2N. \tag{3.6}$$

If this is satisfied, then for the same $\mathbf{m}$

$$\hat{u}_{\mathbf{m}} = (2N)^{-d} \hat{\delta}_{\mathbf{m}}^{-1} \hat{f}_{\mathbf{m}},$$

which we write symbolically as

$$\hat{u} = (2N)^{-d} \hat{\delta}^{-1} \cdot \hat{f},$$

with the understanding that $0^{-1}0 = 0$. Thus $\hat{u}$ is unique, and applying the inverse Fourier transform and using (2.2), it follows that the unique solution $u \in \Pi_{2N}^{(o)}$ is given by

$$u = \overline{F_{2N}}\left(\hat{\delta}^{-1} \cdot \hat{f}\right). \tag{3.7}$$

We summarize the result.

THEOREM 3a. *The system of difference equations* (3.3) *has a unique solution in* $\Pi_{2N}^{(o)}$ *if* δ *is an even sequence whose Fourier image* $\hat{\delta}$ *satisfies* (3.6). *The solution is then given by* (3.7).

We emphasize that the hypotheses concerning δ were made only to be able to guarantee the existence of a unique solution. The Fourier method works, and the solution is represented by (3.7), whenever a solution exists.

Equation (3.7) furnishes the solution of the discretized problem at the expense of three discrete Fourier transforms plus one Hadamard division, and thus in $O(N^d \mathrm{Log}_2 N)$ operations. However, because δ has small support, it is feasible to compute $\hat{\delta}$ analytically and thus to save one discrete Fourier transform. By doing so, one can also verify the condition (3.6).

Our general approach renders unnecessary special treatments of special difference operators such as that by Pickering [12].

Example 3.4: In this and in subsequent examples, if $d = 2$, we let $\mathbf{j} = (j, k)$, $\mathbf{m} = (m, n)$ to avoid double subscripts. For the 5-point approximation to $-\Delta$ considered in Example 3.1 we then obtain for $\hat{\delta} = (\hat{\delta}_{mn})$, noting that $(hN)^2 = 1$,

$$\hat{\delta}_{mn} = (2N)^{-2} \sum_{\mathbf{j} \in Q_{2N}} \delta_{jk} w_{2N}^{-jm-kn}$$

$$= h^{-2}(2N)^{-2}\{4 - w_{2N}^{-m} - w_{2N}^{m} - w_{2N}^{-n} - w_{2N}^{n}\}$$

$$= \left(\sin\frac{m\pi}{2N}\right)^2 + \left(\sin\frac{n\pi}{2N}\right)^2. \tag{3.9}$$

Example 3.5: For the 9-point operator of Example 3.2 we similarly get

$$\hat{\delta}_{mn} = \left(\sin\frac{m\pi}{2N}\right)^2 + \left(\sin\frac{n\pi}{2N}\right)^2 - \frac{2}{3}\left(\sin\frac{m\pi}{2N}\sin\frac{n\pi}{2N}\right)^2.$$

$$(3.10)$$

Example 3.6: In the same notation, the 9-point operator of Example 3.3 has the Fourier image

$$\hat{\delta}_{mn} = \left(\sin\frac{m\pi}{2N}\right)^2 + \left(\sin\frac{n\pi}{2N}\right)^2$$

$$+ \frac{1}{3}\left\{\left(\sin\frac{m\pi}{2N}\right)^4 + \left(\sin\frac{n\pi}{2N}\right)^4\right\}. \qquad (3.11)$$

Condition (3.6) is seen to be satisfied in all three examples.

4. MEHRSTELLENVERFAHREN

Let Lu denote a differential operator of the form

$$Lu = -\Delta u - f(\boldsymbol{\xi}, u). \qquad (4.1)$$

where f now is permitted to depend, and not necessarily linearly so, on u. The difference operators considered in Section 3 all would approximate L by a discrete operator L_h, the value of which at the grid point $\boldsymbol{\xi_j}$ would be given by

$$(L_h u)_{\mathbf{j}} = \sum_{\mathbf{i}} \delta_{\mathbf{i}} u_{\mathbf{j-i}} - f(\boldsymbol{\xi_j}, u_{\mathbf{j}}). \qquad (4.2)$$

The approximation holds in the sense that if u is any sufficiently differentiable function, there holds at a fixed point $\boldsymbol{\xi} = \boldsymbol{\xi_j}$ the relation

$$(L_h u)_{\mathbf{j}} - Lu = O(h^{2p}) \qquad (4.3)$$

as $h \to 0$, where $2p$ is a positive even integer depending only on δ, which is called the order of the finite difference approximation. (For the operators of Example 3.1 and Example 3.2, $p = 1$; for the operator of Example 3.3, $p = 2$.) The relation (4.3) holds whether or not u is a solution of $Lu = 0$.

Collatz [2] observed as early as 1935 that for functions u that are solutions of $Lu = 0$, and only for such functions, the order of approximation can be enhanced by replacing the term $-f(\xi_j, u_j)$ in (4.2) by a suitable average of values of f at nearby meshpoints, that is, by an expression of the form

$$\sum_i \gamma_i f(\xi_{j-i}, u_{j-i}).$$

If, as in Poisson's equation, f is independent of u, this simply takes the form

$$\sum_i \gamma_i f_{j-i},$$

and the finite difference approximation to Poisson's equation appears as

$$\sum_i \delta_i u_{j-i} = \sum_i \gamma_i f_{j-i}. \tag{4.4}$$

Example 4.1: If δ is the operator considered in Example 3.2, then by choosing for γ the stencil

the order of approximation for solutions of $-\Delta u = f$ becomes 4 in place of 2. As a Corollary, it follows that for solutions of Laplace's equation $-\Delta u = 0$, the 9-point operator of Example 3.2 achieves at least the order 4. (Actually, the order then is 6, see Forsythe and Wasow [5], p. 195.)

Numerical methods based on difference operators of the form (4.4) were called "Mehrstellenverfahren" by Collatz. No equivalent term in English appears to be available, which perhaps is indicative of the fact that these operators, although familiar in the case of ordinary differential equations under the name, "linear multistep methods", have not been analyzed with the same degree of abstract understanding that prevails in other areas of numerical analysis. (However, the method of delayed difference corrections (see Fox [5]) in many cases yields discretisations that are similar to the Mehrstellenverfahren, as does the HODIE method of Lynch and Rice [9].) In Section 8, a systematic method for constructing Mehrstellen operators and for appraising their accuracy will be presented. It will emerge, among other things, that in the case of Poisson's equation in a square, and for a comparable amount of work, the Mehrstellenverfahren yields more accurate results than the finite element method.

Here we show that the Mehrstellenverfahren, like the ordinary finite difference method, is amenable to Fourier techniques. In order to take proper care of the boundary conditions, we suppose that the averaging operator γ, like the difference operator δ, satisfies the symmetry condition (3.3). Continuing γ as an even sequence in Π_{2N}, the equation (4.4) may then be written

$$\delta * u = \gamma * f, \tag{4.5}$$

which, by taking Fourier transforms, letting $\hat{\gamma} := F_{2N}\gamma$ and using the convolution theorem, results in

$$\hat{\delta} * \hat{u} = \hat{\gamma} * \hat{f}.$$

If (4.5) has a unique solution in $\Pi_{2N}^{(o)}$, then its Fourier transform thus must satisfy

$$\hat{u} = \hat{\delta}^{-1} \cdot \hat{\gamma} \cdot \hat{f}, \tag{4.6}$$

where again $0^{-1}0 := 0$. From this we then get the solution

$$u = (2N)^d \overline{F_{2N}} \left(\hat{\delta}^{-1} \cdot \hat{\gamma} \cdot \hat{f} \right). \tag{4.7}$$

In addition to the Fourier transform required for $\hat{f}$, only the transform (4.7) needs to be computed numerically, because in all cases of interest, $\hat{\gamma}$ can be evaluated analytically.

Example 4.2: For the stencil of Example 4.1 we get

$$(2N)^2\hat{\gamma}_{mn} = 1 - \frac{1}{3}\left\{\left(\sin\frac{m\pi}{2N}\right)^2 + \left(\sin\frac{n\pi}{2N}\right)^2\right\}. \qquad (4.8)$$

The solution formula (4.7) has the same form as the solution formula (3.8) if we introduce

$$\rho := (2N)^{-2}\hat{\gamma}^{-1}\cdot\hat{\delta}, \qquad (4.9)$$

which is in Π_{2N} if all $\gamma_m \neq 0$. In all examples considered so far,

$$\rho_m = R\left(\left(\sin\frac{m_1\pi}{2N}\right)^2,\ldots,\left(\sin\frac{m_d\pi}{2N}\right)\right), \qquad (4.10)$$

where R is a rational function of its d variables that depends only on the method under consideration; R is a polynomial for pure difference methods. With this abbreviation,

$$\hat{u} = (2N)^{d-2}\rho^{-1}\cdot\hat{f},$$

and the solution itself is given by

$$u = (2N)^{-2}\overline{F_{2N}}\left(\rho^{-1}\cdot\hat{f}\right). \qquad (4.11)$$

5. THE FINITE ELEMENT METHOD

For real or complex valued square integrable functions u and v defined on S, let

$$(u,v) := \int_S u(\boldsymbol{\xi})\bar{v}(\boldsymbol{\xi})\,d\xi_1\cdots d\xi_d$$

denote their scalar product, and let

$$\|u\| := (u, u)^{1/2}.$$

It is well known and easily proved by Green's identity that the solution u of (1.1) minimizes the functional

$$I(u) := \|\|\text{grad } u\|\|^2 + 2(u, f) \qquad (5.1)$$

in the space Ω of all functions u that are continuous on S', zero on ∂S, and at least piecewise differentiable in S. In the finite element method as in the older method of Ritz, u is determined approximately as the function that minimizes $I(u)$ in an appropriate finite-dimensional subspace Ω_1 of Ω. In the finite element method, Ω_1 is usually chosen as a space of piecewise polynomial functions. The pieces (or "elements") on which a function $u \in \Omega_1$ is represented by a single polynomial arise through a triangularization. These spaces Ω_1 are made computationally attractive by the fact that they possess a basis $\{u_i\}$ where each function u_i has small support. In the simplest case, u_i is zero at all nodes of the triangularization with the exception of one, ξ_i say, where it may be assigned the value 1. Any function $u \in \Omega_1$ may then be represented in the form $u = \Sigma \alpha_i u_i$ where $\alpha_i = u(\xi_i)$. If this expression is substituted into (5.1) and the conditions for a minimum as a function of the α_i are written down, a system of equations results which in many cases formally resembles the Mehrstellen equations (4.4). Referring to Schwarz [14] for computational details, we present the results (for $d = 2$) for two well-known choices of Ω_1.

Example 5.1: *Piecewise linear trial functions.* The triangularization used here is sketched in Fig. 1. The space Ω_1 here consists of the functions that are continuous on S', zero on ∂S, and linear in each triangle of the triangularization. It is clear that any function $u \in \Omega_1$ is fully described by its values u_m at the nodes $\xi_m = (m/N, n/N)$ of the triangularization. Carrying out the process of minimization there results a system of linear equations which in

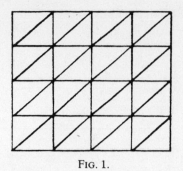

FIG. 1.

stencil form may be written

$$
h^{-2} \begin{array}{|c|c|c|} \hline & -1 & \\ \hline -1 & 4 & -1 \\ \hline & -1 & \\ \hline \end{array} \, u = \frac{1}{12} \begin{array}{|c|c|c|} \hline 0 & 1 & 1 \\ \hline 1 & 6 & 1 \\ \hline 1 & 1 & 0 \\ \hline \end{array} f. \tag{5.2}
$$

These equations clearly have the form (4.4); however, considered as a Mehrstellenverfahren, the local error now is only $O(h^2)$. The solution can again be effected by (4.11); the function ρ in this case is given by

$$
\rho_{mn} = \frac{\left(\sin\dfrac{m\pi}{2N}\right)^2 + \left(\sin\dfrac{n\pi}{2N}\right)^2}{1 - \dfrac{1}{3}\left(\sin\dfrac{m\pi}{2N}\right)^2 - \dfrac{1}{3}\left(\sin\dfrac{n\pi}{2N}\right)^2 - \dfrac{1}{3}\left(\sin\dfrac{[m+n]\pi}{2N}\right)^2}.
$$

$$\tag{5.3}$$

Example 5.2: *Piecewise bilinear trial functions.* Here we subdivide into squares as shown in Fig. 2. The subspace Ω_1 here consists of the functions u which are continuous on S', zero on ∂S, and which, on each square of the subdivision, are represented by a polynomial of the special form $\alpha + \beta\xi + \gamma\eta + \delta\xi\eta$. Again any function $u \in \Omega_1$ is uniquely described by its values u_m at the nodes ξ_m. By minimizing $I(u)$ for $u \in \Omega_1$, the following system of linear

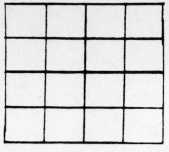

FIG. 2.

equations for the values $u_{\mathbf{m}}$ results:

$$\frac{1}{3h^2}\begin{array}{|c|c|c|} \hline -1 & -1 & -1 \\ \hline -1 & 8 & -1 \\ \hline -1 & -1 & -1 \\ \hline \end{array} \quad u = \frac{1}{36}\begin{array}{|c|c|c|} \hline 1 & 4 & 1 \\ \hline 4 & 16 & 4 \\ \hline 1 & 4 & 1 \\ \hline \end{array} f. \quad (5.4)$$

Once again, the solution of the system is given by (4.11), where now

$$\rho_{mn} = \frac{\left(\sin\dfrac{m\pi}{2N}\right)^2 + \left(\sin\dfrac{n\pi}{2N}\right)^2 - \dfrac{4}{3}\left(\sin\dfrac{m\pi}{2N}\sin\dfrac{n\pi}{2N}\right)^2}{1 - \dfrac{2}{3}\left\{\left(\sin\dfrac{m\pi}{2N}\right)^2 + \left(\sin\dfrac{n\pi}{2N}\right)^2\right\} + \dfrac{4}{9}\left(\sin\dfrac{m\pi}{2N}\sin\dfrac{n\pi}{2N}\right)^2}.$$

$$(5.5)$$

SUMMARY. In all approaches considered so far, a system of linear equations for the elements of a sequence $u \in \Pi_{2N}^{(o)}$ results, the solution of which requires two discrete Fourier transforms F_{2N} in d dimensions. The solution is given explicitly by (4.11) for suitable choice of ρ. If $d = 2$, $\mathbf{m} = (m, n)$, the sequence $\rho = \{\rho_{\mathbf{m}}\}$ is given by

$$\rho_{mn} = R\left(\left(\sin\frac{m\pi}{2N}\right)^2, \left(\sin\frac{n\pi}{2N}\right)^2\right).$$

For the examples given, the functions R are tabulated below.

Example	Method	$R(s,t)$
3.1	naive 5-point	$s + t$
3.2	naive 9-point square stencil	$s + t - \dfrac{2}{3}st$
3.3	naive 9-point long stencil	$s + t + \dfrac{1}{3}(s^2 + t^2)$
4.1	Collatz Mehrstellen	$\dfrac{s + t - \dfrac{2}{3}st}{1 - \dfrac{1}{3}(s + t)}$
5.1	finite element $\triangle$ linear	$\dfrac{s + t}{1 - \dfrac{2}{3}(s + t) + \dfrac{2}{3}st - \dfrac{2}{3}[s(1-s)t(1-t)]^{1/2}}$
5.2	finite element $\square$ bilinear	$\dfrac{s + t - \dfrac{4}{3}st}{1 - \dfrac{2}{3}(s + t) + \dfrac{4}{9}st}$

6. OTHER DIFFERENTIAL OPERATORS

The use of the discrete Fourier transform for solving difference equations is not restricted to discrete models for Poisson's equation under zero boundary conditions. However, for the method to be applicable the following conditions must be satisfied:

(i) It must be possible to write the difference equation as a convolution. This means that the difference operator cannot depend on ξ, i.e., the difference operator must have *constant coefficients*.

(ii) The boundary conditions cannot be taken into account explicitly. Instead, it must be possible to take care of the boundary conditions implicitly, by continuing the solution as a sequence that is periodic in each component of the index vector. It is not necessary that the period be the same in all components. Neither is it necessary that the continued sequence be odd. By continuing u_j as an even sequence, or as a sequence that is even in some components of the index vector, one can treat problems where the condition $\partial u / \partial n = 0$ is prescribed on pairs of opposing faces of S.

(iii) In order to continue the nonhomogeneous term as a periodic function, it had to be assumed that the sequences δ and ρ defining the difference or Mehrstellen approximation are *even* sequences. This implies that the differential operators that are simulated are invariant under the substitutions $\xi_i \to -\xi_i$ $(i = 1, 2, \ldots, d)$. This is the case only if *they do not contain any derivatives of odd order*.

We mention some examples where the foregoing conditions are satisfied.

Example 6.1: *The Helmholtz equation.*   The equation

$$Lu := -\Delta u + \lambda u = f \qquad (6.1)$$

is of the required kind; we suppose $\lambda \geq 0$ for mathematical as well as numerical solvability. If

$$\delta * u = \gamma * f$$

is a Mehrstellen approximation for Poisson's equation, then

$$\delta * u = \gamma * (f - \lambda u)$$

will be a corresponding Mehrstellen approximation, with the same order of the local error term, for the Helmholtz equation (6.1). The equation to be solved in the spectral domain then is

$$(\hat{\delta} + \lambda \hat{\gamma}) \cdot \hat{u} = \hat{\gamma} \cdot \hat{f}. \qquad (6.2)$$

We thus have

$$\hat{u} = (2N)^{d-2} \rho^{-1} \cdot \hat{f},$$

as in (4.9), where

$$\rho := (2N)^{-2}\hat{\gamma}^{-1} \cdot (\hat{\delta} + \lambda\hat{\gamma}).$$

If the rational function R defined by (4.10) is written as

$$R = \frac{P}{Q},$$

where P and Q are polynomials, the corresponding function for the Helmholtz equation is given by

$$R = \frac{P + (2N)^{-2}\lambda Q}{Q}.$$

Any Mehrstellen method for Poisson's equation thus is applicable also to the Helmholtz equation, and any fast Fourier method applicable to the former also works for the latter. Both boundary conditions $u = 0$ and $\partial u / \partial n = 0$ can be treated.

Example 6.2: *The simply supported square plate.* Here we deal with the problem

$$\Delta\Delta u = f(\boldsymbol{\xi}), \qquad \boldsymbol{\xi} \in S, \tag{6.3}$$

describing a square plate loaded at the point $\boldsymbol{\xi}$ with a load proportional to $f(\boldsymbol{\xi})$. Unfortunately, only the boundary conditions

$$u = 0, \qquad \Delta u = 0 \quad \text{on } S \tag{6.4}$$

can be treated by our method; they model a physical situation where the plate is freely supported. If δ is a finite difference approximation to $-\Delta$, (6.3) will be approximated by

$$\delta * \delta * u = f.$$

Using the convolution theorem and applying the discrete Fourier transform, this in view of $d = 2$ becomes

$$(2N)^4\hat{\delta} \cdot \hat{\delta} \cdot \hat{u} = \hat{f},$$

the solution of which can be written symbolically as

$$\hat{u} = (2N)^{-4}\hat{\delta}^{-2}\cdot\hat{f},$$

yielding

$$u = (2N)^{-2}\overline{F_{2N}}(\hat{\delta}^{-2}\cdot\hat{f}). \qquad (6.5)$$

Mehrstellenverfahren for dealing with the plate problem can be treated in a similar way.

7. EIGENFUNCTION EXPANSIONS

A totally different, and theoretically very powerful, approach to the solution of Poisson's equation, which in principle works for any bounded region, is provided by the method of eigenfunction expansions. Let G be a bounded domain in R_d of arbitrary shape and connectivity, and let ∂G be its boundary. As is well known, the eigenvalue problem

$$-\Delta u = \lambda u \text{ in } G$$

$$u = 0 \text{ on } \partial G \qquad (7.1)$$

then has a denumerable set of eigenvalues $0 < \lambda_1 < \lambda_2 \leqq \lambda_3 \leqq \cdots$ and corresponding eigenfunctions $u_1, u_2, u_3, \ldots$ such that

$$-\Delta u_n = \lambda_n u_n \quad \text{in } G,$$

$$u_n = 0 \quad \text{on } \partial G, \qquad (7.2)$$

$n = 1, 2, \ldots$ [If $d = 2$, these eigenfunctions describe the free vibrations of a membrane with fixed boundary ∂G.] The eigenfunctions may be assumed orthonormal, i.e., with the usual definition of the scalar product the relations

$$(u_n, u_m) = \begin{cases} 1, & n = m \\ 0, & n \neq m \end{cases}$$

may be assumed to hold. For this orthonormal set of eigenfunctions there holds an expansion theorem: If g is an arbitrary square integrable function in G, and if

$$\gamma_n := (g, u_n), \qquad n = 1, 2, \ldots, \tag{7.3}$$

then the series

$$\sum_{n=1}^{\infty} \gamma_n u_n$$

converges to g at least in the least squares sense (i.e., its nth partial sums s_n satisfy $\|s_n - g\| \to 0$ for $n \to \infty$). If g is sufficiently smooth and zero on ∂G, this series converges pointwise, even uniformly so, and we then write

$$g = \sum \gamma_n u_n. \tag{7.4}$$

This expansion is unique in the sense that if it represents the zero function,

$$0 = \sum \gamma_n u_n.$$

it necessarily follows that all $\gamma_n = 0$.

If the eigenfunctions are orthogonal but not normalized, (7.4) is replaced by

$$g = \sum \frac{\gamma_n}{\omega_n} u_n, \tag{7.4'}$$

where the γ_n are still given by (7.3), and where

$$\omega_n := (u_n, u_n). \tag{7.5}$$

In order to apply the above to the solution of Poisson's equation $-\Delta u = f$, we assume that the given function f is such that

$$f = \sum \alpha_n u_n,$$

where $\{u_n\}$ is an orthonormal set of eigenfunctions, and

$$\alpha_n := (f, u_n), \qquad n = 1, 2, \dots . \tag{7.6}$$

Since the solution of Poisson's equation is always smoother than the nonhomogeneous term, the solution u then likewise may be expanded,

$$u = \sum \beta_n u_n.$$

Applying $-\Delta$ termwise—we assume that this is permissible—and expressing the fact that u satisfies $-\Delta u = f$ we get, using (7.2),

$$-\Delta u = \sum \beta_n (-\Delta u_n) = \sum \beta_n \lambda_n u_n = f = \sum \alpha_n u_n$$

and thus, by virtue of the uniqueness of the eigenfunction expansions,

$$\beta_n \lambda_n = \alpha_n$$

or

$$\beta_n = \frac{\alpha_n}{\lambda_n}, \qquad n = 1, 2, \dots .$$

That is, the solution of Poisson's equation under zero-boundary conditions is given in the explicit form

$$u = \sum_{n=1}^{\infty} \frac{\alpha_n}{\lambda_n} u_n, \tag{7.7}$$

where the α_n are given by (7.6). If the eigenfunctions are not normalized, (7.6) is replaced by

$$u = \sum_{n=1}^{\infty} \frac{\alpha_n}{\lambda_n \omega_n} u_n, \tag{7.8}$$

where ω_n is defined by (7.5).

The solution formula holds, in principle, for any bounded region G for which the eigenvalue problem (7.2) makes sense. However, if the formula is to be realized algorithmically, the eigenfunctions and eigenvalues must be known. Such is the case for the unit hypercube S. It is convenient to number the eigenvalues and eigenfunctions by means of an index vector $\mathbf{m} = (m_1, \ldots, m_d)$, where the m_i are integers > 0. The eigenfunctions then are

$$u_{\mathbf{m}}(\boldsymbol{\xi}) = i^d \sin(m_1 \pi \xi_1) \cdots \sin(m_d \pi \xi_d); \qquad (7.9)$$

the factor i^d is inserted for later formal convenience. These eigenfunctions are orthogonal but not orthonormal; we have $\omega_{\mathbf{m}} = 2^{-d}$ for all $\mathbf{m}$. The eigenvalue corresponding to $u_{\mathbf{m}}$ is

$$\lambda_{\mathbf{m}} = \pi^2 \left(m_1^2 + \cdots + m_d^2 \right) = \pi^2 \|\mathbf{m}\|^2. \qquad (7.10)$$

The expansion coefficients of the function f are given by

$$\alpha_{\mathbf{m}} = (-i)^d \int_S f(\boldsymbol{\xi}) \sin(m_1 \pi \xi_1) \cdots \sin(m_d \pi \xi_d) \, d\xi_1 \cdots d\xi_d,$$

$$(7.11)$$

and with these coefficients the solution of (1.1) by (7.8) is given by the explicit formula

$$u(\boldsymbol{\xi}) = (2i)^d \sum_{m_i > 0} \frac{\alpha_{\mathbf{m}}}{\lambda_{\mathbf{m}}} \sin(m_1 \pi \xi_1) \cdots \sin(m_d \pi \xi_d). \quad (7.12)$$

This will be called the *eigenfunction solution* of Poisson's equation.

We now wish to show the connection of the eigenfunction solution with the numerical solutions given in Sections 3–5. To this end we define $\alpha_{\mathbf{m}}$ by (7.11) for index vectors $\mathbf{m} = (m_1, \ldots, m_d)$ whose components have arbitrary sign. What we get of course is the continuation of $\alpha_{\mathbf{m}}$ as an *odd* function of the m_i. Expressing the sines by exponentials, the series on the right of (7.12) first becomes

$$\sum_{m_i > 0} \frac{\alpha_{\mathbf{m}}}{\lambda_{\mathbf{m}}} \left(e^{im_1 \pi \xi_1} - e^{-im_1 \pi \xi_1} \right) \cdots \left(e^{im_d \pi \xi_d} - e^{-im_d \pi \xi_d} \right)$$

and then, by virtue of the extended definition of the α_m,

$$\sum_{m \neq 0} \frac{\alpha_m}{\lambda_m} e^{im_1\pi\xi_1} \cdots e^{im_d\pi\xi_d}$$

or

$$u(\xi) = \sum_{m \neq 0} \frac{\alpha_m}{\lambda_m} e^{i\pi(m,\xi)}. \qquad (7.13)$$

(It is possible here to sum with respect to *all* $m \neq 0$, because the terms where at least one component of m is zero are automatically zero.) A truncated form of the representation (7.13) is

$$u(\xi) = \sum_{\substack{m \neq 0 \\ |m| < N}} \frac{\alpha_m}{\lambda_m} e^{i\pi(m,\xi)}, \qquad (7.14)$$

where N is a suitably large integer, and

$$|m| := \max_{1 \leq i \leq d} |m_i|.$$

The truncated form (7.14) of the solution is still exact for Poisson equations such that the expansion coefficients α_m of f satisfy

$$\alpha_m = 0 \quad \text{if } |m| \geq N.$$

This is the case precisely for the functions f that are in the set U_N of linear combinations of the functions

$$\sin(m_1\pi\xi_1) \cdots \sin(m_d\pi\xi_d)$$

where m satisfies $|m| < N$.

We now evaluate the truncated series (7.14) at a point $\xi_j = jh \in S$ where $h := N^{-1}$ and where j is an index vector satisfying $|j| < N$. Letting

$$w := e^{i\pi/N}, \qquad (7.15)$$

the result in view of $(\mathbf{m}, \boldsymbol{\xi}) = \dfrac{1}{N}(\mathbf{m}, \mathbf{j})$ is

$$u(\boldsymbol{\xi_j}) = \sum_{\substack{\mathbf{m} \neq 0 \\ |\mathbf{m}| < N}} \frac{\alpha_{\mathbf{m}}}{\lambda_{\mathbf{m}}} w^{(\mathbf{m},\mathbf{j})}. \qquad (7.16)$$

Interpreting the values $u_j := u(\boldsymbol{\xi_j})$ as elements of a sequence u in $\Pi_{2N}^{(o)}$, and continuing $\alpha_{\mathbf{m}}$ and $\lambda_{\mathbf{m}}$ likewise as sequences $\alpha \in \Pi_{2N}^{(o)}$ and $\lambda \in \Pi_{2N}^{(e)}$, (7.16) expresses the fact that

$$u = (2N)^d \overline{F_{2N}} (\lambda^{-1} \cdot \alpha) \qquad (7.17)$$

where, as usual, we stipulate that $0^{-1} 0 = 0$. We emphasize that the solution formula (7.17) is still exact for functions $f \in U_N$. The formula already strongly resembles the most general numerical solution formula

$$u = (2N)^{d-2} \overline{F_{2N}} (\rho^{-1} \cdot \hat{f}), \qquad (7.18)$$

valid for the numerical methods discussed in Sections 3–5.

The resemblance becomes more pronounced if the Fourier coefficients $\alpha_{\mathbf{m}}$ are evaluated by the trapezoidal rule with step $h = N^{-1}$. Continuing $f(\boldsymbol{\xi})$ as a function that is odd in each ξ_i, this yields

$$\alpha_{\mathbf{m}} = (-i)^d \int_S f(\boldsymbol{\xi}) \sin(m_1 \pi \xi_1) \cdots \sin(m_d \pi \xi_d) \, d\xi_1 \cdots d\xi_d$$

$$= 2^{-d} \int_{|\boldsymbol{\xi}| < 1} f(\boldsymbol{\xi}) e^{-i\pi(\mathbf{m}, \boldsymbol{\xi})} \, d\xi_1 \cdots d\xi_d$$

$$\approx (2N)^{-d} \sum_{|\mathbf{j}| < N} f(\boldsymbol{\xi_j}) w^{-(\mathbf{m},\mathbf{j})}$$

$$= \hat{f}_{\mathbf{m}}.$$

Thus, the expansion coefficients $\alpha_{\mathbf{m}}$ approximately equal the discrete Fourier transform of the sequence $f = \{f_j\}$. It is a basic fact of discrete Fourier analysis that for functions $f \in U_N$ the approximate equality becomes an equality, that is, for all **m** such that $|\mathbf{m}| < N$,

$$\alpha_\mathbf{m} = \hat{f}_\mathbf{m}. \tag{7.19}$$

Thus for functions $f \in U_N$, the exact solution satisfies

$$u = (2N)^d \overline{F_{2N}}(\lambda^{-1} \cdot \hat{f}). \tag{7.20}$$

If the numerical solution given by (7.18) is to approximate this as $h \to 0$ and thus $N \to \infty$, it is clear that, in some sense,

$$(2N)^2 \rho \text{ approximates } \lambda.$$

The sense in which this approximation takes place becomes evident if we consider the special case

$$f(\boldsymbol{\xi}) = \sin(m_1 \pi \xi_1) \cdots \sin(m_d \pi \xi_d)$$

where $0 < m_i < N,\ i = 1, \dots, d$. Here the only nonzero element of the sequence $\alpha = \hat{f}$ in the first octant is $\alpha_\mathbf{m}$, where $\mathbf{m} := (m_1, \dots, m_d)$, and it has the value $(2i)^{-d}$. It follows from (7.12) that the solution is

$$u(\boldsymbol{\xi}) = \frac{1}{\pi^2 (m_1^2 + \cdots + m_d^2)} \sin(m_1 \pi \xi_1) \cdots \sin(m_d \pi \xi_d),$$

$$\tag{7.21}$$

as can be verified directly. The numerical solution at a point $\boldsymbol{\xi} = \boldsymbol{\xi}_\mathbf{j}$, if ρ is expressed in the form (4.10), is given by

$$u(\boldsymbol{\xi}) = \frac{1}{(2N)^2} \frac{1}{R\left(\left(\sin\dfrac{m_1 \pi}{2N}\right)^2, \dots, \left(\sin\dfrac{m_d \pi}{2N}\right)^2\right)}$$

$$\times \sin(m_1 \pi \xi_1) \cdots \sin(m_d \pi \xi_d). \tag{7.22}$$

As $N \to \infty$, this will tend to the exact solution (7.21) at every fixed ξ if and only if

$$(2N)^2 R\left(\left(\sin\frac{m_1\pi}{2N}\right)^2, \ldots, \left(\sin\frac{m_d\pi}{2N}\right)^2\right)$$

$$\to \lambda_{\mathbf{m}} = \pi^2\left(m_1^2 + \cdots + m_d^2\right).$$

It is of interest to verify this fact in some of the examples considered previously, where $d = 2$, and where we use the notation $\mathbf{m} = (m, n)$.

Example 7.1: Five point formula. Here

$$(2N)^2 \rho_{mn} = (2N)^2\left[\left(\sin\frac{m\pi}{2N}\right)^2 + \left(\sin\frac{n\pi}{2N}\right)^2\right]$$

$$= \pi^2 m^2 + \pi^2 n^2 + O(N^{-2})$$

$$= \lambda_{mn} + O(N^{-2}).$$

Example 7.2: Nine point formula. From (3.10) we again get

$$(2N)^2 \rho_{mn} = \lambda_{mn} + O(N^{-2}).$$

Example 7.3: Collatz Mehrstellen operator (Example 4.1). Here we find

$$(2N)^2 \rho_{mn} = (2N)^2 \frac{\left(\sin\frac{m\pi}{2N}\right)^2 + \left(\sin\frac{n\pi}{2N}\right)^2}{1 - \frac{1}{3}\left[\left(\sin\frac{m\pi}{2N}\right)^2 + \left(\sin\frac{n\pi}{2N}\right)^2\right]}$$

$$= \lambda_{mn} + O(N^{-4});$$

the higher order of the error term reflects the increased accuracy of the Mehrstellenverfahren for smooth functions f.

In summary, we see that the numerical solution (7.18) is obtained from the exact solution in terms of eigenfunctions by the following steps:

  (i)   the series of eigenfunctions is truncated;
  (ii) the expansion coefficients are approximated by the trapezoidal rule;
  (iii) the exact eigenvalues

$$\lambda_{\mathbf{m}} = \pi^2 \left(m_1^2 + \cdots + m_d^2 \right)$$

are approximated by $(2N)^2$ times a rational function in the variables

$$s_i := \left(\sin \frac{m_i \pi}{2N} \right)^2, \qquad i = 1, 2, \ldots, d.$$

While there is no way around truncating the eigenfunction series, steps (ii) and (iii) seem somewhat arbitrary. Besides using the trapezoidal rule, there are many other ways for computing Fourier coefficients, among which we mention the use of attenuation factors (see Gautschi [7]), or the method of using general spline functions (see Marti [11]), which is suitable also for functions with discontinuities. And of course, there is no need to approximate the eigenvalues when the exact eigenvalues are known. Thus the use of the eigenfunction expansion appears to be much more flexible than the purely numerical approaches described in the Sections 3–6. Moreover, eigenfunctions can also be used for regions other than the square or the cube if the eigenfunctions are known. Candidates for such regions are the disk and the sphere; however, the computational details remain to be worked out in these cases.

We mention that the use of eigenfunctions for solving Poisson's equation in a square or a rectangle has already been advocated by Sköllermo [15]. She did not discuss the connection between the eigenfunction approach and the strictly numerical approaches discussed earlier.

8. CONSTRUCTION OF MEHRSTELLEN OPERATORS BY PADÉ METHODS

For solving Poisson's equation in a unit hypercube, there is probably no reason why the method of eigenfunction expansions discussed in Section 7 should not be preferred to any method working with finite differences. It yields a representation of the solution that is, in principle, arbitrarily accurate. The two parameters that limit the accuracy in practice—accuracy of the Fourier coefficients α_m and length of the truncated series—are independent of each other and are easily controlled, and because the exact eigenvalues λ_m are known and easily computed, the accuracy that is attainable with a given amount of work is probably higher for the eigenfunction method than for the finite difference method.

However, high order finite difference and Mehrstellen approximations to $-\Delta u = f$ are still useful for problems in non-rectangular domains, and also for nonlinear problems where f itself depends on u. We therefore show how the connection between finite difference operators and eigenvalues that was established in Section 7 can be exploited for the *construction* of such finite difference and Mehrstellen operators.

It emerged from the examples in the preceding sections that any symmetric finite difference operator, if expressed in convolution notation as $\delta * u$ where $\delta \in \Pi_{2N}^{(o)}$, has a discrete Fourier transform that can be expressed as $(2N)^{-2}$ times a polynomial in the d variables

$$s_i := \left(\sin \frac{m_i \pi}{2N} \right)^2, \qquad i = 1, \dots, d.$$

We shall make this connection more explicit. We begin by expressing a symmetric finite difference operator in conventional notation. If such a difference operator acts on u_i, all differences must be centered at ξ_i; moreover, no differences of odd order can appear. Such an operator can therefore be expressed as a polynomial in the basic symmetric difference operator of order two. In place of the conventional definition

$$\delta^2 u_i := u_{i+1} - 2u_i + u_{i-1}$$

it is here preferable to use the operator $\partial^2 := -\frac{1}{4}\delta^2$, that is,

$$\partial^2 u_i := \frac{1}{4}(-u_{i+1} + 2u_i - u_{i-1}).$$

In the case of several variables, if ∂^2 acts on ξ_j, we denote it by ∂_j^2. Thus a symmetric finite difference operator can be expressed as a (not necessarily homogeneous) polynomial

$$P(\partial_1^2, \partial_2^2, \ldots, \partial_d^2). \tag{8.1}$$

Example 8.1: If $d = 2$, the familiar five point approximation to $-\Delta$ is represented by

$$P(\partial_1^2, \partial_2^2) = \frac{4}{h^2}(\partial_1^2 + \partial_2^2).$$

THEOREM 8a. *The sequence $\hat{\delta}$ corresponding to the finite difference operator* (8.1) *is given by*

$$\hat{\delta}_{\mathbf{m}} = P\left(\left(\sin\frac{m_1\pi}{2N}\right)^2, \ldots, \left(\sin\frac{m_d\pi}{2N}\right)^2\right).$$

Proof. Our first task is to express the operator (8.1) in the formalism of sequences in $\Pi_{2N}^{(e)}$. The rule is simply this: Let

$$\delta^{(i)} = \left\{\delta_{\mathbf{j}}^{(i)}\right\} \in \Pi_{2N}^{(e)}$$

be the sequence corresponding to ∂_i^2; it is given by

$$\delta_0^{(i)} = \frac{1}{2}, \qquad \delta_{\mathbf{e}_i}^{(i)} = \delta_{-\mathbf{e}_i}^{(i)} = -\frac{1}{4}, \tag{8.2}$$

where $\mathbf{e}_i$ is the ith unit coordinate vector, and where all other $\delta_{\mathbf{j}}^{(i)}$ in a period cube are zero. Then the operator (8.1) is described by

$$P(\delta^{(1)}, \ldots, \delta^{(d)})^*$$

where $*$ signifies that all multiplications are to be replaced by convolutions. Now from (8.2) it is immediate that

$$\hat{\delta}_{\mathbf{m}}^{(i)} = \left(F_{2N}(\delta^{(i)}) \right)_{\mathbf{m}}$$

$$= (2N)^{-d} \left\{ \frac{1}{2} - \frac{1}{4}(w^{-m_i} + w^{m_i}) \right\}$$

where $w := \exp\left(\frac{2\pi i}{2N} \right)$, that is

$$\hat{\delta}_{\mathbf{m}}^{(i)} = (2N)^{-d} \left(\sin \frac{m_i \pi}{2N} \right)^2.$$

Now by the convolution theorem

$$F_{2N}(\delta^{(i)} * \delta^{(j)}) = (2N)^d \hat{\delta}^{(i)} \cdot \hat{\delta}^{(j)},$$

that is

$$\left(F_{2N}(\delta^{(i)} * \delta^{(j)}) \right)_{\mathbf{m}} = (2N)^{-d} \left(\sin \frac{m_i \pi}{2N} \right)^2 \left(\sin \frac{m_j \pi}{2N} \right)^2.$$

By induction on the number of convolution factors the result thus holds for arbitrary monomials in $\delta^{(1)}, \ldots, \delta^{(d)}$. By the linearity of F_{2N} it thus holds for every polynomial P.

As a consequence of Theorem 8a, a Mehrstellen method of the form

$$(2N)^2 P(\partial_1^2, \ldots, \partial_d^2) u = Q(\partial_1^2, \ldots, \partial_d^2) F \qquad (8.3)$$

in the space of Fourier transforms results in the equation

$$\hat{u} = (2N)^{-2} \rho^{-1} \cdot \hat{f}$$

where $\rho = \{\rho_{\mathbf{m}}\}$ is given by

$$\rho_{\mathbf{m}} = R(\mathbf{s}) = \frac{P(\mathbf{s})}{Q(\mathbf{s})},$$

with $(s_1, \ldots, s_d)$,

$$s_i := \left(\sin \frac{m_i \pi}{2N} \right)^2, \qquad i = 1, \ldots, d. \tag{8.4}$$

In the examples which we have observed,

$$(2N)^2 R(\mathbf{s}) \sim \lambda_{\mathbf{m}} = \pi^2 \|\mathbf{m}\|^2$$

as $N \to \infty$. Let p be the largest integer such that

$$(2N)^2 R(\mathbf{s}) - \lambda_{\mathbf{m}} = O(N^{-p}) \tag{8.5}$$

as $N \to \infty$, uniformly on every set $|\mathbf{m}| < N_0$. This will be called the *spectral order* of the Mehrstellen operator (8.3). We recall that U_M is the set of linear combinations of the functions (7.9) where $|\mathbf{m}| < M$.

THEOREM 8b. *Let the Mehrstellen operator* (8.3) *have spectral order $p > 0$, let $f \in U_M$, and let $\tilde{u}$ denote the numerical solution of* (1.1) *defined by* (8.3). *Then if u denotes the exact solution, there exists a constant κ such that at each $\boldsymbol{\xi} \in S$ where $\tilde{u}$ is defined,*

$$|\tilde{u}(\boldsymbol{\xi}) - u(\boldsymbol{\xi})| \leq \kappa N^{-p} \tag{8.6}$$

for all sufficiently large N.

Proof. The exact solution is given by

$$u(\boldsymbol{\xi}) = \sum_{\substack{\mathbf{m} \neq 0 \\ |\mathbf{m}| < M}} \frac{\alpha_{\mathbf{m}}}{\lambda_{\mathbf{m}}} e^{i\pi(\mathbf{m}, \boldsymbol{\xi})}$$

and the numerical solution is, where defined,

$$\tilde{u}(\boldsymbol{\xi}) = (2N)^{-2} \sum_{\substack{\mathbf{m} \neq 0 \\ |\mathbf{m}| < M}} \frac{\alpha_{\mathbf{m}}}{R(\mathbf{s})} e^{i\pi(\mathbf{m}, \boldsymbol{\xi})}.$$

By virtue of (8.5) there exists μ such that

$$|(2N)^2 R(\mathbf{s}) - \lambda_{\mathbf{m}}| \leq \mu N^{-p}, \qquad |\mathbf{m}| < M,$$

hence if N is such that

$$\mu N^{-p} < \tfrac{1}{2}|\lambda_{\mathbf{m}}| \text{ for all } \mathbf{m} \neq \mathbf{0}$$

then

$$|\tilde{u}(\boldsymbol{\xi}) - u(\boldsymbol{\xi})| \leq \sum_{\substack{\mathbf{m} \neq \mathbf{0} \\ |\mathbf{m}| < M}} \frac{|\alpha_{\mathbf{m}}|}{\lambda_{\mathbf{m}}^2} \cdot 2\mu N^{-p},$$

proving (8.6).

We next show that the concept of spectral order is equivalent to the ordinary concept of order, here called *polynomial order*. The operator (8.3) is said to have polynomial order p_1 if for any polynomial u in the variables $\xi_1, \ldots, \xi_d$ of degree $< p_1 + 2$

$$(2N)^2 P\big(\partial_1^2, \ldots, \partial_d^2\big) u - Q\big(\partial_1^2, \ldots, \partial_d^2\big) \Delta u = 0.$$

For this definition to be unambiguous it is necessary to normalize the operator (8.3). Our normalization consists in setting

$$Q(0, \ldots, 0) = 1.$$

Our definition thus applies only to operators where $Q(0, \ldots, 0) \neq 0$.

THEOREM 8c. *The operator* (8.3), *where* $Q(0, \ldots, 0) = 1$, *has polynomial order* p *if and only if it has spectral order* p.

Proof. Let (8.3) have spectral order p. Then if the s_i are given by (8.4),

$$\frac{P(\mathbf{s})}{Q(\mathbf{s})} - \frac{\pi^2}{(2N)^2}\big(m_1^2 + \cdots + m_d^2\big) = O\big(N^{-p-2}\big), \qquad N \to \infty.$$

Expressing this in terms of the variables

$$\sigma_j := \frac{\pi m_j}{2N}$$

and using the fact that $Q(0,\ldots,0)=1$, this means that the formal power series expansion of

$$P\big((\sin\sigma_1)^2,\ldots,(\sin\sigma_d)^2\big)$$

$$-Q\big((\sin\sigma_1)^2,\ldots,(\sin\sigma_d)^2\big)\big(\sigma_1^2+\cdots+\sigma_d^2\big)$$

begins (at the earliest) with terms of order $p+2$. In this formal expansion we replace σ_j by $-ihD_j$, where $D_j:=\partial/\partial\xi_j$, $h:=N^{-1}$. Then $\sigma_1^2+\cdots+\sigma_d^2=-h^2\Delta$ and

$$\big(\sin\sigma_j\big)^2 = -\tfrac{1}{4}\big(e^{i\sigma_j}-2+e^{-i\sigma_j}\big)$$

$$= \tfrac{1}{4}\big(-e^{-hD_j}+2-e^{hD_j}\big)$$

$$= \partial_j^2.$$

Because the formal expansion now begins with differentiation operators of order $p+2$ at the earliest, zero results if this is applied to a polynomial of order $p+2$. Thus the polynomial order is at least p. The foregoing steps being reversible, it follows that the polynomial order is exactly p.

We are now motivated to look for rational functions $R=P/Q$ that approximate λ_m with high spectral order. This problem may be phrased as an approximation problem of Padé type. Let the s_i be given by (8.4). Then

$$\left(\frac{m_i\pi}{2N}\right)^2 = \left(\arcsin\sqrt{s_i}\right)^2,$$

and the problem is to approximate

$$F(\mathbf{s}) := \left(\arcsin\sqrt{s_1}\right)^2 + \cdots + \left(\arcsin\sqrt{s_d}\right)^2$$

by a rational function R so that the exponent r in

$$R(\mathbf{s}) - F(\mathbf{s}) = O\left(\|\mathbf{s}\|^{r+1}\right) \tag{8.7}$$

becomes as large as possible. If (8.7) holds, then by virtue of $\mathbf{s} = O(N^{-2})$ it is clear that

$$(2N)^2 R(\mathbf{s}) - \lambda_{\mathbf{m}} = O(N^{-2r}),$$

and thus that the Mehrstellen operator defined by R has spectral and polynomial order $2r$.

Because F is a sum of functions of one variable, the multivariate Padé problem (8.7) may be reduced to a one-dimensional problem. By a result of classical analysis (Bailey [1], p. 86)

$$F(s) = \left(\arcsin\sqrt{s}\right)^2$$

$$= s_3 F_2\left(1,1,1; \tfrac{3}{2},2; s\right)$$

$$= s + \tfrac{1}{3}s^2 + \tfrac{8}{45}s^3 + \tfrac{4}{35}s^4 + \cdots. \tag{8.8}$$

By classical methods (for instance, by the quotient-difference algorithm) a Padé approximant

$$R^{(p,q)}(s) = \frac{P_p(s)}{Q_q(s)} \tag{8.9}$$

(deg $P = p$, deg $Q = q$) can be constructed for every pair of integers (p, q) such that $p \geqq 1$, $q \geqq 0$ with the property that

$$F(s) - R^{(p,q)}(s) = O(s^{p+q+1}), \; s \to 0. \qquad (8.10)$$

Clearly, the function

$$F(\mathbf{s}) = \sum_{j=1}^{d} F(s_j)$$

then is approximated by

$$R(\mathbf{s}) := \sum_{j=1}^{d} R^{(p,q)}(s_j) \qquad (8.11)$$

with an error $O(\|\mathbf{s}\|^{p+q+1})$ as $\|\mathbf{s}\| \to 0$. The order of the resulting operator will be $2(p + q)$.

In order to interpret (8.11) as a Mehrstellen or difference operator, it has to be represented as a single fraction. Evidently,

$$R(\mathbf{s}) = \frac{P(\mathbf{s})}{Q(\mathbf{s})} \qquad (8.12)$$

where

$$Q(\mathbf{s}) = \prod_{j=1}^{d} Q_q(s_j), \qquad (8.13)$$

$$P(\mathbf{s}) = \sum_{j=1}^{d} P_p(s_j) \prod_{\substack{k=1 \\ k \neq j}}^{d} Q_q(s_k). \qquad (8.14)$$

In forming the expressions (8.13) and (8.14), terms of order $> p + q$ may be neglected in $P(\mathbf{s})$ and terms of order $\geq p + q$ may be neglected in $Q(\mathbf{s})$, because on expanding these will contribute only $O(\|\mathbf{s}\|^{p+q+1})$ to $R(\mathbf{s})$, and thus will not change the order $2(p + q)$ of the operator.

The following examples illustrate the algorithmic procedure.

Example 8.2: $(p, q) = (1, 0)$, $d = 2$. Here

$$R^{(1,0)}(s) = s,$$

hence

$$R(\mathbf{s}) = s_1 + s_2,$$

which results in the difference operator

$$\frac{4}{h^2}(\partial_1^2 + \partial_2^2)u = f$$

or in stencil notation

$$\frac{1}{h^2} \quad \begin{array}{|c|c|c|} \hline & -1 & \\ \hline -1 & 4 & -1 \\ \hline & -1 & \\ \hline \end{array} u = f.$$

The order of this operator is $2(p + q) = 2$, as is well known.

Example 8.3: $(p, q) = (1, 1)$, $d = 1$. Here

$$R^{(1,1)}(s) = \frac{s}{1 - \frac{1}{3}s},$$

as is easily verified. This directly results in the difference operator

$$\frac{4}{h^2}\partial^2 u = \left[1 - \frac{1}{3}\partial^2\right]f$$

or in conventional notation, using $\partial^2 = -\frac{1}{4}\delta^2$,

$$-\frac{1}{h^2}\delta^2 u = \left[1 + \frac{1}{12}\delta^2\right]f.$$

Written in stencil form this is

$$\frac{1}{h^2} \quad \boxed{-1 \mid 2 \mid -1} \quad u = \frac{1}{12} \quad \boxed{1 \mid 10 \mid 1} \; f$$

which is known as Numerov's linear multistep method (see Henrici [8], p. 292) for solving $-y'' = f(x, y)$.

Example 8.4: $(p, q) = (1,1)$, $d = 2$. By the preceding example,

$$R(s_1, s_2) = \frac{s_1}{1 - \frac{1}{3}s_1} + \frac{s_2}{1 - \frac{1}{3}s_2}$$

$$= \frac{s_1\left(1 - \frac{1}{3}s_2\right) + s_2\left(1 - \frac{1}{3}s_1\right)}{\left(1 - \frac{1}{3}s_1\right)\left(1 - \frac{1}{3}s_2\right)}$$

$$\sim \frac{s_1 + s_2 - \frac{2}{3}s_1 s_2}{1 - \frac{1}{3}(s_1 + s_2)}.$$

Here $\sim$ means that terms $O(\|s\|^2)$ have been neglected. The resulting difference operator is, using the δ^2 notation,

$$-\frac{1}{h^2}\left[\delta_1^2 + \delta_2^2 + \frac{1}{6}\delta_1^2\delta_2^2\right] u = \left[1 + \frac{1}{12}\left(\delta_1^2 + \delta_2^2\right)\right] f.$$

If this is translated into stencil notation, there results the Collatz Mehrstellen operator of Example 4.1. We thus recognize this operator as the two-dimensional analog of the Numerov operator. The order of this operator is $2(p + q) = 4$. For the functions $R(s, t)$ associated with Example 5.1 (triangular finite elements, piecewise linear trial functions) and with Example 5.2 (quadratic elements, piecewise bilinear trial functions) we have only

$$R(s, t) = (s + t)\left(1 + O(\sqrt{st})\right)$$

and

$$R(s,t) = s + t + O(s^2 + t^2),$$

respectively. Although their complexity is comparable to the Collatz Mehrstellen operator of Example 4.1, the order of these methods thus is only 2, and thus they are less accurate for smooth functions f.

Example 8.5: $(p, q) = (2, 1)$, $d = 2$. The quotient-difference algorithm yields

$$R^{(2,1)}(s) = \frac{s - \frac{1}{8}s^2}{1 - \frac{8}{15}s}.$$

This results in

$$R(s_1, s_2) = \frac{s_1 + s_2 - \frac{1}{5}\left(s_1^2 + s_2^2\right) - \frac{4}{15}s_1 s_2 + \frac{8}{75}\left(s_1^2 s_2 + s_1 s_2^2\right)}{1 - \frac{8}{15}(s_1 + s_2) + \frac{64}{225}s_1 s_2}.$$

In stencil notation this becomes

$$\frac{1}{300h^2}$$

	−2	−11	−2	
−2	−4	−228	−4	−2
−11	−228	988	−228	−11
−2	−4	−288	−4	−2
	−2	−11	−2	

$$u = \frac{1}{255}$$

4	22	4
22	121	22
4	22	4

f.

The order of this operator is 6.

Example 8.6: $(p, q) = (1, 1)$, $d = 3$. By virtue of Example 8.3 we have

$$R(\mathbf{s}) = \frac{s_1}{1 - \frac{1}{3}s_1} + \frac{s_2}{1 - \frac{1}{3}s_2} + \frac{s_3}{1 - \frac{1}{3}s_3}$$

$$\sim \frac{s_1 + s_2 + s_3 - \frac{2}{3}\left(s_1 s_2 + s_2 s_3 + s_3 s_1\right)}{1 - \frac{1}{3}(s_1 + s_2 + s_3)}.$$

The translation into a 3-dimensional stencil yields

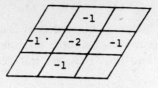

$\frac{1}{6h^2}$ 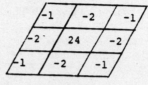 u

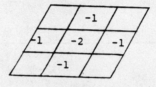

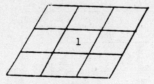

$= \frac{1}{12}$  f .

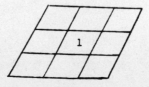

REFERENCES

1. W. N. Bailey, *Generalized Hypergeometric Series*, University Press, Cambridge, 1935.
2. L. Collatz, *The Numerical Treatment of Differential Equations*, Grundlehren **60**, 3rd ed., Springer, Berlin, 1960.
3. R. Courant and D. Hilbert, *Methods of Mathematical Physics*, vol. I, Wiley-Interscience, New York, 1953.
4. F. W. Dorr, "The direct solution of the discrete Poisson equation on a rectangle," *SIAM Rev*, **12** (1970), 248–263.
5. G. E. Forsythe and W. Wasow, *Finite Difference Methods for Partial Differential Equations*, Wiley, New York, 1960.
6. L. Fox, *The Numerical Solution of Two-Point Boundary Problems in Ordinary Differential Equations*, Clarendon Press, Oxford, 1957.
7. W. Gautschi, Attenuation factors in practical Fourier analysis. *Numer. Math*, **18** (1972), 373–400.
8. P. Henrici, *Discrete Variable Methods in Ordinary Differential Equations*, Wiley, New York, 1962.
9. _____, "Fast Fourier methods in computational complex analysis," *SIAM Rev.*, **21** (1979), 481–527.
10. R. E. Lynch and J. R. Rice, "High accuracy finite difference approximation to solutions of elliptic partial differential equations," *Proc. Nat. Acad. Sci*, **75** (1978), 2541–2544.
11. J. Marti, "An algorithm recursively computing the exact Fourier coefficients of *B*-splines with nonequidistant knots," *Z. angew. Math. Phys.*, **29** (1978), 301–5.
12. W. M. Pickering, "Some comments on the solution of Poisson's equation using Bickley's formula and Fast Fourier Transforms," *J. Inst. Math. Appl.* **19** (1977), 337–338.
13. A. J. Oppenheim and R. Schaefer, *Digital Signal Processing*, Prentice-Hall, Englewood Cliffs, 1975.
14. H. R. Schwarz, *Die Methode der finiten Elemente: Eine Einführung unter besonderer Berücksichtigung der Rechenpraxis*, Teubner, Stuttgart, 1980.
15. G. Sköllermo, "A Fourier method for the numerical solution of Poisson's equation," *Math. of Comp.* **29** (1975), 697–711.
16. P. N. Swarztrauber, Fast Poisson solvers. *This vol.*, 1983.

INDEX